The only guides to compare the content and struc...

Engineering

4
SERIES 1

CRAC
Degree Course Guides
2007/08

CRAC
Career Development – FOR LIFE

The CRAC Series of Degree Course Guides

The CRAC Series of Degree
Course Guides
Engineering

**Compilers and series
editors**
Andrew Smith and Marion Owen

This Guide published in 2007 by
Trotman, an imprint of Crimson
Publishing, Westminster House,
Kew Road, Richmond, Surrey
TW9 2ND
www.crimsonpublishing.co.uk

Copyright © Trotman 2007

British Library Cataloguing in
Publication Data
A catalogue record for this book
is available from the British Library

ISBN 978-1-90604-111-3

Printed and bound in Great
Britain by Bell & Bain, Glasgow

This Degree Course Guide is one of a series providing comparative information about first-degree courses offered in the UK. The aim of the series is to help you choose the course most suitable for you.

This Guide is not an official publication, and although every effort has been made to ensure accuracy, the publishers can accept no responsibility for errors or omissions. Changes are continually being made and details have not been given for all courses, so you must also consult up-to-date prospectuses before finally deciding which courses to apply for. The publishers wish to thank the institutions of higher education that have provided information for this edition.

CRAC: The Careers Research and Advisory Centre

CRAC is an independent not-for-profit organisation. CRAC Degree Course Guides are published under exclusive licence by Trotman Publishing. For catalogues or information, write to Trotman Publishing; to buy publications, contact NBN International, Estover Road, Plymouth PL6 7PY. Tel: 0870 900 2665.

Academic consultants:

Introduction and Integrated Engineering and Mechatronics:
Dr James Taylor, University of Lancaster

Chemical Engineering:
Dr Tom Arnot, University of Bath

Civil and Environmental Engineering and Building:
Dr David Richards, University of Southampton

Electrical and Electronic Engineering:
Dr Dave Stone, University of Sheffield

Materials Engineering, Materials Science and Metallurgy:
Dr Claire Davis, University of Birmingham

Mechanical and Manufacturing Engineering:
Dr Peter Willmot, Loughborough University

Engineering

Engineering

Engineering

With the recent changes to student funding and tuition fees, entering higher education can no longer be a decision made lightly by school-leavers. Though overall there has been an increase in the numbers applying to university in the last five years, those numbers have fluctuated and the funding changes have muddied the waters for students who do not have the ready funds to enter, and leave, higher education debt-free.

So is it still worth it? There is evidence that there is still a 'graduate advantage'. Salaries for graduates across the span of a career remain higher than for non-graduates. The demand in the UK for higher skills is great; it is clear that higher levels of skills are associated with higher levels of productivity, and greater productivity coupled with higher-level skills are vital to the growth of the UK economy.

Career choices for graduates are constantly changing. Changes in industry mean that there is an increasing demand for graduates to enter marketing, IT, management accountancy, management consultancy, community and society occupations and other newer professions. Graduate career paths are not necessarily clear-cut, but the majority of graduates find themselves in employment that is related to their long-term career plans. One growing trend is that employers are more and more interested in work experience. The good news here is that universities are increasingly catering for this by providing assessed placements that form part of the degree course and opportunities for volunteering in the local community.

Students and graduates feel that their decision to enter higher education is worthwhile. The 2006 National Student Survey, the official survey of students' feelings about their courses, shows that approximately 80% were satisfied with their higher education compared to their expectations. According to the research project *Seven Years On: Graduate Careers in a Changing Labour Market*, more than two-thirds of graduates say they would do it all over again.

Thus it seems that the prospects for new entrants to higher education are still exciting. Making this decision may be daunting, bewildering or even liberating. It is vital that you take into account all of the factors that you feel are important in making a choice: your career aspirations, your abilities and aptitudes, the teaching facilities, opportunities for placements and, importantly, your passion for the subject. The choice of institutions and degree programmes is vast, and creating a shortlist can prove to be challenging – the biggest challenge being to ensure your choice of course and institution is an informed one. The CRAC *Degree Course Guides* are intended to be accurate, comprehensive and insightful to help equip you to do just this, making them essential reading for anyone considering entry to higher education.

Jeffrey Defries, Chief Executive
The Careers Research and Advisory Centre (CRAC)
www.crac.org.uk

Engineering

The tables provide comparative information about first-degree courses offered at higher education institutions (universities and colleges) in the UK. Each table provides detailed information organised alphabetically by institution and course title. An at-a-glance summary of the material in each main table is given below:

Table 2a: First-degree courses

This gives the degree qualification, duration, foundation year availability, modes of study, whether the course is part of a modular scheme, course type and number of combined courses available.

Table 2b: Subjects in combination

This lists subjects that can be combined with engineering, and where they can be studied.

Tables 3a–3e: Course content

These give detailed information on the content and organisation of the courses, and special features such as study abroad.

Table 4: Assessment methods

This gives the frequency of assessment, years in which there are exams contributing to the final degree mark and percentage assessed by coursework.

Table 5: Entrance requirements

This gives the number of students on each course, typical offer in UCAS points, A-level grades and Scottish Highers, and required and preferred subjects at A-level.

'How to use this Guide' provides an overview of the book's content and more detailed information is provided before each table.

The *Engineering* Guide provides you with comparative information about the honours degree courses in engineering available at higher education institutions in the UK. Naturally, each branch of engineering has its own distinctive style and content, so each is covered in a separate part within the Guide. However, there are certain features common to all types of engineering, so the Guide has a common introduction giving information that applies generally to all the subjects. The *Introduction* has broadly the same structure as each of the individual engineering subject parts, with the same chapter in each giving information about one particular aspect of the courses. However, the *Introduction* does have an additional appendix (see page 27) describing the benefits and obligations of industrial sponsorship, and how to go about choosing and applying to possible sponsors.

The aim of the Guide is to help you answer some of the questions you need to ask if you are going to find the courses that suit you best and interest you most. Some of these questions are shown below, with an indication of which chapter in each part will help you find the answers. You should read the general information in the corresponding chapter in the *Introduction* before reading the chapter in the individual subject part.

What is it like to be an engineer?   The first chapter in the *Introduction* is designed to give you some insight into what engineering is about, what it is like to be an engineer. The first chapter in each of the individual subject parts gives similar information for that specific branch of engineering, and the careers open to graduates.

What courses are offered in these subjects?   Chapter 2 in the *Introduction* begins with a table showing you which part of the Guide to look at for specific subjects, and goes on to give general information about the way courses are organised and what you can find out from TABLE 2a in each of the individual subject parts of the Guide. This table lists the courses in which you can study the branch of engineering covered in that part of the Guide for at least half of your time. TABLE 2b then shows which subjects you can combine with this subject in a two-subject degree. As you might expect for courses that are largely designed to lead to a professional qualification, most of the courses are specialised, so there are relatively few combined courses involving engineering subjects.

At the end of Chapter 2 in each individual subject part of the Guide there is a list of courses in related areas that may also interest you.

What are the courses like?   One of your major concerns will naturally be to find out just what you will study during your undergraduate career. Chapter 3 describes the style and content of the courses, mainly using tables to enable you to compare them easily. However, before you dive into the tables, you should read the surrounding text and Chapter 3 in the *Introduction*, as this will tell you how to interpret the tables and what they can and cannot tell you.

Engineering

1

How will I be taught and examined?   Institutions in higher education do not all use exactly the same teaching and assessment methods, and the balance between them is often different at different institutions. Chapter 4 identifies and distinguishes between the approaches used.

What do I need to get accepted for a course?   Chapter 5 describes the qualifications you will need for admission to degree courses listed in that individual subject part of the Guide. TABLE 5 gives the requirements for individual courses and an indication of the number of places on each course. There is no simple relationship between the number of places on offer and how easy it is to get accepted for a course, but this figure does at least give you some idea of how many fellow students you would have on a course.

Where do I go from here?   This Guide, like the others in the series, aims to provide comparative information about courses to help you decide which ones you want to follow up in more detail, but it is only concerned with the content and organisation of the courses themselves. You will need to look at prospectuses, websites such as www.trotman.co.uk, and the institutions' own sites, shown in TABLE 2a, to find out about such things as tuition fees, accommodation, locality and student life at each institution. Chapter 6 gives a list of sources of further information, including a contact name for each course. In addition, most institutions welcome visits from potential students, and many run open days. Once you have applied, you may be invited for an interview, which will usually include an opportunity to look around the institution's facilities.

Chapter 1: Introduction

The engineer   It is perhaps unfortunate that in English the word 'engineer' is used to describe everyone from the person who comes to your home to repair the washing machine through to the person in charge of the design of multi-million pound projects such as a new supercomputer, the Channel Tunnel, a supertanker or an oil refinery. On the other hand, this link to everyday practical tasks may serve as a reminder that, above all, engineers are concerned with making things that work. A car that turns eyes in the street but breaks down every other week will not sell. A computer system using innovative design and state-of-the-art technology may be cancelled before it is turned on if it is a year late and over budget. An engineer who designs an elegant bridge that falls down will be remembered for the fact that the bridge fell down, not for the elegance of the original construction. This is not to say that aesthetics are not important, just that they are secondary to function.

Perhaps the earliest engineers were the Stone Age people who first produced flint tools. They started a process that has carried on down the ages. The genius who gave civilisation the wheel and the designers of Stonehenge were engineers. Leonardo da Vinci, Sir Christopher Wren and Isambard Kingdom Brunel were all engineers. So what did these people have in common? They all had creative ways of thinking, the ability and energy to convert an idea into a product, and a streak of hard common sense capable of distinguishing the just possible from the merely fantastic.

Modern professional engineers still need all these qualities, but they also need a thorough formal education, usually through a degree course, which will enable them to learn from the successful, and unsuccessful, experiences of previous generations of engineers. Increasingly, as the size and scope of engineering projects grow, engineers work together in teams with other engineers and professionals from other disciplines. Professional engineers also have to deal with non-technical managers, clients and the general public. All this puts a premium on their ability to get on with other people and to communicate with them effectively. It is important to remember that engineering is directly concerned with people and not just with things. Your work as a professional engineer can impinge on many individuals, so you must possess character, integrity and a sense of social awareness.

Science and engineering   Engineering and science have developed in parallel. The engineer often relies on the scientific principles laid out by the scientist, and applies them to practical problems. Apart from the benefit of having their ideas turned into useful results, scientists are often guided by the requirements of engineers as they come up against new problems. The most obvious instances of this are in the area of materials, where, for example, pure research in semiconductors and high temperature superconductors is driven by the potential engineering applications.

You may have the impression that it is the scientist who does the imaginative and inspirational work, while the engineer tidies up and fills in the boring practical details.

3

Nothing could be further from the truth. The problems faced by engineers are just as challenging and demand just as much creativity as any in fundamental science. However, the nature of the problems is different and, more importantly, there are a whole range of other constraints on the solutions that scientists do not need to consider, such as 'how much will it cost?', 'how long will it take?' and 'can I sell the idea to my colleagues, boss and clients?' These constraints are often in conflict, so you have to use judgement in balancing one against the other. Unlike a scientist, who is looking for the answer to a problem, the engineer has to keep looking at several potential answers and must decide which best suits the present conditions.

Given the relationship between science and engineering, it will come as no surprise that most engineering courses use the physical sciences as a starting point from which the individual specialised disciplines grow, and that the most common A-level requirements are for physics and mathematics.

Using models for analysis

Engineering is expanding and developing a greater range of analytical techniques: engineers make increasing use of 'models' to understand problems and possible solutions. These models may be mathematical formulae, computer programs, scaled representations or graphical procedures. What they have in common is that under specified conditions they are known to behave in some respects like the real thing. The skill comes in knowing when the simplifications inherent in a model are appropriate and, perhaps more important, when they are not. Models can be used to explain how circuits work, how stresses are distributed in structures, how machines function or how complex mechanical or human systems work.

Originally, engineers tended to rely more on experiment than analysis. The problems tackled by engineers are much more complex now, so the modern engineer needs to embrace both kinds of expertise. For example, testing all the possible states of a relatively simple microprocessor could take longer than the lifetime of the universe. This means that the engineers have to rely on lengthy analysis at both the design and test stages to break down the required behaviour into smaller and simpler modules.

Research is required not only to discover and investigate new phenomena but also for the creation of more and better models of well-understood but complex systems. These can then be used by other members of the engineering team to design and implement better solutions.

The engineer and society

To a large extent the wealth of a nation depends on its capacity to process and increase the value of its own and imported natural resources. This processing is through the production of manufactured goods and engineering constructions. Engineers thus play a key role in deciding a nation's prosperity. Even with offshore oil, the UK's mineral resources are limited, so engineering is particularly important in ensuring that the standards of living of the population are maintained. The quality of engineers trained in the UK is recognised throughout the world and for many years they have been called upon to provide expert knowledge and advice in design and development for technical activities abroad as well as in this country.

In recent years, people in general, and engineers in particular, have become increasingly conscious of how much society and our environment are affected by the work of engineers. As a result, engineers now have to pay more attention to the social and environmental consequences of their actions. This requires a sensitivity to the issues, as well as particular kinds of expertise, such as communication skills.

Which course is right for you?   As you will see from the lists in Chapter 2 of each of the individual subject parts of this Guide, there is a huge variety of courses on offer, so making a shortlist to apply for is not an easy task. More detailed information about the range of possibilities is given in the rest of this *Introduction* and in the individual subject parts. The following list shows you some of the main questions you will have to answer.

- Do you have the right qualifications or should you take a foundation or access course?
- Do you want to aim for Incorporated or Chartered status?
- Do you want to take a sandwich or full-time course?
- Do you want to spend time abroad?
- Do you want to take a BEng or MEng course?
- Do you want to study engineering alongside another subject?
- Do you want to specialise in a particular branch of engineering or do you want to take a more general course, leaving the possibility of specialisation to a later stage?
- Which branch of engineering do you want to study?
- Are there any particular specialised content features of a course that are important to you?
- What sort of engineering role are you looking for? For example, are you more likely to be interested in research, in the design of solutions to immediate engineering problems, or in playing some part in the implementation of these solutions?

The scope of the Engineering Guide   The *Engineering* Guide divides the field of engineering into a number of groups. Broadly speaking these reflect the traditional divisions between subjects. However, the boundaries are not always sharp, and may not always match the boundaries of your own interests, so if you are interested in engineering, you should at least skim through the course lists in each individual subject part of the Guide to get a feel for the range of courses offered. In particular, you may want to look at *Integrated Engineering and Mechatronics*, as it contains courses that combine the study of two or more engineering subjects or allow specialisation after one or two years' broadly based study. In fact, many of the courses in the parts of the Guide covering specific branches of engineering begin with a broad introduction and often allow scope for specialisation in that particular branch of engineering at a later stage.

The following table summarises the scope of each individual subject part of the Guide, but you should note that there are many variations on the names of courses.

Integrated Engineering and Mechatronics	Engineering, mechatronics, engineering science and other general and combined engineering subjects
Chemical Engineering	Biochemical engineering, chemical engineering, process technology
Civil and Environmental Engineering and Building	Architectural engineering, building, building services engineering, civil engineering, structural engineering, environmental engineering, environmental technology
Electrical and Electronic Engineering	Communication engineering, computer engineering, electrical engineering, electronic engineering, electronics, internet engineering, mobile communications technology, optoelectronics, telecommunications engineering
Materials Engineering, Materials Science and Metallurgy	Materials engineering, materials science, metallurgy, ceramics, glass, polymer technologies
Mechanical and Manufacturing Engineering	Aeronautical engineering, aerospace engineering, agricultural engineering, automotive engineering, computer-aided engineering, manufacturing engineering, marine engineering, mechanical engineering, naval architecture, product design engineering

Table 2a TABLE 2a in each of the individual subject parts of the Guide lists the courses that lead to the award of an honours degree and in which you spend at least half of your time studying engineering. When the tables were compiled they were as up to date as possible, but sometimes new courses are announced and existing courses withdrawn, so before you finally fill in your application you should check the UCAS website, www.ucas.com, to make sure the courses you plan to apply for are still on offer.

Degree Most of the courses covered in these Guides lead to the award of a BEng or MEng, though some lead to a BSc or even a BA. However, nearly all of those accredited by the professional bodies lead to a BEng or MEng (Chapter 6 gives further details, but accreditation status can change so you should check with the institutions offering the courses and the professional bodies before making any final decisions). Many courses have both an MEng and a BEng stream, the MEng normally lasting a year longer. The extra year may be spent in a variety of ways. Some courses concentrate on broadening your understanding of the context in which engineers work in industry, with courses in areas such as law, economics, management and industrial relations, together with a period of industrial experience. Some concentrate on the international context, with language and regional studies courses and a period of time spent abroad. Other courses concentrate on a more specialised study of engineering topics, while others combine aspects of several of these approaches. Chapter 3 gives details of the differences in organisation and content between BEng and MEng courses with the same title; Chapter 5 (in particular, TABLE 5) shows the entrance requirements for each.

Entry to MEng courses varies from course to course. A few MEng courses are a separate stream from the start and you apply directly for admission to that course. For other courses, all students begin on a common BEng/MEng, with the MEng stream separating at a later point. Whether you can gain a BEng or an MEng will also affect how you proceed to professional qualification: see Chapter 6 for further details.

Duration The *Duration* column in TABLE 2a shows the duration of the course including a sandwich year or any time spent abroad, so courses available in both full-time and sandwich variants, or with optional periods abroad, will generally show two possible durations (for example '3, 4', showing that they can involve either three years' continuous study in the UK, or four years if the option for study abroad is taken). Two durations, or more, may also be shown if the course has both a BEng and MEng stream.

Foundation years and franchised courses The *Foundation year* column in TABLE 2a shows whether an optional foundation year is available. These courses have been designed specifically to encourage students who have not studied mathematics or science at A-level to take degrees in science and engineering. The foundation year allows you to acquire the necessary knowledge and skills to begin the main course on a par with students entering the course directly. You can usually apply directly for a degree course including a foundation year, though you should note that the final decision as to whether you need to take a foundation course will be made by the institution to which you are applying.

Foundation years are not the same as Foundation Degrees. These are qualifications in their own right and are currently only offered in England, Wales and Northern Ireland. Courses leading to a Foundation Degree last two years and are particularly related to workplace skills. However, they can provide another route into honours degree courses, especially at the institutions offering them. For more information, see www.foundationdegree.org.uk.

Franchised courses   TABLE 2a also shows whether the foundation year can or must be taken at a franchised institution, which will typically be a college of further education in the same region. Some institutions also franchise complete degree courses, sometimes at colleges outside their immediate locality. Refer to prospectuses for further details of franchising arrangements.

Optional foundation years are not included in the *Duration* column, so you should add one to those figures to give the length including a foundation year.

Direct entry to year 2 in Scotland   In Scotland, most students enter university with a broader, less specialised background than in the rest of the UK, so the first year is often similar in level to a foundation course. If you are well qualified, you may be able to gain exemption from this first year for some courses at Scottish universities. The *Foundation year* column shows courses for which this is possible, and in these cases the *Duration* figures should be reduced by one for direct entry to the second year.

Modes of study   The *Modes of study* column in TABLE 2a shows whether a course is available in full-time, part-time, sandwich or time abroad modes. The Guide does not include courses that are only available part time, but TABLE 2a does show if courses are available part time in addition to one of the other modes. Chapter 3 in this *Introduction* and in the individual subject parts of the Guide gives further information about sandwich courses.

Courses may be shown as involving time abroad for a number of reasons. The most straightforward is if the named course has an optional or compulsory period of study or work experience overseas. The period spent abroad will usually be about a year; courses in which you spend less than six months abroad do not qualify for the 'time abroad' description. However, to avoid duplication of information, a time abroad entry may also mean that there is a variant of the course with a slightly different title (usually including a phrase such as 'with European studies', 'International' or 'with French') and involving study or a work placement abroad. Chapter 3 in this *Introduction* and in the individual subject parts of the Guide gives further information about work and study abroad. Note that a course is shown as including time abroad only when that opportunity arises *as part of the engineering course*. If time can be spent abroad only if you combine the study of engineering with a modern language, the course is not shown as a time abroad course in TABLE 2a.

Modular schemes   Many institutions offer modular schemes in which engineering can be studied alongside a wide range of other, often quite unrelated, subjects, though the requirements of the professional institutions mean that the number and choice of modules that can be taken from outside the engineering subject may be very restricted. Note that although many courses are described as 'modular', this usually means that the engineering content itself is organised in modules; a modular *course*, in this sense, may or may not be part of a modular *scheme*.

Modular schemes can give you much greater flexibility in choosing what you study and when, and can have the particular benefit of allowing you to delay specialisation until you know more about the subject. However, there is a great deal of variation in the way modular schemes are organised, and there is always some restriction on the combination of modules you can take. If you feel strongly that you want to take a particular combination of subjects in a modular scheme, you should check prospectuses carefully, as some institutions do not guarantee that all advertised combinations will be available.

Course type   This Guide describes courses in which you can spend at least half of your time studying engineering. Specialised courses, in which you spend substantially more than half your time studying engineering, are shown as ● in TABLE 2a. The *No of combined courses* column in TABLE 2a shows how many subjects can be combined with engineering in a combined course. In these you spend up to two-thirds of your time studying engineering, and the rest studying another subject. You can use TABLE 2b to find out where you can study a specific subject in combination with engineering (you should use the UCAS website or prospectuses if you want to know what combinations are available at a specific institution).

The validation requirements of the various professional institutions mean that there are not many combined courses involving engineering subjects.

Subjects available in   TABLE 2b shows those subjects that can make up
combination with engineering   between a third and a half of your degree programme alongside engineering in the combined degrees listed in TABLE 2a. For example, if you are interested in combining engineering with physics, first look up physics in the list to find the institutions offering physics in combination with engineering. You can then use the index number given after the institution name to find which of the courses at the institution can be combined with physics – there is no index number if there is only one course in TABLE 2a.

It is not possible to describe in the space available here the many different ways in which combined courses may be organised, so you should read prospectuses carefully. For example, you should find out whether the subjects are taught independently of each other or are integrated in any way. Combined courses in modular schemes often provide considerable flexibility, allowing you to vary the proportions of the subjects and

9

include elements of other subjects. However, this means that you may lose some of the benefits of more integrated courses.

The names given in the table for the combined subjects have been standardised to make comparison easier, so the name used at a particular institution may not be exactly the same as that given in the table. However, in nearly all cases it will be very similar, so you shouldn't find much difficulty in identifying a particular course combination when you look at the prospectus.

Naturally, the content of engineering courses depends very much on the branch of engineering, so you will need to look at the individual subject parts of the Guide for most of the detailed information. However, there are a number of general points about the way courses are organised that are common to most of the subjects, and it is convenient to gather them together in this *Introduction*.

Sandwich courses and other industrial experience   As you can see from TABLE 2a in the individual subject parts of the Guide, there are a large number of sandwich courses available in all the engineering subjects. Because of the importance of sandwich courses, Chapter 3 in each part of the Guide gives detailed information about how they are organised and of any other opportunities for industrial experience. As the range of possibilities is similar in all the subjects, this section of the *Introduction* explains what they are and gives general background information.

On many sandwich courses you can spend the industrial experience period abroad, so you should also look at the *Study and work abroad* section later in this chapter.

Sandwich or full time?   The industrial training periods in sandwich courses mean that you should be better able to relate theoretical concepts to industrial practice, and you will have a much better feel for the social and cultural aspects of working in industry. Hopefully, you will also have proved that you can work productively in an industrial environment, and you may have made some valuable contacts for when you are looking for a job. Some institutions also offer a Diploma in Industrial Studies for successful completion of the industrial component of a sandwich course.

Sandwich courses are longer than full-time courses: BEng sandwich courses usually last four years, though there are some exceptions (see TABLE 2a); MEng courses may also have a sandwich component, and will then generally last five years. However, the sandwich period can usually be counted towards the period of industrial training required for professional recognition by the professional institutions, so you will be at no disadvantage compared with students on full-time courses in the time it takes to gain, for example, Chartered Engineer status. Taking a sandwich course can, however, affect your social life, as the extra time taken means that you will get 'out of phase' with other students on full-time courses, and you may also lose contact with them during industrial periods.

One further possible advantage of a sandwich course arises when you come to project work, which is a major component of all engineering courses. An idea for a project can often emerge from an industrial training period, and it may be possible to carry it out in co-operation with the company. In these cases, your progress on the project might be supervised jointly by a college tutor and a tutor from industry.

However, you should remember that many full time courses also give a variety of opportunities for industrial experience, which can give some of the advantages of sandwich courses.

Introduction

Structure of sandwich courses   There are several different ways of organising sandwich courses, the commonest being described as 'thick' or 'thin'. Thick sandwich courses have one or two industrial periods lasting about a year; thin sandwich courses have shorter periods, often of about six months, spent alternately in industry and university or college. Thin sandwich courses can offer the advantage of greater integration of academic work and industrial training, but there are more periods of disruption as you move from one environment to the other.

The vast majority of sandwich courses now on offer have the thick sandwich pattern. These can be divided into two main types: 2–1–1, in which there is a 12-month period of industrial training between the second and final years of the academic course, and 1–3–1, in which a three-year full-time degree course is 'sandwiched' between two one-year periods in industry.

The diagram shows a number of different ways in which a four-year sandwich course can be organised (MEng courses usually have an extra year at the end, and courses in Scotland an extra year at the beginning). There are, however, a number of variations on these patterns, so you should check with institutions if the precise details are important to you.

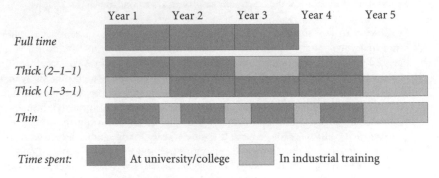

University-based or industry-based?   Students on sandwich courses can be either industry- or university-based. Students are industry-based if a sponsoring firm undertakes at the outset to meet all the industrial training requirements of the course, which is usually, though not necessarily, carried out within the sponsoring firm itself. Students normally receive student apprentice rates of pay during the periods of training, and may also receive some financial support from the firm during the academic periods.

University-based students have no formal link with any one firm, and the industrial training is arranged by the university, though, subject to the university's approval, it may also be possible to make your own arrangements. The industrial period(s) may be spent at a number of different firms, or one or more training periods could be spent in the university workshops.

Other forms of industrial experience   All institutions encourage training in industry during the long vacation, and at some places it may be compulsory. Some institutions also recommend students to spend a year working in industry before they take up the course, though they may have different recommendations for sandwich and full-time students. Another possible way of gaining industrial experience, which some institutions allow, is to take a year out (an 'intercalated year') in the middle of a full-time course. In some ways this is similar to a sandwich course, except that you are unlikely to be supervised by the university/college, and although the period will usually count as industrial training for the purposes of professional qualification, you should check that this will be the case.

Some large firms run 1–3–1 schemes, which incorporate a full-time course. In these, you make an agreement with the firm for five years, in return for which the firm usually provides some funding while you are at university or college, with a year's training before the degree course and a further year after graduation. This training may provide exemption from a compulsory long vacation training period, but some firms expect their 1–3–1 students to work for them during the summer anyway as part of the agreement. Not all companies require students to make an agreement and they may operate quite separate pre-university and postgraduate schemes. For more information on industrial sponsorship in general, see the Appendix to this *Introduction*.

TABLE 3a in all the individual subject parts of the Guide gives information about sandwich courses.

Study and work abroad   **The international context**   The increasing recognition of the importance of European and global markets for engineering products has led to a rapid increase in the number of opportunities for spending time abroad during degree courses. The steady reduction of barriers to the movement of goods, services and professionals within the European Union creates new challenges and opportunities for graduate engineers and the engineering industry at large. National borders are becoming increasingly irrelevant to large companies, but the opportunities for small companies are just as important, as they can now address a much larger market with specialist products or services. No company can ignore these developments, as their traditional markets are also being opened up to much wider competition.

The academic world is becoming equally international, and many universities have formed close links with institutions abroad. Today, many research projects are international, involving close co-operation between institutions in different countries. In some instances, collaboration has extended to the development of joint courses, with parts of the total study programme taken in each of the participating institutions.

European engineers   To exploit the opportunities and meet the challenges, industry needs innovative engineers and engineering managers with a pan-European vision, a command of key European languages and an insight into the European Union and its institutions. Only then can they compete effectively within the single market and against US, Japanese and East Asian rivals.

Engineers are now having to think of their careers in a European context. It is becoming even more important to attain the title of Chartered Engineer (CEng), which can lead to the title of European Engineer (EurIng). This title, which under EU rules is recognised throughout the Union, signifies that the holder has been judged to have achieved full professional status as an engineer. It is likely that in future, proficiency in a European language will be necessary for qualification as a EurIng.

TABLE 3a gives details of where study or work abroad is possible, together with further details, which are explained in the following sections.

Industrial experience abroad   Many institutions allow or encourage students to spend the industrial period of sandwich courses working abroad. Institutions may have special arrangements with foreign companies, or you may have to make your own arrangements, though, naturally, this will be subject to approval by the institution. The EU strongly favours the principle of students spending time working in other countries, and runs a co-ordinating scheme, Leonardo da Vinci, to encourage this (see Chapter 7).

Study abroad   Another option is to spend time abroad studying rather than in industrial experience. Many courses have links with overseas academic institutions with whom they arrange exchanges, usually through Erasmus (European Community Action Scheme for the Mobility of University Students), which is part of the broader Socrates programme. See the website www.erasmus.ac.uk for up-to-date information.

The European Credit Transfer Scheme (ECTS) also operates within the EU Erasmus initiative. Students at participating institutions can move from one to another, accumulating credits towards a degree, which is awarded by the institution at which the final credits are earned.

If you are thinking of spending some of your time studying abroad, you should be aware that the university system and styles of teaching vary between countries, so it is important to find out how what you study in the overseas university fits in with the work you do in this country, and what support is available while you are abroad.

Language training   Whilst the language of technology and commerce may well be English, in selling goods and ideas and in intricate negotiations, the maxim 'I sell in any language, but I buy in my own' remains true. To be effective in Europe, engineers will need a good command of one or more European languages. Many courses include some language teaching, which may be an optional or compulsory element and may or may not contribute to the overall assessment of the course: see TABLE 3a for details.

Special courses   The increasing importance of including an international dimension in engineering studies has been recognised by a number of universities and colleges, who are providing a variety of European and other international programmes through named variants of engineering courses. In general, the Guides do not give separate entries for these, as the engineering content is usually the same as that in the normal course. However, TABLE 3a shows where a course has a named variant that includes a period of time abroad as an integral component of the course. Examples of such names are 'European mechanical engineering' and 'Civil engineering with a year abroad', as well as the more traditional minor combinations with a foreign language.

The extent and depth of language teaching can vary widely, ranging from a modest 60 hours a year to extended degree courses that also provide instruction in the relevant 'technical' languages, with some of the engineering subjects and seminars given in French or German. In the majority of cases, the minimum entrance requirements in languages are at GCSE level and not at A- or AS-level, so the development of language skills remains modest. A few courses do, however, take language teaching to a higher level and require an A-level language for admission.

Some courses also contain a significant element of more general European studies. This may include a study of the European institutions and the structure and regulations of the single market. It may also include the socio-economic history of the region and the variety of cultures in different areas of Europe. An understanding of these latter issues can be just as important as the more technical legal issues, as they can affect the impression you create and the way you relate to other people.

Some of the special degree programmes can lead to dual engineering qualifications, such as an MEng and a European diploma.

Teaching methods   Most courses are taught using a mixture of practicals, lectures, and classes or seminars (discussion groups of varying sizes). The total time taken up each week is around 20–24 hours for almost all the courses, though the proportion of time devoted to each activity may vary, as may the size of the study groups. A typical distribution is 9–13 hours of lectures a week, with 2–8 hours a week allowed for teaching through small study groups. Practical work usually occupies 5–9 hours a week in the early years.

The balance between different teaching methods also varies between the early and late stages of a course, with the emphasis switching from lectures to smaller group teaching and project work. In particular, project work entirely replaces the formal laboratory programme in the later years of some courses.

Other teaching methods   As well as the traditional teaching methods, many institutions use a variety of other techniques. Amongst these are technology-based teaching methods such as computer-aided learning, film, video and CCTV.

Many courses also make use of case studies, design studies, modelling and simulation, especially in the final year. These act as a focal point, bringing subject material together from various disciplines and from industry. Many courses use business games to give a context to their management studies component. This usually happens in the second or third year. All departments emphasise the teaching of engineering design and open-ended problem-solving, which can include group working and presentation. This can occur in any, or more usually all, years of the course.

Practical work   Practical work is a major component of all the courses and can be organised in a variety of ways, from formal laboratory or computer exercises to long-term project work. In many courses the practical work component is compulsory, and often your laboratory notes and reports are used as the basis for continuous assessment. You will be able to find out more about the organisation of practical work if you visit the institution and look around their teaching laboratories: this can be a good topic for discussion at an interview.

All courses include basic workshop training, which is usually completed in the first year. In some courses it takes the form of a concentrated training period lasting three or four weeks; in others one day a week may be allocated to it.

Projects   All courses include a large amount of project work, particularly towards the end of the course. Projects vary from short exercises that may be carried out within a week, to a major project involving many aspects of research, design, manufacture and testing, which may extend throughout a whole year. Projects can play a part in all years of the course, but the final-year project is normally much more extensive than those in earlier years, and is often a major component of the course assessment.

Design is the main objective of engineering and is thus an important component of all courses. Design examples will be given in lectures, but it is in the final project that you will be able to pull all the threads of the course together in a design context. You will usually have considerable freedom in the planning and execution of projects.

Engineers in industry normally work as part of a team, so some institutions see group projects as an important preparation for work beyond the degree course. There are, however, difficulties associated with group projects when they have to be used as the basis of the individual's assessment. TABLE 4 shows whether group projects are used in the course, and an estimate of how much time is spent on project work.

Assessment methods   Most institutions use a variety of assessment methods such as formal written examinations, continuous assessment of coursework and extended projects or dissertations. TABLE 4 gives information about the balance between these methods. It shows in which years there are written examinations, and if they contribute to the final degree classification. However, the contribution from examinations in earlier years is often less than that from the examinations in the final year. You should also note that although an examination may not contribute to the final result, passing it may be a condition for continuing with the course.

Many courses allow a wide range of options, which are often assessed in different ways, so it is difficult to give precise figures for the contributions of different assessment methods. For this reason, TABLE 4 shows the possible maximum and minimum contributions from coursework and projects or dissertations.

Oral examinations are used in some courses, though sometimes only for borderline candidates. In particular, you may have to give an oral presentation of the results of your project as well as a written report. TABLE 4 shows what part oral examinations form in the assessment of the course.

Frequency of assessment   On many types of course, especially modular courses, assessment is carried out more frequently than in the traditional pattern of end-of-year examinations. Often, each module is assessed independently, soon after it has been completed. The precise details of when assessments are carried out vary from course to course: TABLE 4 shows if courses are assessed every term, semester (there are two semesters in a year) or year. Note that even if all modules are assessed, those occurring early in the course may not count as much in the final result as later ones.

The mix of assessment methods   All assessment systems have advantages and drawbacks: for example, reducing the significance of final examinations may simply mean that short periods of high stress are replaced by a series of deadlines and continuous low-level stress throughout the course. Which of these you prefer will depend on your temperament. Because students vary in their response to different assessment methods, institutions usually employ a combination of methods, which also allows them to match the assessment method to the skill being tested. In some cases you may be able to change the make-up of your assessment regime, for example by choosing a dissertation or project instead of a formal examination.

How to apply   Applications for courses in this Guide must be made online at www.ucas.com, the website of the Universities and Colleges Admissions Service (UCAS).

The UCAS tariff   The UCAS points scheme, or 'tariff', combines GCE Advanced levels, GCE Advanced Subsidiary levels, AVCE and Scottish qualifications in a single points system. For example, a GCE A-level with an A grade or a Scottish Advanced Higher with an A grade will score 120 points, while an AS-level with an A grade, an A-level with a D grade, or a Scottish Higher with a B grade will score 60 points.

For some courses, entrance requirements are expressed simply in terms of points, but many also make offers in terms of particular grades in A-levels or Highers, and some require minimum grades in specific A-levels or Highers. TABLE 5 summarises these requirements. However, for many courses the requirements are more complex, so you should take the information in the table only as a starting point, and an indication of the relative demands of the courses. It is vital to check prospectuses and university and college websites, to find out the full details of requirements for any courses you are considering.

Other qualifications   All institutions accept certain other qualifications for their general requirement. For example, the International Baccalaureate (IB) and the European Baccalaureate (EB) are generally accepted. The publications listed in Chapter 6 give information about some other qualifications and about the acceptability of the IB and the EB for course-specific requirements. Again, if in doubt, you should consult institutions direct.

Institutions are becoming increasingly flexible about entrance requirements, so you should not assume that you will not be accepted if you lack standard qualifications. For example, in some franchising schemes you can begin at a level below the starting point of a degree course by taking the early stages of a course at a college of further education. Provided you are successful, you will then move on to an institution of higher education for the later stages of the course. You may also be interested in the Access courses offered at some universities and colleges, some of which are taught in the evenings. Successful completion of an Access course, while not guaranteeing you a place on a degree course, will give you an advantage when you apply.

Open days and interviews   Many institutions and departments run open days, when you can look round their facilities and meet staff and students. In some cases you may be invited to visit only after you have been offered a place. For some courses, interviews play an important part in the selection process. They are usually conducted in as relaxed and friendly a manner as possible; the intention is to find out more about your personal qualities than can be learned from your application form alone.

Introduction

Student numbers   TABLE 5 also gives an estimate of the number of students on each course (the figure is shown in brackets if it is the total for a larger group, such as a faculty, or if it includes other students, such as those on a combined degree scheme, not just those studying engineering). This number will give you some idea of how many other students you are likely to be taught with, but you should remember that some lectures may be common to several courses, and students on the same course may choose different specialised options, particularly in later years.

You should bear in mind that there is no simple equation between the number of places and the probability of acceptance, since a course with a large number of places may have those places because it is popular and oversubscribed. You cannot even use the ratio of applicants to places as a reliable guide, as one course may have relatively few applicants because it is known to have high entrance requirements, while another may attract a large number of applicants who view it as an insurance policy in case they fail to gain entry to another course.

Entrance requirements   The information in TABLE 5 is for general guidance only, since admissions tutors consider applicants individually, and may take many factors into account other than examination grades.

Introduction

19

The engineering institutions and the Engineering Council

Each branch of engineering has its own professional institution. However, the boundaries of what each institution covers are not always sharply defined and two people with the same educational and professional experience, and doing essentially the same job, might belong to two different institutions. Because of this, the names and addresses of the relevant professional institutions are given in Chapter 7 of this *Introduction*, rather than in the individual Engineering Guides.

The overall controlling authority for the engineering profession is the Engineering Council, which is responsible both for setting the standards of qualification for engineers and technicians, and for maintaining the register of those qualified. In practice, however, most of the day-to-day administration is delegated to a number of nominated chartered engineering institutions or other affiliated institutions, and in obtaining professional recognition, your main contact will be with the relevant professional institution for your branch of engineering.

Professional engineering status

Obtaining a degree in engineering does not of itself make you an engineer. Becoming a professional engineer includes three phases: education (a degree), followed by a period of initial professional development, and finally a professional review. The education base is normally an accredited BEng or MEng degree; the initial professional development phase is a combination of training and experience, which together develop the professional competence necessary for registration as a professional engineer.

Levels of professional recognition

The Council maintains the registers for all levels within the engineering profession, from Engineering Technicians (EngTech), through Incorporated Engineers (IEng) to, at the highest level, the Chartered Engineer (CEng), the last two being the two levels of professional engineer. The Engineering Council have issued the following definitions of the responsibilities of these two types of engineer:

<u>Chartered Engineers</u>   are concerned primarily with the progress of technology through innovation, creativity and change. They develop and apply new technologies, promote advanced designs and design methods, introduce new and more efficient production technologies and marketing and construction concepts, and pioneer new engineering services and management methods. They may be involved with the management and direction of high-risk and resource-intensive projects. Professional judgement is a key feature of their role, allied to the assumption of responsibility for the direction of important tasks, including the profitable management of industrial and commercial enterprises.

Incorporated Engineers act as exponents of today's technology and, to this end, they maintain and manage applications of current technology at the highest efficiency. Incorporated Engineers require a detailed understanding of a recognised field of technology, so that they can exercise independent judgement and management in that field. They provide independently, and as leaders, a significant influence on the overall effectiveness of the organisation in which they work, often in key operational management roles.

When you become a corporate member of one of the engineering institutions and gain the status of IEng or CEng, you will be recognised as having finally qualified in your profession.

Requirements for Chartered and Incorporated status Before you can use either of these titles, you must satisfy certain stringent requirements, which are given below.

Chartered Engineer Education Successful completion of a four-year accredited education programme, typically an MEng degree or an honours BEng degree plus one year's further postgraduate learning.

Initial professional development Successful completion of an appropriate monitored programme of training and the gaining of experience, which is likely to take at least four years. You will normally be able to count a period of industrial training and experience taken during a sandwich course towards this requirement.

Professional review You must satisfy a review panel composed of members of the appropriate engineering institution that you are fit and competent to become a recognised member of the profession.

Membership of an institution You must be accepted for corporate membership by one of the nominated chartered engineering institutions or one of their affiliated institutions.

Age The time taken to satisfy the preceding conditions will almost certainly mean that you satisfy the final requirement, which is that you be at least 25 years old.

Incorporated Engineer Education Successful completion of a three-year accredited education programme, typically a BSc or BEng degree or Higher National qualification plus one or two years' additional matching learning.

Initial professional development Successful completion of an appropriate monitored programme of training and the gaining of experience, which is likely to take at least three years. You will normally be able to count a period of industrial training and experience taken during a sandwich course towards this requirement.

<u>Professional review</u> You must satisfy a review panel composed of members of the appropriate engineering institution that you are fit and competent to become a recognised member of the profession.

<u>Membership of an institution</u> You must be accepted for corporate membership by one of the nominated chartered engineering institutions or one of their affiliated institutions.

<u>Age</u> The time taken to satisfy the preceding conditions will almost certainly mean that you satisfy the final requirement, which is that you be at least 23 years old.

The educational requirement

The *Engineering* Guide is concerned principally with helping you find a course that will enable you to satisfy the educational requirement for professional recognition. The usual educational route begins from a sound GCSE base, followed by a course of A-levels in mathematics and physics or engineering science, leading to entry to an honours degree course. However, many universities and colleges offer foundation or access courses (see Chapter 2 in this *Introduction* – page 6), enabling students without qualifications in the appropriate subjects to get the necessary background to start the main course.

Accredited courses

The engineering profession has an established procedure for accrediting individual courses at universities. Typically, four-year accredited MEng will permit registration for CEng, while three-year accredited degree courses will lead either to CEng (following additional training following graduation), or to IEng registration. The MEng degree is becoming established as the preferred course for those with sufficient academic ability and ambition. In many cases the MEng is a separate course, but in some cases, a course can lead to the award of either an MEng or BEng degree, with the two streams separating at some point. A table in Chapter 3 of each of the *Engineering* Guides gives information about these courses, indicating where the MEng variant differs from the BEng.

The issue is further complicated by the fact that the standards for accreditation change occasionally. However, once accreditation has been granted to a course, it lasts for five years, so during the life of this Guide you may be able to enter a course that will lead to registration, despite the fact that it may no longer qualify for accreditation at today's standards.

Graduates with degrees from courses that have not received full accreditation from the relevant professional institutions are eligible for registration as Chartered Engineers through the 'individual application' route. However, those qualifying through this route are in the minority, and the procedure usually takes longer and is less straightforward than for graduates of accredited courses.

The individual subject parts of the Guide give an indication of which courses are accredited and by whom, but the process of accreditation is ongoing, so you should confirm with the professional institution whether it has approved a particular course you are considering. In particular, you should note that new courses cannot be accredited in advance, but this does not mean that the qualification will not be recognised for professional purposes if you start the course before it is accredited. You should also note that approval by a particular institution may depend on your taking certain options within a course. This is particularly likely to be the case for integrated engineering courses and those, such as mechatronics, that combine two or more branches of engineering.

Publications   Unless indicated otherwise, all items in the following list are available
from Trotman Publishing; phone 0870 900 2665, or buy online at
www.trotman.co.uk (follow the link to 'Careers Portal', then to 'Your Bookshop').
The Big Guide 2008   UCAS, 2007, £29.50
How to Complete Your UCAS Application 2008 Entry   Trotman, 2007, £11.99
Scottish Guide 2008 Entry   UCAS, 2007, £14.95
Trotman's Green Guides: Engineering Courses 2007   Trotman, 2006, £9.99
The Ultimate University Ranking Guide   C Harris. Trotman, 2004, £14.99
Student Book 2008   K Boehm & J Lees-Spalding. Trotman, 2007, £16.99
Experience Erasmus: The UK Guide   Careerscope Publications, 2006, £16.95
Taking a Year Off   Margaret Flynn. Trotman, 2002, £11.99
Students' Money Matters 2007   G Thomas. Trotman, 2007, £14.99
Mature Students' Directory   Trotman, 2004, £19.99
Disabled Students' Guide to University   E Caprez. Trotman, 2004, £21.99
Making the Most of University   K van Haeften. Trotman, 2003, £9.99

Websites

Universities and Colleges Admissions Service   www.ucas.com
Higher Education Funding Council for England   www.aimhigher.ac.uk
Higher Education and Research Opportunities in the United Kingdom   www.hero.ac.uk
Teaching Quality Information   www.tqi.ac.uk; the site gives a range of information on
courses, such as drop-out rates, student satisfaction ratings and graduate destinations
Student Loans Company   www.slc.co.uk
Student Support in Scotland   www.student-support-saas.gov.uk
Student Support for Northern Ireland   www.education-support.org.uk/students
Erasmus   www.erasmus.ac.uk
European Commission Erasmus pages   http://ec.europa.eu/; use the A–Z to select Edu-
cation, then (from the menu) Programmes and Actions, then Erasmus
Leonardo da Vinci Programme   www.europa.eu.int/comm/education/leonardo.html
The European Choice: A Guide to Opportunities for Higher Education in Europe
www.eurochoice.org.uk
Prospects Occupational Information (Higher Education Career Services Unit)
www.prospects.ac.uk; from the 'Jobs and Work' menu, select 'Explore types of jobs'.
National Bureau for Students with Disabilities   www.skill.org.uk

Engineering careers and sponsorship information

Getting into Engineering   Neil Harris. Trotman, 2002, £9.99
Get 2007: Engineering   Trotman, 2007, £9.99

Professional institutions   Information about the general regulations concerning the
academic   qualifications   required   for   entry   to   the

Introduction

24

engineering profession and the basic requirements for education and training can be obtained from the Engineering Council, 10 Maltravers Street, London WC2R 3ER (www.engc.org.uk). Detailed information can be obtained from the relevant engineering institutions; careers information is also available from them.

The Science Engineering and Manufacturing Technologies Alliance (SEMTA) publishes a number of leaflets and booklets on education, training and careers for engineers; these can be obtained from 14 Upton Road, Watford WD18 0JT (www.semta.org.uk).

Nominated chartered engineering institutions

The following professional bodies are regarded by the Engineering Council as the principal qualifying bodies in the areas covered by this Guide. They all supply information about training and careers. Other more specialised institutions are listed on the Council's website.

British Computer Society 1 Sanford Street, Swindon SN1 1HJ; www.bcs.org.uk

Chartered Institution of Building Services Engineers 222 Balham High Road, London SW12 9BS; www.cibse.org

Chartered Institution of Water and Environmental Management 15 John Street, London WC1N 2EB; www.ciwem.org

Energy Institute 61 New Cavendish Street, London W1G 7AR; www.energyinst.org.uk

Institute of Marine Engineering, Science and Technology 80 Coleman Street, London EC2R 5BJ; www.imarest.org

Institute of Materials, Minerals and Mining Library and Information Service, 1 Carlton House Terrace, London SW1Y 5DB; www.iom3.org

Institute of Measurement and Control 87 Gower Street, London WC1E 6AA; www.instmc.org.uk

Institute of Physics 76 Portland Place, London W1B 1NT; www.iop.org

Institute of Physics and Engineering in Medicine Fairmount House, 230 Tadcaster Road, York YO24 1ES; www.ipem.org.uk

Institution of Agricultural Engineers West End Road, Silsoe, Bedford MK45 4DU; www.iagre.org

Institution of Chemical Engineers Careers Liaison Department, Davis Building, 165–189 Railway Terrace, Rugby CV21 3HQ; www.icheme.org.uk or www.whynotchemeng.com

Institution of Civil Engineers Careers Department, 1 Great George Street, London SW1P 3AA; www.ice.org.uk

Institution of Engineering and Technology Educational Activities, Michael Faraday House, Six Hills Way, Stevenage, Hertfordshire SG1 2AY; www.theiet.org

Institution of Engineering Designers Courtleigh, Westbury Leigh, Westbury, Wiltshire BA13 3TA; www.ied.org.uk

Institution of Mechanical Engineers Schools and University Liaison Service, 1 Birdcage Walk, London SW1H 9JJ; www.imeche.org.uk

Institution of Nuclear Engineers 1 Penerley Road, London SE6 2LQ; www.inuce.org.uk

Institution of Structural Engineers 11 Upper Belgrave Street, London SW1X 8BH; www.istructe.org

Royal Aeronautical Society 4 Hamilton Place, London W1V 7BQ; www.raes.org.uk

Royal Institution of Naval Architects 10 Upper Belgrave Street, London SW1X 8BQ; www.rina.org.uk

Society of Operations Engineers 22 Greencoat Place, London SW1P 1PR; www.soe.org.uk

Introduction

Other professional bodies and trade organisations

Chartered Institute of Building Englemere, Kings Ride, Ascot, Berkshire SL5 7TB; www.ciob.org.uk

Chartered Institute of Housing Octavia House, Westwood Way, Coventry, Warwickshire CV4 8JP; www.cih.org

Chartered Society of Designers 5 Bermondsey Exchange, 179–181 Bermondsey Street, London SE1 3UW; www.csd.org.uk

Chemical Industries Association Kings Buildings, Smith Square, London SW1P 3JJ; www.cia.org.uk

Confederation of Paper Industries 1 Rivenhall Road, Swindon SN5 7BD; www.paper.org.uk

Institute of Operations Management University of Warwick Science Park, Sir William Lyons Road, Coventry CV4 7EZ; www.iomnet.org.uk

Institute of Wood Science Stocking Lane, Hughenden Valley, High Wycombe, Buckinghamshire HP14 4NU; www.iwsc.org.uk

Society of British Aerospace Companies Ltd Salamanca Square, 9 Albert Embankment, London SE1 7SP; www.sbac.co.uk

The Textile Institute 1st Floor, St James's Buildings, Oxford Street, Manchester M1 6FQ; www.texi.org

Sponsors can provide you with two types of assistance during your degree course. The first is industrial experience. If you are taking a sandwich course, you will probably spend the industrial period(s) working within your sponsoring company, who will organise and supervise your training. Sponsoring companies also provide industrial training during long vacations for both full-time and sandwich students.

The second benefit of sponsorship is financial. While you are working for the company you will be paid at student apprentice rates of pay. As well as income while you work for the company, some sponsors also offer bursaries to help support you during the academic period.

The benefits to the sponsor The cost of sponsoring a student is considerable, so companies expect some benefit in return, though to some extent they do feel responsibility to support higher education in general as the source of their future workforce. Many also benefit through the technical and research co-operation with universities that can grow out of contacts made through sponsorship arrangements. However, the main benefit comes from being able to identify and encourage students of high ability, train them in the company's technologies, prepare them for working in an industrial environment, and see at first hand how they respond to the training they are given. In this context, the cost of sponsorship can seem small compared with the cost of making a bad appointment at a later stage. However, it should be said that when times are hard, short-term financial pressures can mean that sponsorships are cut back.

Working for the company after graduation There is no obligation for your sponsoring company to offer you a job once the sponsorship period has been completed. However, the company will have devoted considerable resources to your training, and they will have offered you the sponsorship in the first place in the expectation that you will be a valuable asset to them in the future. If they do offer you a job, you are usually under no obligation to accept it, though you may feel there is some moral obligation to work for them for at least a year or two. By the time you have graduated, you will know far more about the sponsoring company than about any other potential employer. Of course, they will also know far more about you than about any other potential employee. So provided the experience has been reasonably successful, there is much less risk on both sides if you do take up an offer of employment.

Types of course There are more opportunities for sponsorship in engineering than in other degree subjects. Some sponsorship is tied to specific courses at specific universities, but other sponsors may allow you to choose, though this will probably be subject to their approval.

Introduction

Sponsors support full-time courses and all the varieties of sandwich course, though an individual sponsor may only support one type. Probably the commonest sponsored course is the 1–3–1 thick sandwich (or 1–4–1 for MEng courses). Sponsors also support 2–1–1 and 2–1–2 courses, but in some cases they will only begin interviewing students for sponsorship when they are in their first or second year. You should, however, bear this possibility in mind if you do not find sponsorship before the course begins.

Choosing a sponsor   Picking the sponsorships to apply for is an important decision, but you should also consider the effect it has on other choices you have to make. Usually, the best approach is to decide first which courses you want to apply for, and then investigate who might sponsor you for those courses. Given the long-term implications of a choice of course and sponsor, it is probably wise to give the short-term financial benefit the least weight in your considerations unless your financial position means that you could not take a course without this support.

If you have strong feelings about working for a particular company, it may be sensible to place getting sponsorship from them at the top of your priorities, but you should bear in mind that degree courses last at least three years, and your feelings and preferences may have changed by the time you graduate and have had more experience.

Applying for sponsorship   Sponsoring companies all have their own procedures, but in general you should be thinking and finding out about potential sponsors at the same time as you are deciding about which courses to apply for. In some cases the decisions are linked, since some sponsors only sponsor students on particular courses, and some courses may only accept sponsored students. You basically have three choices:

- You can write first to the university or college; they may help you to find an employer to sponsor you.
- You can write first to the employer; the employer may suggest a university with which there is an established relationship.
- You can apply simultaneously to both university and employer; then if you are turned down by an employer you may still be accepted by the university, and the university may try to find you another sponsor.

To find out about what companies offer, you should begin by talking to your careers teacher or adviser and to your specialist subject teachers. They should be able to tell you about previous students' experiences as well as having information about individual companies. You should also look in your careers room or library for special booklets or leaflets produced by companies and giving details of their sponsorship schemes. Once you have identified a number of companies you are interested in, write to them for further information and application forms. As for how many you should write to, remember that, apart from the time involved, there is no limit to how many sponsorships you can apply for, and that you will be facing strong competition.

Part 2

Chapter 1: Introduction

See the *Introduction* to this Guide (page 3) for an overview of how to use each chapter in the individual subject parts, including this one, together with general information about engineering and the structure of engineering courses.

The scope of this Guide   This Guide gives specific information about broadly based engineering courses. This includes courses that combine the study of more than one engineering discipline (especially mechatronics – see below) as well as those with titles such as 'Engineering', 'General engineering' and 'Integrated engineering'. There is no rigid definition of what the different titles mean, so in any particular case you should look carefully at the course description. In particular, it is important to distinguish between courses that are designed to provide an integrated cross-disciplinary approach throughout, and others that provide a general introduction allowing you to delay the choice of which specialisation to follow (see overleaf). One way that the courses in this part of the Guide may provide extra breadth is by giving a fuller understanding of underlying principles and by integrating the various engineering disciplines. Alternatively, they may incorporate studies traditionally outside the scope of engineering courses. In some cases, both of these paths are followed.

You should note that some specialised courses that include aspects of a number of technologies appear in the other parts of this Guide. For example, courses in avionics appear in *Mechanical and Manufacturing Engineering*.

Specialised and non-specialised courses   Traditionally, engineers have concentrated on a particular branch of engineering. Most engineering courses reflect this, so the other parts of this Guide describe courses devoted to the specialist engineering subjects. However, most large and complex engineering projects require expertise that crosses the traditional boundaries between the disciplines. Specialists certainly have their place within project teams, but there is also a need for team members who can integrate the contributions supplied by the different fields of engineering.

Engineering in areas of advanced technology is rapidly evolving and draws upon knowledge and techniques from many other disciplines. This means that an engineer trained in the fundamentals underlying all branches of engineering, and having a broad appreciation of diverse technologies, may well be more valuable than one who has developed a narrow technical expertise in one area. New and creative solutions in one branch of engineering are often produced by adapting techniques developed in other areas. For example, the theory developed for the analysis of electrical circuits can be used for traffic flow analysis and in the study of the structure of buildings. Novel solutions are likely to come from people with a broad understanding of engineering who can view problems from a fresh perspective and apply new techniques to them.

Delayed choice   Another reason you may opt to take an integrated engineering course is that it enables you to delay choosing which branch of engineering you are going to specialise in. In most cases you can delay both your selection of a specialised engineering area, such as mechanical, electrical, electronic or manufacturing, and the individual topics you study within the area. Each course will vary in the range of topics offered as well as the emphasis given to a particular topic (see the section *Specialisation*, page 39, and TABLE 3b for more details). You should, however, note that many of the specialised courses in the other parts of the Guide start with a general engineering introduction, and it may be possible to switch specialisations at an early stage of those courses too.

Being able to delay the point at which you specialise in a particular branch of engineering can be very valuable since you may only truly find out where your interests lie once you have started the course in earnest. However, you should not use the possibility of delaying specialisation as an excuse for not finding out as much as you can about the individual subjects: by making flexibility your highest priority, you may miss out on a specialised course that actually suits your needs and interests better.

Mechatronics   Mechatronics is the integrated study of mechanical engineering, low-power electrical and electronic engineering and computer control systems with the aim of producing sophisticated machines in a wide variety of application areas. A modern hi-fi sound system is a classic mechatronics device: precision electro-mechanical systems deal with the disc and tapes, electronics is obviously heavily involved, and microcomputers are used to interface between the control buttons on the front panel and the internal systems. Most modern devices and systems, such as cameras, video recorders, microwave ovens and personal computers, not to mention many of the systems in modern cars, could be classified as mechatronic, which all makes mechatronics a growing field within engineering.

A broad perspective   In making a decision about whether to choose a general or more specialised engineering degree, bear in mind that engineers often work in multidisciplinary teams, so they need a broad view of their subject as well as a sound knowledge of several specialisms. For example, as goods are manufactured, they may be machined, moved along belts, lifted, rotated, packaged and wrapped by mechanical equipment, so a knowledge of mechanical engineering is important. This equipment is often driven by electrical motors and controlled by electronic equipment, and the goods are weighed and measured, and their progress sensed electronically, so electronic and electrical engineering is also required. Many manufacturing operations are controlled by computer, a process known as computer-aided manufacture, so there is also a need for computer programming and for an understanding of how computers work. This is just one example; a broad engineering knowledge is required in many other fields, such as transport, chemicals, oil exploration and engineering equipment production. As a result, there are many occasions when an employer may prefer an engineer who has a broad outlook rather than particular specialist knowledge.

What do engineering graduates do? Most integrated engineering graduates go straight into employment, the large majority working in industry or commerce. However, some engineering graduates go on to further study – such as pursuing Master's courses in specialist engineering areas or commencing PhDs.

Typical employers are: engineering research and development companies; engineering design consultancies; the computing industry, including computer manufacturers and software firms; and the electrical and electronics manufacturing and telecommunications industries. Rather smaller numbers are employed in the public sector, such as local and national government and the health service.

Engineers can fill a number of different roles with these employers, including research, design and development, production engineering, and quality assurance. For the more commercially minded, there are technical sales and marketing, and for those who can write well, there is patent work or technical writing. You may also be able to make a career in engineering consultancy or in contracting – organising the logistics of major projects, managing subcontractors and supervising the installation of equipment on site.

After graduation, industrial training, and a period in a responsible engineering post, you can apply for membership of one of the engineering institutions as a chartered engineer, which confers the full status of a professional engineer. Chartered engineer status is becoming ever more important as the profession becomes increasingly internationalised, since it is fully recognised within Europe and the wider world. Chartered engineers are the leaders of the profession – the managers, creators and instigators. For further details on professional qualification, see Chapter 6 in the *Introduction* (page 20) and later in this part of the Guide (page 47).

However, in taking an engineering degree you are not committed to a career in engineering. Around 40% of graduate vacancies are open to people from any degree background. This is because many employers are interested in personal skills, intellectual ability and experience, rather than specific subject knowledge. An engineering degree is recognised as teaching a wide variety of sought-after skills, such as analytical problem-solving, teamwork, project management and organisational skills. A wide range of organisations are therefore happy to employ engineering graduates in posts where they make no direct use of engineering knowledge. For example, engineers find work in a variety of financial organisations such as banks and building societies, accountancy firms and insurance companies. Some work in the wholesale and retail commercial sector, and others are employed in sales and buying activities.

You may worry that an integrated engineering degree might be considered too generalist for you to understand any engineering subject in sufficient depth, but good graduates from these degree courses are welcomed by employers. However, technical knowledge is not all that you will need, so you should take every opportunity to develop your personal skills through your course, work experience, or participation in student sports or societies – it is often these skills and experience that can give you the edge in competing for jobs both within and outside engineering.

Integrated Engineering and Mechatronics

31

Chapter 2: The courses

TABLE 2a lists the specialised and combined courses at universities and colleges in the UK that lead to the award of an honours degree in integrated engineering or mechatronics. When the table was compiled it was as up to date as possible, but sometimes new courses are announced and existing courses withdrawn, so before you finally fill in your application you should check the UCAS website, www.ucas.com, to make sure the courses you plan to apply for are still on offer.

See Chapter 2 in the *Introduction* (page 6) for advice on how to use TABLE 2a and for an explanation of what the various columns mean.

Table 2a — First-degree courses in **Integrated Engineering and Mechatronics**

Foundation year: ● at this institution; ○ at franchised institution; ◐ second-year entry
Modes of study: ● full time; ▼ part time; ○ time abroad; ◑ sandwich
Course type: ● specialised; ◑ combined

Institution / Course title	⊙⊙⊙ see combined subject list – Table 2b	Degree	Duration (Number of years)	Foundation year	Modes of study	Modular scheme	Course type	No of combined courses
Aberdeen www.abdn.ac.uk								
Engineering		BEng/MEng	4, 5	● ◐	● ▼ ○	🌀	●	0
Engineering (integrated)		BEng	4	◐	● ▼ ○		●	0
Mechanical and electrical engineering		MEng	5	◐	●		●	0
Aston www.aston.ac.uk								
Electromechanical engineering		BEng	3, 4	●	● ○ ◑		●	0
Birmingham www.bham.ac.uk								
Engineering with business management		BEng/MEng	3, 4	●	● ○ ◑		●	0
Mathematical engineering		MSci	4	●	● ○ ◑		●	0
Blackpool and The Fylde C www.blackpool.ac.uk								
Mechatronics		BEng	3	●	● ▼		●	0
Bolton www.bolton.ac.uk								
Computer-aided engineering		BEng	3, 4	●	●		●	0
Bradford www.bradford.ac.uk								
Engineering and management		BEng	3, 4	●	● ◑		● ◑	1
Bristol www.bris.ac.uk								
Engineering design with study in industry		MEng	5		○ ◑		●	0
Engineering mathematics ⊙		MEng	4	●	○		◑	1
Bristol UWE www.uwe.ac.uk								
Engineering		BSc	3, 4	● ○	● ▼ ◑		●	0
Cambridge www.cam.ac.uk								
Engineering		BA/MEng	3, 4	●	○		●	0
Cardiff www.cardiff.ac.uk								
Integrated engineering		BEng/MEng	3, 4, 5	●	● ○ ◑		●	0
Central Lancashire www.uclan.ac.uk								
Computer-aided engineering		BEng	3		● ▼ ○ ◑		●	0
Robotics and mechatronics		BEng	3, 4		● ▼ ○ ◑		●	0
City www.city.ac.uk								
Engineering and energy management		BEng/MEng	3, 4	●	● ○		●	0
Coventry www.coventry.ac.uk								
Engineering studies		BSc	3, 4	●	● ◑		●	0
De Montfort www.dmu.ac.uk								
Engineering		BSc	3, 4	● ○	● ▼ ○ ◑		●	0

First-degree courses in **Integrated Engineering and Mechatronics**												
Institution Course title	① ② ③ see combined subject list – Table 2b	**Degree**	**Duration** Number of years	**Foundation year** ● at this institution	○ at franchised institution	◐ second-year entry	**Modes of study** ● full time; ➤ part time	○ time abroad	◑ sandwich	✿ **Modular scheme**	**Course type** ● specialised; ◑ combined	**No of combined courses**
Durham www.durham.ac.uk												
General engineering		BEng/MEng	3, 4				●	○			●	0
Edinburgh www.ed.ac.uk												
Electrical and mechanical engineering		BEng/MEng	4, 5			◐	●	○			●	0
Exeter www.exeter.ac.uk												
Engineering		BEng/MEng	3, 4				●	○		✿	● ◑	1
Glamorgan www.glam.ac.uk												
Mechatronic engineering		BEng/MEng	3, 4	●			●		◑		●	0
Glasgow Caledonian www.gcal.ac.uk												
Mechanical electronic systems engineering		BEng	4, 5			◐	● ➤	○	◑		●	0
Mechatronics		BSc	4			◐	●	○			●	0
Greenwich www.gre.ac.uk												
Mechatronics		BEng	3	●			●		◑		●	0
Harper Adams UC www.harper-adams.ac.uk												
Engineering design and development		BSc/BEng	3, 4				●		◑		●	0
Hertfordshire www.herts.ac.uk												
Computer-aided engineering		BEng	4, 5				●		◑		●	0
Huddersfield www.hud.ac.uk												
Computer-aided engineering		BEng/MEng	3, 4, 5	●			●	○	◑		●	0
Engineering with technology management		BEng	3, 4				● ➤	○	◑		●	0
Technology with business management		BSc	3, 4				●	○	◑		●	0
Hull www.hull.ac.uk												
Computer-aided engineering		BEng	3		○		●				●	0
King's College London www.kcl.ac.uk												
Mechatronics		BEng/MEng	3, 4				●				●	0
Lancaster www.lancs.ac.uk												
Engineering		BEng	3, 4				●	○	◑		●	0
Mechatronics		BEng/MEng	3, 4				●		◑		●	0
Leeds www.leeds.ac.uk												
Mechatronics and robotics		BEng/MEng	3, 4	●			●	○			●	0
Leicester www.le.ac.uk												
General engineering		BEng/MEng	3, 4, 5	●			●	○	◑		●	0
Liverpool www.liv.ac.uk												
Engineering ①		BEng/MEng	3, 4	●			●	○			● ◑	1
Mechatronics and robotic systems		BEng/MEng	3, 4	●	○		●				●	0
Liverpool John Moores www.ljmu.ac.uk												
Mechatronics		BEng	3, 4	●	○		●		◑	✿	●	0
London South Bank www.sbu.ac.uk												
Computer-aided engineering		BSc	3, 4				●		◑		●	0
Engineering		BEng	3, 4				●		◑		●	0
Mechatronics		BEng	3, 4				●		◑		●	0
Manchester www.man.ac.uk												
Mechatronic engineering		BEng/MEng	3, 4, 5	●			●		◑		●	0
Manchester Metropolitan www.mmu.ac.uk												
Engineering		BEng	3, 4	●			● ➤	○	◑		●	0
Napier www.napier.ac.uk												
Engineering with management ①		BEng	4			◐	●	○	◑		● ◑	1
Mechatronics		BEng	4				●				●	0

(continued) Table 2a — First-degree courses in **Integrated Engineering and Mechatronics**

Institution / Course title	①②③ see combined subject list – Table 2b	Degree	Duration (Number of years)	Foundation year (● at this institution; ○ at franchised institution; ◐ second-year entry)	Modes of study (● full time; ▼ part time; ○ time abroad; ◐ sandwich)	Modular scheme	Course type (● specialised; ◐ combined)	No of combined courses
Newcastle www.ncl.ac.uk								
Mechanical engineering and mechatronics		MEng	4	●	● ○		◐	1
Northampton www.northampton.ac.uk								
Engineering		BSc	3		● ▼ ○ ◐		●	0
Northumbria www.northumbria.ac.uk								
Computer-aided engineering		BEng	3, 4	●	● ◐		●	0
Engineering with business studies		BSc	3, 4	●	○ ◐		●	0
Nottingham www.nottingham.ac.uk								
Integrated engineering		BEng/MEng	3, 4	●	●		● ◐	3
Oxford www.ox.ac.uk								
Engineering		MEng	4		●		● ◐	1
Queen Mary www.qmul.ac.uk								
Engineering		BSc(Eng)	3	●	● ○		●	0
Engineering science		BEng	3	●	● ○		●	0
Richmond www.richmond.ac.uk								
Computer-aided engineering		BSc	4	●			●	0
Robert Gordon www.rgu.ac.uk								
Mechanical and electrical engineering		BEng/MEng	4, 5	◐	● ▼			0
Sheffield www.sheffield.ac.uk								
Mechatronics		BEng/MEng	3, 4	●	●		● ◐	2
Southampton www.soton.ac.uk								
Electromechanical engineering		BEng/MEng	3, 4	● ○	●	○ ◐		0
Southampton Solent www.solent.ac.uk								
Engineering with business		BEng	3		●		●	0
Staffordshire www.staffs.ac.uk								
Mechatronics		MEng	3, 4, 5	● ○	● ▼ ○ ◐		●	0
Strathclyde www.strath.ac.uk								
Electrical and mechanical engineering		BEng/MEng	4, 5	◐	● ○		●	0
Surrey www.surrey.ac.uk								
Sustainable systems engineering		BEng/MEng	3, 4, 5	●	● ○ ◐		●	0
UCL www.ucl.ac.uk								
Engineering with business finance		BEng/MEng	3, 4, 5		● ○ ◐		● ◐	1
Ulster www.ulster.ac.uk								
Engineering		BEng/MEng	3, 4		● ▼ ◐		● ◐	1
Warwick www.warwick.ac.uk								
Combined technology		BEng	3		●		●	0
Engineering ①		BEng/MEng	3, 4, 5		● ○ ◐		● ◐	1
Systems engineering		BEng/MEng	3, 4, 5		● ○ ◐		●	0
Wolverhampton www.wlv.ac.uk								
Mechatronics		BEng	3, 4	○	●	◐ ✿	●	0

Subjects available in combination with integrated engineering

TABLE 2b shows those subjects that can be taken in roughly equal proportions with integrated engineering or mechatronics in the combined degrees listed in TABLE 2a. See Chapter 2 in the *Introduction* (page 6) for general information about combined courses and for an explanation of how to use TABLE 2b.

Subjects to combine with **Integrated Engineering or Mechatronics**	
Business studies Sheffield, Ulster, Warwick ①	**Materials engineering** Nottingham
Computer science Oxford	**Mathematics** Bristol ①
Finance UCL	**Mechanical engineering** Newcastle,
Management studies Bradford, Exeter,	Nottingham
Napier ①, Sheffield	**Product design** Liverpool ①
Manufacturing engineering Nottingham	

Other courses that may interest you   The following courses are not included in any of the Guides in the series, but are related to the courses in this part of the Guide. These courses may include those where you spend less than half your time studying engineering, as well as courses providing a more intensive study of a specialised aspect of engineering.

- Technology management (Bradford)
- Medical/biomedical engineering (Birmingham, Cardiff, Leeds, Queen Mary and Surrey).

You may also find courses of interest in the other parts of this Guide and in the following Guides:

- *Biological Sciences* (particularly for biotechnology)
- *Physics and Chemistry* (for applied physics)
- *Mathematics, Statistics and Computer Science.*

Industrial experience and time abroad Many courses provide a range of opportunities for spending a period of industrial training in the UK or abroad or of study abroad. TABLE 3a gives information about the possibilities for the courses in this Guide. See Chapter 3 in the *Introduction* (page 11) for further information.

Table 3a — Time abroad and sandwich courses

Institution / Course title ① ②: see notes after table	Named 'international' variant of the course	Location: ● Europe; ○ North America; ▼ industry; ◐ academic institution	Maximum time abroad (months)	Time abroad assessed	Language study: ○ optional; ● compulsory; * contributes to assessment	Socrates–Erasmus	Sandwich courses: ● thick; ○ thin	Arranged by: ● institution; ○ student
Aberdeen								
Engineering		● ○ ▼ ◐	12	●	○*	●		
Engineering (integrated)	●	● ○ ▼ ◐	12	●	○*	●		
Aston								
Electromechanical engineering ① ② ④	●	● ○ ▼ ◐	15	●	●	●		●
Birmingham								
Engineering with business management ① ③		● ○ ▼ ◐	12	●			● ●	●
Mathematical engineering ① ③		● ○ ▼ ◐	12	●			● ●	●
Bolton								
Computer-aided engineering						●		
Bradford								
Engineering and management							●	● ○
Bristol								
Engineering design with study in industry		● ○ ▼	12	●	○*		○	● ○
Engineering mathematics	●	● ◐	12	●	○*		●	○
Bristol UWE								
Engineering ③							●	○
Cambridge								
Engineering ① ③ ⑤		○ ◐	8	●	○*		●	○
Cardiff								
Integrated engineering							●	○
Central Lancashire								
Computer-aided engineering		● ▼ ◐	12		○*		● ●	○
Robotics and mechatronics ③		● ○ ▼ ◐	12	●	○*		● ●	○
De Montfort								
Engineering							●	○
Durham								
General engineering ③	●	● ○ ▼ ◐	12	●	○			
Edinburgh								
Electrical and mechanical engineering						●		
Exeter								
Engineering ③		● ○ ◐	6	●	○*	●		
Glamorgan								
Mechatronic engineering ③	●	▼	12		○		● ●	
Glasgow Caledonian								
Mechanical electronic systems engineering ③	●	● ○ ▼ ◐	14	●	○		● ●	● ○
Mechatronics	●		12					

Integrated Engineering and Mechatronics

Time abroad and sandwich courses

Institution / Course title ① ②: see notes after table	Named 'international' variant of the course	Location: ● Europe; ○ North America; ◗ industry; ◑ academic institution	Maximum time abroad (months)	Time abroad assessed	Language study: ○ optional; ● compulsory; * contributes to assessment	Socrates–Erasmus	Sandwich courses: ● thick; ○ thin	Arranged by: ● institution; ○ student
Greenwich								
Mechatronics ③							●	●
Hertfordshire								
Computer-aided engineering		● ○ ◗ ◑	12	●		●	●	● ○
Huddersfield								
Computer-aided engineering		● ○ ◗	12	●		●	● ●	●
Engineering with technology management		● ○ ◗ ◑	12		○	●	● ●	● ○
Technology with business management ③		●	12			●	● ●	●
King's College London								
Mechatronics ②					○			
Lancaster								
Engineering		○ ◑	12	●				
Leeds								
Mechatronics and robotics ③		● ○ ◑	12	●	○*	●		
Leicester								
General engineering ③	●	● ○ ◗ ◑	12	●	○	●		○
Liverpool								
Engineering ③					○			
Mechatronics and robotic systems ③		● ◗	12		○			
Liverpool John Moores								
Mechatronics ③						●	●	
Manchester								
Mechatronic engineering ③						●	●	● ○
Manchester Metropolitan								
Engineering	●	● ◗ ◑	12		○*	●	●	●
Napier								
Engineering with management		● ◗	6		○		○ ●	
Newcastle								
Mechanical engineering and mechatronics ① ③		● ◑	9	●	○*	●		
Northampton								
Engineering ③		◗	12				●	
Northumbria								
Engineering with business studies		●	12				●	○
Nottingham								
Integrated engineering ③								
Oxford								
Engineering ③			0		○*			
Queen Mary								
Engineering	●	● ○	12			●		
Engineering science	●	● ○	12		○	●		
Robert Gordon								
Mechanical and electrical engineering ③								
Southampton								
Electromechanical engineering ③		● ◑	8	●	○*	●		
Staffordshire								
Mechatronics ③		● ◗ ◑	18	●	○*	●	● ●	● ○
Strathclyde								
Electrical and mechanical engineering		● ○ ◑	6	●	○*	●		

Integrated Engineering and Mechatronics

(continued) Table 3a Time abroad and sandwich courses

Institution / Course title ① ⑤: see notes after table	Named 'international' variant of the course	Location: ● Europe; ○ North America; ◗ industry; ◐ academic institution	Maximum time abroad (months)	Time abroad assessed	Language study: ○ optional; ● compulsory; * contributes to assessment	Socrates–Erasmus	Sandwich courses: ● thick; ○ thin	Arranged by: ● institution; ○ student
Surrey								
Sustainable systems engineering	●	○ ◗ ○	12	●		●	●	●
UCL								
Engineering with business finance ① ② ③		◗	12		○*		●	○
Ulster								
Engineering	●	◗	18	●	○*	●	●	●
Warwick								
Combined technology					○*			
Engineering ④	●	○ ◗ ○	12	●	○*	●	●	○
Wolverhampton								
Mechatronics							●	

① A year of industrial experience before the course starts is recommended for students on non-sandwich courses
② A year of industrial experience before the course starts is recommended for students on sandwich courses
③ Students on non-sandwich courses can take a year out for industrial experience
④ Other patterns of sandwich course : Aston <u>Electromechanical engineering</u> 1–3–1 also available
⑤ Minimum period of vacation industrial experience for non-sandwich students (weeks) : Cambridge <u>Engineering</u> 8

Course content   Courses in engineering build directly from a study of mathematics and the physical sciences. However, the teaching of these subjects is placed in an engineering context. From the start of the course, the aim is to give a thorough knowledge of the practice of engineering, and to develop an understanding of how machines, engines, circuits, structures, systems and materials work and behave.

For the first year or two, you will study topics in the physical sciences including electricity (and electronics), mechanics, thermodynamics, fluid mechanics, structures (or structure of materials) and materials science – a range of topics similar to those found in the specialist engineering degrees. Most courses also include some systems engineering and management science in their first two years, and many courses, especially in mechatronics, include a good deal of computing.

All courses consider engineering applications, but the emphasis given to them varies considerably between courses that could be described as applications-based and others that focus more on the basic engineering science.

Drawing, or 'design communication', is an important part of many courses, and is mostly done using computers (computer-aided design, or CAD) rather than more traditional methods. It is usually taught in the first year, though the amount of time devoted to it varies greatly between courses, so if you are particularly interested in this area, you should check how these techniques are taught in the courses.

Generally speaking, in the first year, and to some extent in the second year, the courses in integrated engineering show greater similarities than differences, though this is less true of the combined courses, which can be highly individual.

Specialisation   Most courses begin with a general engineering introduction, which is often shared with other engineering courses. TABLE 3b shows where this is the case and at what point it is possible to specialise in a particular branch of engineering. TABLE 3b shows where a particular branch of engineering is available as a final-year specialisation. It shows where a subject is a compulsory (●) or optional (O) component, where there are both compulsory and optional components (◑) and where a particular engineering branch can form a major specialisation occupying at least 50% of your time in the final year (●* O* ◑*). The subject headings may not exactly reflect those given in prospectuses but indicate the engineering area covered.

Table 3b — Final-year course content

O optional; ● compulsory; ◑ both; ① ② : see notes after table; * major area of specialisation

Institution	Course title	Chemical	Civil	Structural	Environmental	Electrical	Electronic	Control	Systems	Materials	Mechanical	Manufacturing	Aeronautical	Start of specialisation
Aberdeen	Engineering ①		O*	O*	O	O*	O*	O		O	O*			Year 3
	Engineering (integrated)		●	O		●	●	O		O	●			Year 3
Aston	Electromechanical engineering ①				O	O	O	O			O			Year 2
Birmingham	Engineering with business management ①										●*		●*	Level 1
	Mathematical engineering ②							O*			●*	●*		
Blackpool and The Fylde C	Mechatronics					●	●	●	●	●	●	●		
Bolton	Computer-aided engineering ①													Year 2
Bristol	Engineering design with study in industry ①		O*	O*		O*	O*	O*		O*	O*		O*	
	Engineering mathematics		O	O		◑	◑	◑	O		◑		◑	
Bristol UWE	Engineering ①				O	O	O	O		O	O	O	O	Year 2
Cambridge	Engineering ①	O*	O*	O*	O	O*	O*	O*	O	O*	O*	O*	O*	Year 3
Cardiff	Integrated engineering				O	●	●	●	●	O	●	●		Semester2
Central Lancashire	Computer-aided engineering ①							O			◑	●		Year 2
City	Engineering and energy management ①										●			Year 3
De Montfort	Engineering ①					●	●				●	●		
Durham	General engineering		◑	◑		◑	◑	◑		◑	◑	◑	◑	Year 3
Edinburgh	Electrical and mechanical engineering ①			O		●	O	O	O	O	●	O		Year 2
Exeter	Engineering ①		O*	O	O		O*	O	O	O	O*	O		Year 2
Glamorgan	Mechatronic engineering					●	●	●	●		●			
Glasgow Caledonian	Mechanical electronic systems engineering		●				●	●	●		●	●		Year 3
	Mechatronics					●	●	●			●			Year 3
Greenwich	Mechatronics						O	O	O		O	O		Semester 2
Hertfordshire	Computer-aided engineering ①						●*		●*			●*		
Huddersfield	Computer-aided engineering					●	●	●	●		●	●	●	Year 2
	Engineering with technology management					◑	◑*	◑*	◑		◑	◑*		Final year
	Technology with business management ①									●	●	●	●	Year 1
Hull	Computer-aided engineering ①			●*							●*	●*		
King's College London	Mechatronics						◑		◑	◑	◑	◑	◑*	Year 2
Leeds	Mechatronics and robotics						◑	◑	◑		◑			
Leicester	General engineering ①				O	O	◑	◑	◑	O	O	◑		Semester 4
Liverpool	Engineering					◑*	◑*	O		◑	◑*	O		Semester 2
Liverpool John Moores	Mechatronics					●	●	●		O	●	O		Level 2
Manchester	Mechatronic engineering					◑*	◑*	◑*	◑*					

(continued) Table 3b — Final-year course content

Key: ○ optional; ● compulsory; ◐ both; * major area of specialisation; ① ②: see notes after table

Institution	Course title	Chemical	Civil	Structural	Environmental	Electrical	Electronic	Control	Systems	Materials	Mechanical	Manufacturing	Aeronautical	Start of specialisation
Manchester Metropolitan	Engineering ①					○	○	○			●	●	○	Term 2
Newcastle	Mechanical engineering and mechatronics ①					●	●	●	●	●	●	●	●	Year 3
Northampton	Engineering					●	●	●	●		●	●		
Northumbria	Engineering with business studies							○			○	○		
Nottingham	Integrated engineering									◐	◐	◐		Year 2
Oxford	Engineering ①	○*	◐*	◐*		◐*	◐*	◐*		◐	◐*	○*		Year 3
Queen Mary	Engineering		○*		○*			○*		○*	○*		○*	
	Engineering science		○*					○*		○*	○*		○*	Year 2
Robert Gordon	Mechanical and electrical engineering					●*	●*	●*	●*		●*			Semester 2
Southampton	Electromechanical engineering ①					●	●	●	●	◐	●	●		Year 2
Staffordshire	Mechatronics				○			●	●	●	●	○		Year 2
Strathclyde	Electrical and mechanical engineering					○	○	○	○	○	○			
Surrey	Sustainable systems engineering ①	●*			●*					●*				
UCL	Engineering with business finance					○	○	○			●*	○		
Ulster	Engineering					●	○	●	●		●	○		
Warwick	Combined technology		○	○	○	○	○	○	○	○	○	○		Year 1/2
	Engineering		○*	○		○	○*	○	○*	○	○*	○*		Year 1/2
Wolverhampton	Mechatronics ①							○*	○*			○		Semester 2

Aberdeen ① Safety and reliability engineering
Aston ① Nuclear power; thermodynamics; finite element analysis
Birmingham ① Business/management; CAD; computer-integrated manufacturing ② Thermofluids; plasticity; material forming; mechatronics; Lagrangian and Hamiltonian mechanics; geometric modelling
Bolton ① CAD/CAM; modelling
Bristol ① Communications; design information systems; dynamic systems modelling process; software; water resources
Bristol UWE ① Music technology
Cambridge ① Information engineering
City ① Energy management
De Montfort ① Mechatronics; IT; business studies
Edinburgh ① Project management; CAD/CAM; finite element analysis; computational fluid dynamics
Exeter ① Engineering and management
Hertfordshire ① Human and financial resource management; reliability engineering

Huddersfield ① Quality management; marketing; customer care programmes
Hull ① Engineering management; CAD/CAM
Leicester ① Quality and reliability
Manchester Metropolitan ① Robotics and automation; mechatronics; instrumentation; heat transfer; fluids and aerodynamics; management of quality; stress analysis
Newcastle ① Automatic control; distributed control systems; robotics; mechanical power transmissions; image processing and machine vision; high performance embedded systems
Oxford ① Information engineering
Southampton ① Power systems technology; automation and robotics; acoustics and noise control
Surrey ① Transport systems; business and environmental management
Wolverhampton ① Communication networks; expert systems

BEng/MEng courses TABLE 2a lists many courses that can lead to the award of either a BEng or an MEng degree. For these courses, most of the tables in this Guide give information specifically for the BEng stream (the tables show information for an MEng course if the course is only available as MEng). Much of the information will also apply to the MEng stream, but TABLE 3c shows you where there are differences for the MEng stream. It shows at what point the MEng course separates from the BEng (for those courses where they are not separate from the start), and what proportion of students are expected to leave with an MEng degree.

Table 3c	MEng course differences						
Institution	Course title	MEng separates from BEng	MEng proportion %	More engineering	More management	More languages	Other differences
Aberdeen	Engineering	Year 3	30	●	●		More extensive group project; engineering analysis and methods; engineering and project management
Aston	Electromechanical engineering						
Birmingham	Engineering with business management	Level 3	60	●	●		
Cambridge	Engineering		100				Exit after 3 years with BA exceptional
Cardiff	Integrated engineering	Year 2	30	●	●		
City	Engineering and energy management	Year 3	20	●	●		
Durham	General engineering	Year 3					MEng: focus on theory, innovation and management; BEng: focus on high quality engineering design
Edinburgh	Electrical and mechanical engineering	Year 4	75	●			Industrial/European placement; group working exercise
Exeter	Engineering	Year 3	20	●	●		Major group design project; more independent learning
Huddersfield	Computer-aided engineering	Year 4	5	●	●		Major group project
King's College London	Mechatronics	Year 2	15	●			
Leicester	General engineering	Year 3	25	●	●		Major group design project
Liverpool	Engineering	Year 3	30	●	●		Optional assessed industrial project placement in semester 2 of year 3; group project work in years 3 and 4
Manchester	Mechatronic engineering	Year 2	25	●	●		Year 4 group project; enterprise studies
Nottingham	Integrated engineering	Year 3	50	●	●		More project work
Robert Gordon	Mechanical and electrical engineering	Year 3	10	●	●		Multidisciplinary project year 5
Sheffield	Mechatronics	Year 3					
Southampton	Electromechanical engineering	Year 3	30	●	●	●	MEng do individual project in year 3 and group projects in year 4
Staffordshire	Mechatronics	Year 4	20	●			
Strathclyde	Electrical and mechanical engineering	Year 3	60	●	●		Year 5 group project
Surrey	Sustainable systems engineering			●	●		More project work at later levels
UCL	Engineering with business finance	Year 3	40	●			Year 4 group design project
Ulster	Engineering	Year 2	20	●	●		Students can obtain German Diplom Ingeneur
Warwick	Engineering	Year 3	60	●			Optional year overseas; multidisciplinary group project in final year; intercalated year or year in research; wide range of final-year electives

General studies and additional subjects   The courses in this part of the Guide are very broad, and as well as a wide range of engineering subjects, you will almost certainly study some management and business studies. If you want to combine your study of engineering with an unrelated subject in more or less equal proportions, look up the combined subject in TABLE 2b. A few engineering courses form part of a modular degree scheme (see TABLE 2a), on which there are more opportunities to take a wide range of other subjects, though the requirements for professional qualification may mean that the choice is more restricted than it would be if you were following a course in a non-engineering subject.

See Chapter 4 in the *Introduction* (page 16) for general information about teaching and assessment methods used in all types of engineering course, including those covered in this part of the Guide. It also explains how to interpret TABLE 4, which gives information about the teaching and assessment methods used on individual courses.

Table 4 — Assessment methods

Key for frequency of assessment column: ◑ term; ◒ semester; ○ year

Institution	Course title	Frequency of assessment	Years of exams contributing to final degree (years of exams not contributing to final degree)	Coursework: minimum/maximum %	Project/dissertation: minimum/maximum %	Time spent on projects in: first/intermediate/final years %			Group projects: ● compulsory; ○ optional	Orals: ◑ if borderline; ● everyone; ○ for projects; ○ everyone
Aberdeen	Engineering	◑	(1),(2),3,4	15/20	39/42	20	20	50	●	○
	Engineering (integrated)	◑	(1),(2),3,4	15/20	39/42	20	20	50	●	○
Aston	Electromechanical engineering	○	1,2,3,4,5	20/20	20/20	40	15	30	●	◐○
Birmingham	Engineering with business management	◐	(1),2,3,4	30/40	14/14	20	25	50	●	○
	Mathematical engineering	◐	(1),2,3,4	10/15	14/14	10	15	30	●	◐○
Blackpool and The Fylde C	Mechatronics	◑	1,2,3		20/33	0	33	33	●	○
Bolton	Computer-aided engineering	◑	(1),2,3	30/40	40/50	10	25	30	●	○
Bradford	Engineering and management	◑	(1),2,3,4							
Bristol	Engineering design with study in industry	○	(1),2,3,4,5	15/25	25/25	10	30	30	●	◐○
	Engineering mathematics	○	(1),2,3,4	0/50	25/30	10	15	30	●	◐○
Bristol UWE	Engineering	◑	(1),2,3	0/40	25/40	17	17	25	○	○●
Cambridge	Engineering	○	(1),(2),3,4	0/20	50/50	5	20	50	●	○
Cardiff	Integrated engineering	◑	(1),2,3	8	16	15	15	25	●	◐○
Central Lancashire	Computer-aided engineering	○	(1),2,3	65/75	10/10	30	60	80		○●
	Robotics and mechatronics	○	(1),2,4	20/20	20/20	10	25	34		○
City	Engineering and energy management	○	(1),2,3	20/25	20/25	5	10	25	●	○
De Montfort	Engineering	◑	1,2,3,4	20/50	50/50	12	25	50		◐○
Durham	General engineering	○	(1),2,3,4	5/10	25/25	5	10	50	●	◐○
Edinburgh	Electrical and mechanical engineering	◑	(1),(2),3,4,5	22/22	26/26	0	15	30	●	◐○
Exeter	Engineering	◑	(1),2,3,4	15/30	25/30	0	25	25	●	○●
Glamorgan	Mechatronic engineering	◑	(1),2,4,5	30/50	25/30	10	20	30	●	◑
Glasgow Caledonian	Mechanical electronic systems engineering	◑	(1),(2),3,(4),5	20/30	33		10	33		◐○
	Mechatronics	◐	1,2,3,4		20/30	10	10	20		
Greenwich	Mechatronics	◐	(1),2,3	20/20	25/25	15	15	25		◐○
Hertfordshire	Computer-aided engineering	◐	(1),2,4,5	20/25	35/40	10	20	25		○
Huddersfield	Computer-aided engineering	○	(1),2,4	30/50	15/15	50	50	60	●	○●
	Engineering with technology management	○	(1),(2),3,4,5	22/22	33/33	0	16	33	●	◐○●
	Technology with business management	◑	(1),2,3,4	30/40	25/30	25	25	35	●	○●
Hull	Computer-aided engineering	◑	(1),2,3,4	30/50	20/35	10	30	50	●	◐○●
King's College London	Mechatronics	◑	1,2,3,4	32/38	21/21	6	10	30	●	
Leeds	Mechatronics and robotics	◑	(1),2,3,4	5/10	25/33	25	25	25	○	◐○
Leicester	General engineering	◑	1,2,3,4	10/20	10/20	10	20	35	●	○
Liverpool	Engineering	◑	(1),2,3,4	15/25	15/25	15	15	30	●	○
	Mechatronics and robotic systems	◑	(1),2,3	15/15	25/25	20	20	25		◐○

Integrated Engineering and Mechatronics

43

(continued) Table 4

Assessment methods

Institution	Course title	Frequency of assessment (Key: ◑ term; ◐ semester; ○ year)	Years of exams contributing to final degree (years of exams not contributing to final degree)	Coursework: minimum/maximum %	Project/dissertation: minimum/maximum %	Time spent on projects in: first/intermediate/final years %			Group projects: ● compulsory; ○ optional	Orals: ◐ if borderline; ○ for projects; ● everyone
Liverpool John Moores	Mechatronics	◐	(1),**2,3**	20/**50**	30/**30**	20	20	30	○	
Manchester	Mechatronic engineering	◐	(1),**2,3,4**	20/**20**	15/**50**	10	20	25		
Manchester Metropolitan	Engineering	◑	(1),**2,3**	40/**60**	/**25**	3	5	20		○
Napier	Engineering with management	◐	(1),(2),**3,4**	30/**30**	25/**25**	10	15	25	○	○●
Newcastle	Mechanical engineering and mechatronics	◐	(1),**2,3,4**	7/**7**	24/**24**	5	25	33	●	○
Northampton	Engineering	○	(1),**2,3**			0	0	33		
Northumbria	Engineering with business studies	◐	(1),**2**,(3),**4**							
Nottingham	Integrated engineering	◐	(1),**2,3,4**	20/**30**	33/**33**	10	20	33		
Oxford	Engineering		(1),**3,4**	13/**13**	25/**25**	10	15	50	●	◐
Queen Mary	Engineering	○	**1,2,3**	20						
	Engineering science	○	**1,2,3**	10	20	0	25	25		●○
Robert Gordon	Mechanical and electrical engineering	◐	(1),(2),**3,4,5**	**20**	**25**	33	20	25	●	○◐●
Sheffield	Mechatronics	◐	(1),**2,3,4**							
Southampton	Electromechanical engineering	◐	(1),**2,3,4**	10/**10**	20/**20**	5	15	25	●	◐○
Staffordshire	Mechatronics	◐	(1),**2,3**	20/**30**	40/**40**	10	20	40	●	○●
Strathclyde	Electrical and mechanical engineering	◐	(1),(2),**3,4,5**	0/**10**	25/**25**	0	15	25	●	◐○
Surrey	Sustainable systems engineering	◐	(1),**2,3,4**			10	10	25	●	
UCL	Engineering with business finance	○	**1,2,3,4**	25/**33**	20/**20**	10	10	25	●	○
Ulster	Engineering	◐	(1),(2),**4,5**	17/**25**	25/**42**	16	16	25	●	◐
Warwick	Combined technology	○	(1),**2,3**	18/**24**	18/**18**	12	12	12	●	
	Engineering	○	(1),**2,3**	25/**50**	18/**18**	12	12	25	●	○●
Wolverhampton	Mechatronics	◐	(1),**2,3**	30	16			25		

See Chapter 5 in the *Introduction* (page 18) for general information about entrance requirements that applies to all types of engineering course, including those covered in this part of the Guide. It also explains how to interpret TABLE 5, which gives information about entrance requirements for courses in integrated engineering and mechatronics.

Table 5 — Entrance requirements

Institution	Course title	Number of students (includes other courses)	Typical offers (BSc/BEng)			Typical offers (MEng)			● compulsory, ○ preferred A-level Mathematics	A-level Physics
			UCAS tariff points	A-levels	SQF Highers	UCAS tariff points	A-levels	SCQF Highers		
Aberdeen	Engineering	(170)		CCD	ABBBC		BCC	ABBBC	●	●
	Engineering (integrated)	(170)		CCD	ABBBC		BCC	ABBBC	●	●
	Mechanical and electrical engineering						BC	AABB		
Aston	Electromechanical engineering	10	240–280		BBBBC				●	○
Birmingham	Engineering with business management	20	280	ABB	ABBBB	320	ABB	AABBB	●	○
	Mathematical engineering	15					ABB	ABBBB	●	●
Blackpool and The Fylde C	Mechatronics	20	160						○	○
Bolton	Computer-aided engineering	15	80						○	○
Bradford	Engineering and management	(80)	200–240						○	○
Bristol	Engineering design with study in industry	25					AAB	AAAAB	●	●
	Engineering mathematics	(20)					ABB	AAAAB	●	○
Bristol UWE	Engineering	25	140–180						●	○
Cambridge	Engineering	(300)		AAA			AAA		●	●
Cardiff	Integrated engineering	20	240	BBC		240	ABB		●	○
Central Lancashire	Computer-aided engineering	(60)	180						●	
	Robotics and mechatronics	25	200						●	○
City	Engineering and energy management	10		BCD			BBC		●	●
Coventry	Engineering studies	25	160							
De Montfort	Engineering	200	120–180						○	○
Durham	General engineering	(140)		RRR	AAAA		AAR	AAAA	●	○
Edinburgh	Electrical and mechanical engineering	10		BBB	BBBB		BBB	BBBB	●	○
Exeter	Engineering	25	240			300			●	○
Glamorgan	Mechatronic engineering	16	260			260			●	○
Glasgow Caledonian	Mechanical electronic systems engineering	10		BC	BBBC				●	●
	Mechatronics	15		CD	BBC				○	○
Greenwich	Mechatronics	20	240	CCC	CCCC				●	○
Harper Adams UC	Engineering design and development		240		BBBC					
Hertfordshire	Computer-aided engineering		240–280							
Huddersfield	Computer aided engineering	5	220		DDDD	260		AADD	●	○
	Engineering with technology management	40	160						○	○

Integrated Engineering and Mechatronics

(continued) Table 5

Entrance requirements

| Institution | Course title | Number of students (includes other courses) | Typical offers (BSc/BEng) | | | Typical offers (MEng) | | | ● compulsory; O preferred A-level Mathematics | A-level Physics |
			UCAS tariff points	A-levels	SQF Highers	UCAS tariff points	A-levels	SCQF Highers		
Huddersfield (continued)	Technology with business management	15	220		BBBB					
Hull	Computer-aided engineering	10	220–280						●	O
King's College London	Mechatronics	20		BBB	AABBB		BBB	AABBB	●	O
Lancaster	Engineering	(90)		BBC	AAAB				●	O
	Mechatronics	(90)		BBC	AAAB		ABB	AABBB	●	O
Leeds	Mechatronics and robotics	20		BBB	BBBBB		ABB		●	●
Leicester	General engineering	(90)	260			300			●	O
Liverpool	Engineering	20	280	BBC	AAAB	340	ABB	AAAAA	●	●
	Mechatronics and robotic systems	6	280		AAAB	320		AAAAB	●	●
Liverpool John Moores	Mechatronics	15	140–200						●	●
London South Bank	All courses			CC	BBB					
Manchester	Mechatronic engineering	30	320	ABB	AABBB	320	ABB	AABBB	●	●
Manchester Metropolitan	Engineering	30	240						●	O
Napier	Engineering with management	20	220	CD	BCC				●	O
	Mechatronics		220	CD	BCC					
Newcastle	Mechanical engineering and mechatronics	(10)				320	ABB		●	O
Northampton	Engineering		180–220							
Northumbria	Computer-aided engineering		240							
	Engineering with business studies		240		CCC					
Nottingham	Integrated engineering	30		BBB			ABB		●	●
Oxford	Engineering	(170)					AAA	AAAAB	●	●
Queen Mary	Engineering	20	260–340		BBBCC				●	O
	Engineering science	5	260–340		BBBCC				●	●
Richmond	Computer-aided engineering		260							
Robert Gordon	Mechanical and electrical engineering	100	220–240	CCD	BBCC		CCD	BBCC	●	O
Sheffield	Mechatronics			AAB	AABB		AAB	AABB	●	O
Southampton	Electromechanical engineering	20	350	ABB		370	ABB		●	●
Southampton Solent	Engineering with business		120							
Staffordshire	Mechatronics	50	280			280			●	●
Strathclyde	Electrical and mechanical engineering	30		BBC	AABBC		BBB	AAABC	●	●
Surrey	Sustainable systems engineering	20	260	CCC	BBBB	300	BBB	BBBB	●	
UCL	Engineering with business finance	30		BBB			AAB		●	O
Ulster	Engineering	15	260			300			●	O
Warwick	Combined technology	10		BBC					O	O
	Engineering	60		BBB			AAB		O	O
	Systems engineering			BBC			ABB			
Wolverhampton	Mechatronics	30	160–220						O	

See Chapter 6 in the *Introduction* (page 20) for general information about professional qualification that applies to all types of engineering course, including those covered in this part of the Guide.

The breadth of the courses in this part of the Guide means that a single course may be accredited by several institutions, though the accreditation by an individual institution may be conditional on your having taken certain options, which may not coincide with those required by another institution. TABLE 6 shows which institutions have accredited individual courses. The Institution of Engineering and Technology accredits courses at both Chartered and Incorporated level, so the table shows these as ● and ○, respectively.

Remember that the process of accreditation goes on continuously, so you should check with the professional institutions for the current position on specific courses. Information has not been supplied for courses that are not included in the table. You should also bear in mind that two courses with similar names at the same university may have different accreditation, so it is vital to check the details for the precise course you are interested in.

See Chapter 7 in the *Introduction* (page 24–26) for the professional institutions' addresses.

Integrated Engineering and Mechatronics

Table 6 — Accreditation by professional institutions

Inst = Institution; CI = Chartered Institution; IOM3 = Institute of Materials, Minerals and Mining

Institution	Course title	Inst of Chemical Engineers	Institute of Energy	Inst of Civil Engineers	Inst of Structural Engineers	Inst of Engineering & Technology	Inst of Measurement and Control	IOM3	Inst of Mechanical Engineers	Royal Aeronautical Society
Aberdeen	Engineering			●	●	●			●	
	Engineering (integrated)					●			●	
Aston	Electromechanical engineering					●			●	
Birmingham	Engineering with business management					●				
	Mathematical engineering								●	
Blackpool and The Fylde C	Mechatronics					○				
Bristol	Engineering design with study in industry			●	●	●			●	●
Cambridge	Engineering			●	●	●			●	●
Cardiff	Integrated engineering		●			●			●	
City	Engineering and energy management								●	
Durham	General engineering			●	●	●			●	
Edinburgh	Electrical and mechanical engineering					●			●	
Glamorgan	Mechatronic engineering					●				
Glasgow Caledonian	Mechanical electronic systems engineering					○			●	
	Mechatronics									
King's College London	Mechatronics								●	
Lancaster	Engineering			●		●				
Leeds	Mechatronics and robotics					●			●	

(continued) Table 6

Accreditation by professional institutions

Institution	Course title	Inst of Chemical Engineers	Institute of Energy	Inst of Civil Engineers	Inst of Structural Engineers	Inst of Engineering & Technology	Inst of Measurement and Control	IOM3	Inst of Mechanical Engineers	Royal Aeronautical Society
Leicester	General engineering					●	●		●	
Liverpool	Engineering					●			●	
Manchester	Mechatronic engineering					●				
Manchester Metropolitan	Engineering						●		●	
Newcastle	Mechanical engineering and mechatronics					●			●	
Northampton	Engineering					○				
Nottingham	Integrated engineering					●		●	●	
Oxford	Engineering	●		●	●	●	●		●	
Robert Gordon	Mechanical and electrical engineering								●	
Sheffield	Mechatronics					●	●			
Southampton	Electromechanical engineering					●				
Strathclyde	Electrical and mechanical engineering					●			●	
UCL	Engineering with business finance								●	
Warwick	Engineering					●	●		●	

Inst = Institution
CI = Chartered Institution
IOM3 = Institute of Materials, Minerals and Mining

See Chapter 7 in the *Introduction* (page 24) for a list of sources of information that apply to all types of engineering course, including those covered in this part of the Guide. It also lists a number of books that can give some general background to engineering and the work of engineers. Visit the websites of the professional institutions for information on careers, qualifications and course accreditation, as well as information about publications, many of which you can download.

Chapter 7 in each of the other individual subject parts of the Guide contains further suggestions for background reading in particular engineering disciplines.

The courses   This Guide gives you information to help you narrow down your choice of courses. Your next step is to find out more about the courses that particularly interest you. Prospectuses cover many of the aspects you are most likely to want to know about, but some departments produce their own publications giving more specific details of their courses. University and college websites are shown in TABLE 2a.

You can also write to the contacts listed below.

Aberdeen Student Recruitment and Admissions Service (sras@abdn.ac.uk), University of Aberdeen, Regent Walk, Aberdeen AB24 3FX

Aston Dr Geof Carpenter (g.f.carpenter@aston.ac.uk), School of Engineering and Applied Science, Aston University, Aston Triangle, Birmingham B4 7ET

Birmingham Engineering with business management Dr R J Cripps (mfg.mech.admissions@bham.ac.uk), School of Engineering; Mathematical engineering Dr Richard Kaye (maths@bham.ac.uk), School of Manufacturing and Mechanical Engineering; both at University of Birmingham, Edgbaston, Birmingham B15 2TT

Blackpool and The Fylde C Admissions Office (admissions@blackpool.ac.uk), Blackpool and The Fylde College, Ashfield Road, Bispham, Blackpool FY2 0HB

Bolton Admissions Tutor (sw5@bolton.ac.uk), Engineering and Design, Bolton University, Deane Road, Bolton BL3 5AB

Bradford Mr Jack Bradley (ug-eng-enquiries@bradford.ac.uk), Admissions Tutor, School of Engineering, Design and Technology, University of Bradford, Bradford BD7 1DP

Bristol Engineering design with study in industry Admissions Tutor (engineering-design@bristol.ac.uk), Engineering Design Programme; Engineering mathematics Admissions Officer (enm-admissions@bristol.ac.uk), Department of Engineering Mathematics; both at University of Bristol, Queen's Building, University Walk, Bristol BS8 1TR

Bristol UWE Pat Cottrell (admissions.cems@uwe.ac.uk), Faculty of Computing, Engineering and Mathematics, University of the West of England Bristol, Coldharbour Lane, Frenchay, Bristol BS16 1QY

Cambridge Cambridge Admissions Office (admissions@cam.ac.uk), University of Cambridge, Fitzwilliam House, 32 Trumpington Street, Cambridge CB2 1QY

Cardiff Dr T O'Doherty (odoherty@cardiff.ac.uk), School of Engineering, Cardiff University, PO Box 917, Cardiff CF24 1XH

Central Lancashire <u>Computer-aided engineering</u> Admissions Office (uadmissions@uclan.ac.uk); <u>Robotics and mechatronics</u> Martin Varley (mrvarley@uclan.ac.uk), Department of Technology; both at University of Central Lancashire, Preston PR1 2HE

City Undergraduate Admissions Office (ugadmissions@city.ac.uk), City University, Northampton Square, London EC1V 0HB

Coventry Recruitment and Admissions Office (info.rao@coventry.ac.uk), Coventry University, Priory Street, Coventry CV1 5FB

De Montfort Admissions Unit (cse@dmu.ac.uk), Faculty of Computing Sciences and Engineering, De Montfort University, The Gateway, Leicester LE1 9BH

Durham Dr Tim Short (engineering.admissions@durham.ac.uk), School of Engineering, University of Durham, South Road, Durham DH1 3LE

Edinburgh Undergraduate Admissions Office (sciengug@ed.ac.uk), College of Science and Engineering, University of Edinburgh, The King's Buildings, West Mains Road, Edinburgh EH9 3JY

Exeter Admissions Secretary (eng-admissions@exeter.ac.uk), Department of Engineering, School of Engineering, Computer Science and Mathematics, University of Exeter, North Park Road, Exeter EX4 4QF

Glamorgan Alex Beaujean, School of Electronics, University of Glamorgan, Pontypridd, Mid Glamorgan CF37 1DL

Glasgow Caledonian <u>Mechanical electronic systems engineering</u> Angela Geddes (age@gcal.ac.uk); <u>Mechatronics</u> George Cullen (gcu@gcal.ac.uk); both at School of Engineering, Science and Design, Glasgow Caledonian University, Cowcaddens Road, Glasgow G4 0BA

Greenwich Admissions Co-ordinator (eng-courseinfo@gre.ac.uk), School of Engineering, University of Greenwich, Medway Campus, Pembroke, Chatham Maritime, Kent ME4 4TB

Harper Adams UC Admissions Officer (admissions@harper-adams.ac.uk), Harper Adams University College, Newport, Shropshire TF10 8NB

Hertfordshire Admissions Officer (admissions@herts.ac.uk), University of Hertfordshire, Hatfield, Hertfordshire AL10 9AB

Huddersfield Department of Engineering and Technology (engtech@hud.ac.uk), University of Huddersfield, Queensgate, Huddersfield HD1 3DH

Hull Admissions Tutor (engineering-admissions@hull.ac.uk), Department of Engineering, University of Hull, Hull HU6 7RX

King's College London Admissions Tutor (ugadmissions.engineering@kcl.ac.uk), Division of Engineering, King's College London, Strand, London WC2R 2LS

Lancaster Dr R V Chaplin, Engineering Department, Lancaster University, Lancaster LA1 4YR

Leeds Admissions Secretary (ug-admissions@mech-eng.leeds.ac.uk), Department of Mechanical Engineering, University of Leeds, Leeds LS2 9JT

Leicester Mr I M Jarvis (ms263@le.ac.uk), Department of Engineering, University of Leicester, University Road, Leicester LE1 7RH

Liverpool <u>Engineering</u> Admissions Tutor (ugeng@liv.ac.uk), Department of Engineering, University of Liverpool, Brownlow Hill, Liverpool L69 3GH; <u>Mechatronics and robotic systems</u> Admissions Tutor (admiss.ug.eee@liv.ac.uk), Department of Electrical Engineering and Electronics, University of Liverpool, Brownlow Hill, Liverpool L69 3GJ

Liverpool John Moores Student Recruitment Team (recruitment@ljmu.ac.uk), Liverpool John Moores University, Roscoe Court, 4 Rodney Street, Liverpool L1 2TZ

London South Bank Admissions Office, London South Bank University, 103 Borough Road, London SE1 0AA

Manchester Admissions Tutor (ug.elec.eng@manchester.ac.uk), School of Electrical and Electronic Engineering, University of Manchester, PO Box 88, Manchester M60 1QD

Manchester Metropolitan Dr W J McCann (j.mccann@mmu.ac.uk), Department of Engineering and Technology, Manchester Metropolitan University, John Dalton Building, Chester Street, Manchester M1 5GD

Napier Caroline Turnbull, School of Engineering, Napier University, 10 Colinton Road, Edinburgh EH10 5DT

Newcastle Admissions Office (enquiries@ncl.ac.uk), University of Newcastle upon Tyne, 6 Kensington Terrace, Newcastle upon Tyne NE1 7RU

Northampton Admissions Office (admissions@northampton.ac.uk), University of Northampton, Park Campus, Broughton Green Road, Northampton NN2 7AL

Northumbria Admissions (er.educationliaison@northumbria.ac.uk), University of Northumbria, Trinity Building, Northumberland Road, Newcastle upon Tyne NE1 8ST

Nottingham Dr S J Pickering (stephen.pickering@nottingham.ac.uk), School of Mechanical, Materials & Manufacturing Engineering, University of Nottingham, University Park, Nottingham NG7 2RD

Oxford Deputy Administrator (Academic), Department of Engineering Science, Oxford University, Parks Road, Oxford OX1 3PJ

Queen Mary Marian Langbridge (m.langbridge@qmul.ac.uk), Department of Engineering, Queen Mary University of London, Mile End Road, London E1 4NS

Richmond Admissions Office (enroll@richmond.ac.uk), Richmond, The American International University in London, Queens Road, Richmond, Surrey TW10 6JP

Robert Gordon Admissions Office (admissions@rgu.ac.uk), The Robert Gordon University, Schoolhill, Aberdeen AB10 1FR

Sheffield Admissions (ugacse@sheffield.ac.uk), Department of Mechanical and Process Engineering, University of Sheffield, Mappin Street, Sheffield S1 3JD

Southampton Admissions Secretary (ucas@ecs.soton.ac.uk), Department of Electronics and Computer Science, University of Southampton, Southampton SO17 1BJ

Southampton Solent Faculty of Technology Admissions, Southampton Solent University, East Park Terrace, Southampton SO14 0RD

Staffordshire Ann Grainger, School of Engineering and Advanced Technology, Staffordshire University, Beaconside, Stafford ST18 0AD

Strathclyde Mr Dougie Grant (d.grant@eee.strath.ac.uk), Department of Electronic and Electrical Engineering, University of Strathclyde, Royal College Building, Glasgow G1 1XW

Surrey Miss Natasha Baines (eng-admissions@surrey.ac.uk), School of Engineering, University of Surrey, Guildford GU2 7XH

UCL Dr Chris Nightingale (ugadmissions@meng.ucl.ac.uk), Department of Mechanical Engineering, University College London, Torrington Place, London WC1E 7JE

Ulster A E Walbridge (ae.walbridge@ulst.ac.uk), University of Ulster, Shore Road, Newtownabbey BT37 0QB

Warwick Director of Undergraduate Admissions, Department of Engineering, University of Warwick, Coventry CV4 7AL

Wolverhampton Admissions Unit (enquiries@wlv.ac.uk), University of Wolverhampton, Compton Road West, Wolverhampton WV3 9DX

Part 3

Chapter 1: Introduction

See the *Introduction* to this Guide (page 3) for an overview of how to use each chapter in the individual subject parts, including this one, together with general information about engineering and the structure of engineering courses.

With many subjects, your school experience will give you a fair idea of what to expect if you study them at degree level. This is much less true of chemical and biochemical engineering, so you may not have a clear picture in your mind of what it might involve. While it is true that chemical and biochemical engineering are based on the science subjects, the particular problems that chemical engineers have had to handle have led to the development of many ideas and skills that are specific to the subject itself.

What do chemical engineers do? We are all familiar with the products of chemical and biochemical engineering: we wear clothes of man-made fibres; our food is grown with the aid of chemical fertilisers; we use soaps and detergents; we cure diseases with the help of antibiotics and pharmaceuticals; we use toothpaste and toiletries; we drive petrol-powered cars; we use energy; and we eat and drink many forms of processed food and drinks products. New opportunities in the chemical area include advanced materials for computing and electronics; clean synthesis via better-controlled or new chemistry; clean energy via the development of the hydrogen economy; alternative sources of fuel and portable power such as fuel cells. In biochemical engineering there are opportunities for the development of better healthcare products; smart drug delivery; tissue engineering and biomedical applications; and environmental treatment for contaminated land and water.

The basic materials and chemical compounds we use are all made in production facilities designed and built using the science and technology that chemical and biochemical engineering provides. These facilities range hugely in size from vast industrial plant at one extreme to the nano-scale, or 'lab on a chip', at the other. Chemical and biochemical engineers learn how to recognise common factors in very different complex situations, and to use fundamental principles and insight to analyse them. This means that they also need to have a good understanding of control systems and analytical tools such as computer modelling.

The skills required Modern chemical engineering is both an intellectually satisfying discipline and an intensely practical subject, with these two aspects reinforcing each other. You may believe that science and technology are impersonal, but this is very far from the truth, as all engineering involves learning how to work with others in a team, usually as the team leader, with all that that implies. Being able to create and maintain communications at an individual and group level is essential to good engineering, and no engineer can avoid the social and economic consequences of his or her decisions and actions.

What do chemical engineering graduates do?   Chemical and biochemical engineering are subjects that have clearly identified areas of employment associated with them, such as chemicals, plastics and polymers, food, beverages, pharmaceuticals, oil, water, energy, business and finance. Most chemical and biochemical engineering graduates go on to jobs where they employ the specific knowledge and skills learned during their degree courses. Around 1,000 students graduate each year in chemical or biochemical engineering and traditionally there is a very high rate of employment, with starting salaries around £24,000.

Further study   Many graduates go on to further study, mostly taking Master's (MSc) courses to gain specialist knowledge, or to open up different employment opportunities. Popular areas include biotechnology and energy technology and management. Others embarked upon research through the pursuit of a PhD. Such research qualifications are advantageous and are sometimes required for research and development jobs, for example, in the chemical or pharmaceutical industries. A good class of degree is required and funding can take time to obtain, so it is wise to investigate further study, particularly PhD opportunities, at the beginning of your final year.

Who employs chemical engineers?   The knowledge and skills of a chemical or biochemical engineer can be put to use in a wide range of process industries, including the chemicals industry, food and drink manufacturing, the electricity, gas and water supply industries, and the oil industry in organisations concerned with oil extraction or the manufacture of petrochemicals. All of these organisations need chemical engineers to design and manage modern plant, and to develop novel processes. Other openings arise in the electronics, materials, pharmaceuticals and cosmetics industries, and in the nuclear and mining industries. Here chemical engineers work on the processing of nuclear materials and the reclamation of minerals from ores. Engineering contractors specialising in the design, construction and commissioning of chemical plant for their clients are also regular recruiters. Graduates are also employed by environmental and technical consultancies and the manufacturers and designers of process plant and equipment.

A variety of opportunities   The variety of roles that chemical engineers undertake can be illustrated by their employment in the pharmaceutical industry, and their importance is shown by the many key positions they hold. For example, in the construction of a new plant, chemical engineers carry out initial feasibility studies, prepare design plans, and then supervise and co-ordinate the construction and start-up of the plant. Once the plant is running, chemical engineers are involved in its operation, management and maintenance. They are also employed to market and sell the products, and manage financial aspects of what is a multi-million pound business.

You can see that chemical and biochemical engineers can find themselves working in a number of roles: research and development, plant management, economic analysis, design, construction, commissioning and operation, safety analysis, consultancy, pollution control or prevention, technical sales, marketing and management. As these skills and roles are important in many sectors of industry, chemical and biochemical engineers can be very valuable individuals in a wide range of companies.

Organisations large and small   Many of the organisations regularly recruiting chemical engineers are national or international companies with household names. They include Esso, BP and Chevron in the oil industry; Birds Eye and Sara Lee in the food and drinks sector; AstraZeneca and GlaxoSmithKline in pharmaceuticals; npower in energy; Thames Water and Wessex Water in the water utility business; Procter & Gamble and Unilever in healthcare; and ICI and Dow in the chemicals industry.

However, increasing numbers of graduates work for small companies. Anywhere materials are manufactured, processed or formed into components or products may employ chemical engineers, and many of these places are relatively small concerns. In a small company you may be the only chemical engineer, so you will have much greater responsibility at an earlier stage than in a big company. Your role may also extend beyond pure chemical engineering, giving the job greater variety than in a larger company. All this puts a premium on your flexibility and ability to communicate and work with people from a wide range of backgrounds. The broad engineering training provided by chemical engineering degree courses is a good preparation for these challenges.

For further information on opportunities in chemical engineering, see the website of the Institution of Chemical Engineers at www.icheme.org.

Opportunities outside chemical engineering   Although most chemical engineering graduates wish to find employment where they can put their degree subject to use, around 40% of all graduate opportunities are open to graduates from any degree discipline: they are recruited on the basis of their intellectual and personal skills and experience, rather than subject knowledge. A degree course in chemical or biochemical engineering develops many skills, such as numeracy, analytical thinking, teamwork, communication skills and problem-solving, that are required by a wide range of employers. It is therefore not surprising that financial institutions such as banks and accountancy firms, amongst others, regard chemical engineers as excellent recruits. Sandwich placements and other work experience, and participation in student sports and societies, offer additional opportunities for skills development. Chemical engineers can therefore complete their degrees with technical and personal skills valued by both industrial and commercial organisations.

TABLE 2a lists the specialised and combined courses at UK universities that lead to the award of an honours degree in chemical engineering and a number of closely related courses, including biochemical engineering and process engineering.

When the table was compiled it was as up to date as possible, but sometimes new courses are announced and existing courses withdrawn, so before you finally fill in your application you should check the UCAS website, www.ucas.com, to make sure the courses you plan to apply for are still on offer.

See Chapter 2 in the *Introduction* (page 6) for advice on how to use TABLE 2a and for an explanation of what the various columns mean.

Table 2a — First-degree courses in Chemical Engineering

Chemical Engineering

Institution / Course title	①②③ see combined subject list – Table 2b	Degree	Duration (Number of years)	Foundation year (● at this institution / ○ at franchised institution / ◐ second-year entry)			Modes of study (● full time / part time / ○ time abroad / ◑ sandwich)				Modular scheme	Course type (● specialised / ◑ combined)		No of combined courses
Aberdeen www.abdn.ac.uk														
Chemical engineering		BEng/MEng	4, 5			◐	●			◑		●		0
Aston www.aston.ac.uk														
Chemical engineering		BEng/MEng	3, 4, 5	●	○		●		○	◑		●		0
Chemical engineering and applied chemistry ①		MEng	4, 5	●	○		●		○	◑		●	◑	1
Chemistry, technology and design		BEng	3, 4	●	○		●			◑		●		0
Bath www.bath.ac.uk														
Biochemical engineering		MEng	4, 5	●			●		○	◑		●		0
Chemical and bioprocess engineering		BEng	3, 4	●			●			◑		●		0
Chemical engineering		MEng	3, 4, 5	●			●		○	◑		●		0
Belfast www.qub.ac.uk														
Chemical engineering		BEng/MEng	3, 4, 5	●	○		●		○	◑		●		0
Birmingham www.bham.ac.uk														
Chemical engineering		BEng/MEng	3, 4	●			●		○	◑	🕏	●	◑	1
Cambridge www.cam.ac.uk														
Chemical engineering		BA/MEng	3, 4				●					●		0
Edinburgh www.ed.ac.uk														
Chemical engineering		BEng/MEng	4, 5			◐	●		○			●	◑	2
Heriot-Watt www.hw.ac.uk														
Chemical engineering		BEng/MEng	4, 5, 6			◐	●			◑		●		0
Imperial College London www.imperial.ac.uk														
Chemical engineering		MEng	4				●		○			●		0
Leeds www.leeds.ac.uk														
Chemical engineering ①		BEng/MEng	3, 4	●			●		○	◑		●	◑	1
Pharmaceutical chemical engineering		BEng/MEng	3, 4	●			●		○	◑		●		0
London South Bank www.sbu.ac.uk														
Chemical and process engineering		BEng	3, 4				●			◑		●		0
Loughborough www.lboro.ac.uk														
Chemical engineering ①		BEng/MEng	3, 4, 5	●			●		○	◑		●	◑	1
Chemical engineering with environmental protection		BEng	3, 4	●			●		○	◑		●		0
Process technology and management		BSc	3, 4				●			◑		●		0

First-degree courses in **Chemical Engineering**

Institution / Course title	combined subject — Table 2b	Degree	Duration (Number of years)	Foundation year (● at this institution; ○ at franchised institution; ◑ second-year entry)			Modes of study (● full time; ▼ part time; ○ time abroad; ◑ sandwich)			Modular scheme	Course type (● specialised; ◑ combined)		No of combined courses
Manchester www.man.ac.uk													
Chemical engineering		BEng/MEng	3, 4	●			●	○	◑		●	◑	4
Newcastle www.ncl.ac.uk													
Chemical and process engineering		BEng/MEng	3, 4, 5	●			●	○	◑	✿	●		0
Nottingham www.nottingham.ac.uk													
Chemical engineering		BEng/MEng	3, 4	●			●	○			●	◑	2
Oxford www.ox.ac.uk													
Chemical engineering		MEng	4	●			●				●		0
Paisley www.paisley.ac.uk													
Chemical engineering		BEng	4, 5			◑	● ▼	○	◑		●	◑	1
Sheffield www.sheffield.ac.uk													
Chemical engineering		BEng/MEng/MChem	3, 4	●			●	○		✿	●	◑	4
Strathclyde www.strath.ac.uk													
Chemical engineering ①		BEng/MEng	4, 5			◑	● ▼	○			●	◑	1
Chemical engineering with process biotechnology		BEng	4			◑	● ▼	○			●		0
Surrey www.surrey.ac.uk													
Chemical and bio-systems engineering		BEng/MEng	3, 4, 5	●			●	○	◑		●		0
Chemical engineering		BEng/MEng	3, 4, 5	●			●	○	◑		●		0
Swansea www.swan.ac.uk													
Chemical and biochemical engineering		MEng	4	●			●		◑		●		0
Chemical and bioprocess engineering		BEng	3, 4	●			●	○	◑		●		0
Teesside www.tees.ac.uk													
Chemical engineering		BEng	3, 4	●			● ▼		◑		●		0
UCL www.ucl.ac.uk													
Biochemical engineering ①		BEng/MEng	3, 4, 5	●			●	○	◑		●	◑	1
Chemical engineering ②		BEng/MEng	3, 4, 5				●	○	◑		●	◑	1

Subjects available in combination with chemical engineering

The requirements for professional accreditation mean that in general there is little scope for combining chemical engineering with other subjects. However, TABLE 2b shows the subjects that can make up between a third and half of a degree course when studied with chemical engineering in the combined degrees listed in TABLE 2a. See Chapter 2 in the *Introduction* (page 6) for general information about combined courses and for an explanation of how to use TABLE 2b.

Table 2b — Subjects to combine with Chemical Engineering

Applied chemistry Aston①, Strathclyde①
Biochemical engineering UCL②
Biotechnology Manchester
Business studies Manchester
Chemical engineering UCL①
Chemistry Leeds①, Manchester, Paisley, Sheffield

Environmental engineering Edinburgh, Nottingham
Environmental science Manchester
Fuel engineering Sheffield
Management studies Birmingham, Edinburgh, Loughborough①, Sheffield
Modern languages Nottingham, Sheffield

Other courses that may interest you   The following courses are closely related to the courses in this Guide: they include courses where you spend less than half your time studying chemical engineering, as well as courses providing a more intensive study of a specialised aspect of the subject.

- Chemistry with chemical engineering (Huddersfield)
- Fire engineering (Central Lancashire)
- Fire and explosion science (Leeds).

Other parts of this Guide that may particularly interest you are *Integrated Engineering and Mechatronics* and *Civil and Environmental Engineering and Building*.

You may also find courses in the following Guides of interest:

- *Physics and Chemistry:* for applied chemistry courses
- *Biological Sciences:* for biotechnology courses
- *Agricultural Sciences and Food Science and Technology:* for food science and food technology courses.

Industrial experience and time abroad Many courses provide opportunities for spending a period of industrial training in the UK or abroad, or for study abroad. This industrial experience allows you to use the skills and knowledge gained from the academic training, earn a salary and enhance your employment opportunities. TABLE 3a gives information about the possibilities for the courses in this Guide. See Chapter 3 in the *Introduction* (page 11) for further information.

Table 3a — Time abroad and sandwich courses

Institution / Course title (① ② : see notes after table)	Named 'international' variant of the course	Location: ● Europe; ○ North America; ◖ industry; ◐ academic institution	Maximum time abroad (months)	Time abroad assessed	Language study: ○ optional; ● compulsory; * contributes to assessment	Socrates–Erasmus	Sandwich courses: ● thick; ○ thin	Arranged by: ● institution; ○ student
Aston								
Chemical engineering ① ③	●	● ◖ ◐	12	●	○*	●	● ●	●
Chemical engineering and applied chemistry ③	●	● ◖ ◐	12	●	○	●	● ●	●
Chemistry, technology and design							●	●
Bath								
Biochemical engineering ①	●	● ○ ◖ ◐	12	●	○*	●	● ○	●
Chemical engineering ①	●	● ○ ◖ ◐	12	●	○*	●	● ○	●
Belfast								
Chemical engineering ③ ④		● ○ ◖ ◐	12		○	●		
Birmingham								
Chemical engineering ① ② ③		● ○ ◖ ◐	12	●	○	●	● ●	●
Edinburgh								
Chemical engineering	●	○ ◐	12	●	○			
Heriot-Watt								
Chemical engineering ③		● ◐	3	●	○*	●		● ○
Imperial College London								
Chemical engineering ③	●	● ○ ◖ ◐	12	●	○*	●		
Leeds								
Chemical engineering ③		● ○ ◐	12	●	○*	●		
Loughborough								
Chemical engineering		● ○ ◖ ◐	12	●	○*	●	● ●	●
Chemical engineering with environmental protection		● ○ ◖ ◐	12	●	○*	●	● ●	●
Manchester								
Chemical engineering ③	●	● ○ ◖ ◐	12	●	○*	●	● ●	●
Newcastle								
Chemical and process engineering ③	●	● ○ ◖ ◐	12	●	○*	●	● ●	●
Nottingham								
Chemical engineering ③	●	● ○ ◐	12		○*			
Oxford								
Chemical engineering ③			0		○			
Paisley								
Chemical engineering ③	●	◖	12				●	● ○
Strathclyde								
Chemical engineering		● ○ ◖ ◐	12		○	●		

Chemical Engineering

Time abroad and sandwich courses

Institution / Course title (① ② : see notes after table)	Named 'international' variant of the course	Location: ● Europe; ○ North America; ◑ industry; ◐ academic institution	Maximum time abroad (months)	Time abroad assessed	Language study: ○ optional; ● compulsory; * contributes to assessment	Socrates–Erasmus	Sandwich courses: ● thick; ○ thin	Arranged by: ● institution; ○ student
Strathclyde (continued)								
Chemical engineering with process biotechnology		● ○ ◑ ◑	12		○	●		
Surrey								
Chemical and bio-systems engineering		● ○ ◑ ◑	12	●		●	●	●
Chemical engineering		● ○ ◑ ◑	12	●		●	●	●
Swansea								
Chemical and biochemical engineering ③		● ◑	3	●	○	●	●	●
Chemical and bioprocess engineering ① ③	●	● ○ ◑ ◑	12	●	○*	●	●	● ○
Teesside								
Chemical engineering	●	◑ ◑	12	●		●	●	● ○
UCL								
Biochemical engineering ③	●	● ○ ◑ ◑	9	●	○*	●	●	● ○
Chemical engineering ③	●	● ○ ◑ ◑	12	●	○*	●	●	● ○

① A year of industrial experience before the course starts is recommended for students on non-sandwich courses
② A year of industrial experience before the course starts is recommended for students on sandwich courses
③ Students on non-sandwich courses can take a year out for industrial experience
④ Minimum period of vacation industrial experience for non-sandwich students (weeks) : Belfast Chemical engineering 12

Course content   If you were to compare the syllabuses for the chemical engineering courses listed in this Guide, you would find them all very similar. The subjects covered are almost identical, the courses differing only in the depth of treatment of some topics and in the inclusion or exclusion of a few others. This family likeness is not accidental, as it reflects the nature of the problems that all chemical engineers have to tackle. The academic training you will receive is designed to develop an understanding of both the principles and theories of the conceptual and mathematical models that you will use to solve these problems.

A typical course   TABLE 3b shows a three-year schedule of lectures in a typical BEng chemical engineering degree course (the figures give the typical time spent on each topic in hours per week). In practice, courses may vary a little from this pattern and may use slightly different names for topics, or they may group the content together differently. In particular, four-year MEng courses allow you time to study a wider range of topics in greater depth.

One other element common to all departments is teaching in the use of computers, which is often integrated into the main coursework. Most departments have extensive facilities, with access to both microcomputers and powerful workstations.

The content of a typical **Chemical Engineering** course					
Year 1	Time (hours per week)	**Year 2**	Time (hours per week)	**Year 3**	Time (hours per week)
Chemistry	3	Chemistry	2	Advanced transport phenomena including fluid and particle dynamics	2
Mathematics	2	Applied mathematics and process modelling	2	Separation processes and advanced unit operations	2
Fluid mechanics and heat transfer	2	Thermodynamics	1	System analysis and control	1
Engineering drawing and design	2	Fluid mechanics and heat and mass transfer (transport phenomena)	2	Research project	4
Chemical process principles	2	Engineering of chemical reactors and separators	3	Business and management topics	1
Treatment of experimental data and communication skills	1	Plant design and materials	2	Pollution control and environmental awareness	2
Laboratory class	4	Instrumentation and systems analysis	1	Specialised options	2
Tutorial	1	Laboratory class	1	Design project	4

The following sections describe the content covered in each year in a little more detail.

The first year   In the first year, your school chemistry and mathematics will be reinforced and extended, while in fluid mechanics and heat transfer you will begin to learn about how liquids and gases behave in motion, how they can be pumped and their flow measured, and how heat is transferred to them from, say, the walls of a pipe. Initially, quite simple situations will be looked at, more complex and more realistic situations being tackled later in the course. In chemical process principles, you begin to learn about the basic book-keeping necessary to build a plant, how to keep account of the materials entering and leaving, and the heat flows required in each part of the process. This will also include the fundamental ideas necessary for the more advanced design techniques. Some courses also cover biology and biochemistry in addition to chemistry.

Laboratory classes will often be used to illustrate the theory covered in lectures in a practical way.

The second year   In the second year, thermodynamics (often taught first as part of first-year chemistry) covers the production of work from heat and refrigeration, and the theory required for the design of distillation columns and other separation processes. In engineering of chemical reactors and separators you will cover the principles behind the design and operation of processing steps which can be used in many different industries. For example, distillation is common to the refining of petrol, to the production of perfumes and essences, to the making of alcohol and to very many other processes. Some of the courses will also cover the design of biochemical reactors and bio-separations.

Instrumentation and systems analysis is about how you find out what is happening in a process: the measurement of pressure, temperature, fluid flow and composition on the industrial scale; and how your plant will react to changes in the operating conditions.

Again, laboratory classes will often be a key component.

The final year(s)   In the final year of a BEng course you will build on the knowledge gained in the first two years. For example, in systems analysis and control, the second-year work will be extended to include the control of a complete plant. Options allowing you to consider a more specialised topic in detail are quite common in the final year and both design and research projects will be key. The third year of an MEng course and the final year of a BEng course often share some common material, with the final year of the MEng introducing further depth of study in various areas. The research and design project activities in an MEng course are typically more substantial than for a BEng programme.

Course orientation   Although the content of courses is broadly similar, institutions often lay emphasis on different elements of the course to give them a distinct orientation. In particular, it is possible to distinguish courses that might be described as process-oriented from those that are more science-oriented. Process-oriented courses place greater emphasis on the more commercial aspects of the subject, such as process design, cost engineering, operational research and process optimisation, while science-oriented courses emphasise the physico-chemical and fluid-dynamic aspects of the subject.

Some courses cover both chemical and biochemical engineering in an integrated manner, whereas others run these topics as separate programmes. A particular course orientation may also show itself in the range of optional subjects provided, which may closely reflect the interests of those who run the course or the interests of industrial partners.

Although chemical engineering courses are on the whole very similar, some courses have specific features that distinguish them from the others. TABLE 3c shows some unusual or distinctive features of course content and organisation.

General engineering introduction   At many institutions the introductory stages of several engineering courses are taught together. The rather different science foundation of chemical engineering means that it is fairly unusual for chemical engineering to be included with the other engineering courses in this introductory period, but TABLE 3c shows where this occurs and at what point it is possible to specialise in chemical engineering.

Institution	Course title	Point of specialisation after common introductory period	Unusual or distinctive features
	Distinctive features and joint introductory periods		
Aston	Chemical engineering		Integrated department of chemical engineering and applied chemistry giving students a greater insight into chemical aspects; flexibility between the range of courses; extensive team project work including joint project with chemists
	Chemical engineering and applied chemistry		Contains core content required by IChemE and RSC and can lead to Chartered Engineer and Chartered Chemist status after appropriate experience
	Chemistry, technology and design		Combines chemistry and chemical engineering; possible progression to MEng
Cambridge	Chemical engineering	Year 2	In year 1, students read Natural sciences, Computer science or Engineering, moving to chemical engineering from year 2. Year 4: broad range of optional courses and research project
Edinburgh	Chemical engineering	Term 2	First year offers flexibility to go on to chemical engineering, other engineering branches or chemistry
Heriot-Watt	Chemical engineering		Years 3 and 4 options: offshore hydrocarbon processing; liquid effluent treatment; food and beverage processing; polymer technology
Imperial College London	Chemical engineering		Design taught in all years; link projects fuse supervised vacation training with course work project in Department
Manchester	Chemical engineering		Large pilot plant enables experimental work and projects to be done on full-scale equipment
Newcastle	Chemical and process engineering		Choices between MEng and BEng and whether to spend a year in industry can be made at end of year 2
Oxford	Chemical engineering	Year 3	Gives a broad introduction to all aspects of engineering with later specialisation in chemical engineering
Paisley	Chemical engineering	Year 3	Has a heavier weighting of chemistry than many chemical engineering courses
Strathclyde	Both courses		Emphasis on design, pollution control, personal skills and biotechnology
Surrey	Chemical engineering	Year 2	Extensive training in personal skills development throughout the degree programme; co-operative exchange programme for industrial year with countries including: USA; Europe; Singapore; Australia; New Zealand; Kenya
Swansea	Chemical and bioprocess engineering	Year 1	
Teesside	Chemical engineering		Year 1 largely based on industrial case studies with partial delivery in industry
UCL	Biochemical engineering		Design project in state-of-the-art bioprocess pilot plant
	Chemical engineering		2-day teamworking workshop in year 2 with external management organisation

Specialisation   All courses include options in later years, which allow you to develop a particular interest in some branch of the subject. The options on offer often reflect the research interests of the department or individual members of staff, so they tend to change more rapidly than other parts of the course. This has the advantage of introducing you to the latest developments in the field, and in some cases you may even be taught by the people who were responsible for them.

The range of options offered varies widely. They are usually in specialised technical areas that would probably mean little to you at this stage, so details are not included here. However, if you do have a particular interest, you should consult prospectuses or contact departments via their website or direct.

General studies and additional subjects   Specialised engineering courses run the risk of focusing your attention entirely

Chemical Engineering

on technical subjects. To prevent this, some institutions include topics on liberal, complementary or general studies in their specialised courses. TABLE 3a contains information about language teaching; you will need to look at prospectuses or ask institutions direct for information about other subjects.

Where a course is part of a modular degree scheme, there will probably be more opportunities to take a wide range of other subjects, though the requirements for professional qualification may mean that the choice is more restricted than it would be if you were following a course in a non-engineering subject. TABLE 2a shows where a course is part of a modular degree scheme allowing a relatively free choice of modules from a large number of subjects.

Research   If you already have an ambition to do postgraduate research work, you may be hesitant about taking an applied science degree, as you may think that a pure science degree will give you a better chance of going on to do research. However, around one in five graduate chemical engineers go on to further study or training, with the majority of these working on research projects. You can enter courses leading to an MSc or PhD degree immediately after graduating, or you may prefer to spend a year or so in industry first.

You will be able to do research at all the institutions mentioned in this chapter, although their laboratory facilities and research reputation do vary. Many people stay on at the same institution, but many also move on to another department to pursue a particular specialisation or just for the stimulation of a new environment.

As an undergraduate, you will benefit from the department's research activities, which often find expression in the teaching within the department, and through the meetings of your departmental chemical engineering society.

BEng/MEng courses   TABLE 2a lists many courses that can lead to the award of either a BEng or an MEng degree. For these courses, most of the tables in this Guide give information specifically for the BEng stream (the tables show information for MEng courses that are MEng only). Much of the information will also apply to the MEng streams, but TABLE 3d shows you where there are differences for the MEng stream. It shows at what point the MEng course separates from the BEng (for those courses where they are not separate from the start), and what proportion of students are expected to leave with an MEng degree.

It is important to note that currently the minimum academic requirement towards becoming a Chartered Engineer is at the Master's level – this can be achieved either by taking an MEng degree, or by completing a BEng, followed by an MSc at a later date.

MEng course differences

Institution	Course title	MEng separates from BEng	MEng proportion %	More engineering	More management	More languages	Other differences
Aston	Chemical engineering	Year 2		●	●	●	
Belfast	Chemical engineering	Year 2		●	●		
Birmingham	Chemical engineering	Year 3		●	●		More project work
Cambridge	Chemical engineering	Year 3/4	95	●			BA awarded after 3 years
Edinburgh	Chemical engineering	Year 4		●		●	Advanced study; projects; extended projects incorporating further communication and management skills
Heriot-Watt	Chemical engineering	Year 3	10	●	●		Specialises in industrial market sector; enter dependent on performance after year 2 BEng course
Leeds	Chemical engineering	Year 3		●			Further design and research projects; time abroad is only available for MEng course
Loughborough	Chemical engineering	Year 3		●	●		Includes a semester of professional development in industry or another university; this means MEng students can spend up to 18 months abroad; student must achieve high grades in examinations and coursework
Manchester	Chemical engineering	Year 3	75	●		●	More specialised subjects in greater depth; research project
Newcastle	Chemical and process engineering	Year 3	60	●			Specialisation in bioprocess engineering, process control, sustainable engineering, intensified processing or European language
Sheffield	Chemical engineering	Year 3		●		●	'MEng with language' students spend full year abroad studying chemical engineering in a foreign language (BEng students only spend a summer abroad)
Strathclyde	Chemical engineering	Year 4		●	●		More substantial project work
Surrey	Chemical and bio-systems engineering			●	●		More project work at later levels; greater breadth and depth
	Chemical engineering	Year 3		●	●		More depth and breadth; research project; interdisciplinary design project
Swansea	Chemical and bioprocess engineering						
UCL	Biochemical engineering	Year 3	60	●	●		Research projects
	Chemical engineering	Year 3		●			

Chapter 4: Teaching and assessment methods

See Chapter 4 in the *Introduction* (page 16) for general information about teaching and assessment methods used in all types of engineering course, including those covered in this part of the Guide. That chapter also explains how to interpret TABLE 4, which gives information about projects and assessment methods used on individual courses.

The large scale of chemical plant has practical implications for the way in which projects are carried out. In particular, group project work is a compulsory element of all courses. Nearly all departments require you to undertake two large projects: a design project and a research project. The large scale of chemical plant has practical implications for the way in which projects are carried out. In particular, group project work is a compulsory element of all courses. Nearly all departments require you to undertake two large projects: a design project and a research project. These projects are usually carried out in the third or fourth years.

Design projects   In the design project you will be expected to work in a small group to carry out the design of a complete process starting from raw material and ending with a product. You will be expected to participate in the overall planning and design work of the group, and to take responsibility for the design of a particular section of the plant, such as a chemical reactor with its associated equipment. The project will take from one to two terms and may involve the use of computers. (If your course does not include a design project, you will have to take the Institution of Chemical Engineers design project examination before you can become a chartered engineer.) In some departments you may also do small design exercises in earlier years.

Research projects   A research project usually requires both literature and experimental work. This is often genuine research and not just an artificial example. For instance, you might be asked to work on a fundamental topic such as mass transfer as part of an investigation into the behaviour of catalysts, or to carry out a preliminary study of a new design of bioreactor for making pharmaceuticals.

Table 4

Assessment methods

Institution	Course title	Frequency of assessment	Years of exams contributing to final degree (years of exams not contributing to final degree)	Coursework: minimum/maximum %	Project/dissertation: minimum/maximum %	Time spent on projects in: first/intermediate/final years %			Group projects: ● compulsory; ○ optional	Orals: ◐ if borderline; ● everyone ○ for projects;
Aston	Chemical engineering	○	(1),**2,3,4**	**5/5**	36/**36**	35	20	30	●	○
	Chemical engineering and applied chemistry	○	(1),**2,3,4**	10/10	40/**40**	35	20	30	●	○
	Chemistry, technology and design	○	**2,3,4**	10/10	30/**30**	20	30	40	●	◐○
Bath	Biochemical engineering	◑	(1),**2,3,4**	**15**	**35**	10	30	50	●	◐○
	Chemical engineering	◑	(1),**2,3,4**	**15**	**35**	10	30	50	●	◐○

Key for frequency of assessment column: ● term; ◑ semester; ○ year

Institution	Course title	Frequency of assessment	Years of exams contributing to final degree (years of exams not contributing to final degree)	Coursework: minimum/maximum %	Project/dissertation: minimum/maximum %	Time spent on projects in: first/intermediate/final years %			Group projects: ● compulsory; ○ optional	Orals: ◐ if borderline; ○ for projects; ● everyone
Belfast	Chemical engineering	◐	(1),**2,3,4**		15/**25**	10	20	20	●	○
Birmingham	Chemical engineering	◐	(1),**2,3**,4	5/**10**	30/**40**	8	10	30	●	○
Cambridge	Chemical engineering	○	(1),(2),**3,4**	5/**15**	10/**20**		25	30	●	
Edinburgh	Chemical engineering	◐	(1),(2),**3,4,5**	5/**15**	35/**50**	20	40	40	●	○
Heriot-Watt	Chemical engineering	◑	(1),(2),**3,4,5**	0/**0**	39/**46**	20	25	40	●	◐
Imperial College London	Chemical engineering	○	**1,2,3,4**		**40**	40	40	50	●	○
Leeds	Chemical engineering	◐	(1),**2,3,4**	25/**25**	25/**25**	0	20	30	●	○
Loughborough	Chemical engineering	◐	(1),**2,3,4,5**	30/**30**	20/**40**	30	30	30	●	◐○
	Chemical engineering with environmental protection	◐	(1),**2,3,4**	30/**30**	20/**40**	30	30	30	●	◐○
Manchester	Chemical engineering	◐	**1,2,3,4**	10/**15**	35/**40**	33	20	40	●	○
Newcastle	Chemical and process engineering	◐	**1,2,3,4**	5/**15**	15/**30**	20	20	30	●	◐○
Nottingham	Chemical engineering	◐	(1),**2,3,4**	10/**20**	20/**30**	8	15	25	●	◐○
Oxford	Chemical engineering		(1),**3,4**	13/**13**	25/**25**	10	15	50	●	◐
Paisley	Chemical engineering	◐	(1),(2),**3,5**	15/**30**	20/**20**	5	5	20	●	◐
Sheffield	Chemical engineering	◐	(1),**2,3,4**	5/**5**	40/**55**	20	20	40	●	
Strathclyde	*Both courses*	◐	(1),**2,3,4**		**20**	0	5	42	●	◐
Surrey	Chemical and bio-systems engineering	◐	(1),**2,3,4**			10	10	25	●	○
	Chemical engineering	◐	(1),**2,3,4**	15/**35**	0/**50**	0	20	30	●	
Swansea	Chemical and biochemical engineering	◐	(1),**2,3,4**	15/**15**	20/**30**	30	40	40	●	
	Chemical and bioprocess engineering	◐	(1),**2,3**	15/**15**	15/**20**	15	20	25	●	○
Teesside	Chemical engineering	○	(1),**2,3,4**	10/**15**	25/**30**	10	15	30	●	◐○
UCL	Biochemical engineering		**1,2,3,4**	5/**15**	10/**15**	10	15	20	●	○
	Chemical engineering	○	**1,2,3,4**	5/**15**	10/**15**	10	15	20	●	○

Chemical Engineering

See Chapter 5 in the *Introduction* (page 18) for general information about entrance requirements that applies to all types of engineering course, including those covered in this part of the Guide. It also explains how to interpret TABLE 5, which gives information about entrance requirements for individual courses.

Chemical Engineering

Table 5 — Entrance requirements

Institution	Course title	Number of students (includes other courses)	Typical offers (BSc/BEng) UCAS tariff points	A-levels	SQF Highers	Typical offers (MEng) UCAS tariff points	A-levels	SQF Highers	A-level Chemistry	A-level Mathematics	A-level Physics
Aberdeen	Chemical engineering		280	BBC	BBCCC	300	BBB	ABBBC		●	●
Aston	Chemical engineering	50	240–280	BCC	BBBBC	300	BBB	AABBB	●	○	
	Chemical engineering and applied chemistry	5				320	BBB	AABBB	●	○	
	Chemistry, technology and design	5	240–280	BCC	BBBBC				●	●	
Bath	Biochemical engineering	(60)				300			●	●	○
	Chemical and bioprocess engineering		260						●	●	○
	Chemical engineering	(60)				300			●	●	○
Belfast	Chemical engineering	50		BCC	BBBC		BBB		●	●	
Birmingham	Chemical engineering	(100)	300	ABB	ABBBB	300	ABB	ABBBB	●	●	
Cambridge	Chemical engineering	40		AAA			AAA		●	●	○
Edinburgh	Chemical engineering	35		BBB	BBBB		BBB	BBBB	●	●	○
Heriot-Watt	Chemical engineering	45		CCD	BBBB		CCD	BBBB	●	●	○
Imperial College London	Chemical engineering	100					ABB		●	●	○
Leeds	Chemical engineering	40		BBB			BBB		●	○	○
	Pharmaceutical chemical engineering			BBB	BBBBB		BBB	BBBBB			
London South Bank	Chemical and process engineering			CC	BBB						
Loughborough	Chemical engineering	60	260–280			300–320			○	●	○
	Chemical engineering with environmental protection	15	260–280						○	●	○
	Process technology and management		240								
Manchester	Chemical engineering	100		AAB			AAB		●	●	○
Newcastle	Chemical and process engineering	40		BBB	BBBBB		ABB	AABBB	●	●	
Nottingham	Chemical engineering	50		ABB			ABB		●	●	○
Oxford	Chemical engineering	(170)					AAA	AAAAB	●	●	●
Paisley	Chemical engineering	(20)		DD	BBC				○	●	○
Sheffield	Chemical engineering	10		ABB	AABB		ABB	AAAB	●	●	
Strathclyde	Chemical engineering	75		BCC	BBBB		ABB	AAAB	●	●	○
	Chemical engineering with process biotechnology	75		BCC	BBBB				●	●	○

● compulsory; ○ preferred

| Institution | Course title | Number of students (includes other courses) | Typical offers (BSc/BEng) | | | Typical offers (MEng) | | | A-level Chemistry | A-level Mathematics | A-level Physics |
			UCAS tariff points	A-levels	SQF Highers	UCAS tariff points	A-levels	SQF Highers	● compulsory; ○ preferred		
Surrey	Chemical and bio-systems engineering	20	260	BCC	BBBB	300	BBB	BBBB	○	●	
	Chemical engineering	30	260	BCC	BBBB	300	BBB	BBBB	○	●	
Swansea	Chemical and biochemical engineering	20				300			●	●	
	Chemical and bioprocess engineering	20	260						●	●	
Teesside	Chemical engineering	20	180–240						○	○	○
UCL	Biochemical engineering	35		ABB			ABB		○	○	○
	Chemical engineering	50		BBB			BBB		○	○	○

See Chapter 6 in the *Introduction* (page 20) for general information about professional qualification that applies to all types of engineering course, including those covered in this part of the Guide.

Some courses are accredited by several institutions, but the accreditation by an individual institution may be conditional on your having taken certain options, which may not coincide with those required by another institution. Remember that the process of accreditation goes on continuously, so you should check with the professional institutions for the current position on specific courses.

Teesside states that 'the Institution of Chemical Engineers invites students from this course to apply individually for accreditation'. All the other institutions appearing in TABLE 2a offer at least one course accredited by the Institution. In addition, Newcastle's course is accredited by the Energy Institute.

The Royal Society of Chemistry has accredited Chemical engineering and applied chemistry at Aston.

Chapter 7 of the *Introduction* (page 24–26) lists the professional institutions' addresses and websites.

Chemical Engineering

See Chapter 7 in the *Introduction* (page 24) for a list of sources of information that apply to all types of engineering course, including those covered in this part of the Guide. It also lists a number of books that can give some general background to engineering and the work of engineers. Visit the websites of the professional institutions for information on careers, qualifications and course accreditation, as well as information about publications, many of which you can download. Chapter 7 in each of the other individual subject parts of the Guide contains further suggestions for background reading in particular engineering disciplines.

The courses This Guide gives you information to help you narrow down your choice of courses. Your next step is to find out more about the courses that particularly interest you. Prospectuses cover many of the aspects you are most likely to want to know about, but some departments produce their own publications giving more specific details of their courses. University and college websites are shown in TABLE 2a.

You can also write to the contacts listed below.

Aberdeen Student Recruitment and Admissions Service (sras@abdn.ac.uk), University of Aberdeen, Regent Walk, Aberdeen AB24 3FX

Aston Dr Geof Carpenter (g.f.carpenter@aston.ac.uk), School of Engineering and Applied Science, Aston University, Aston Triangle, Birmingham B4 7ET

Bath Dr T C Arnot (t.c.arnot@bath.ac.uk), Admissions Tutor, Department of Chemical Engineering, University of Bath, Claverton Down, Bath BA2 7AY

Belfast Admissions Officer, The Queen's University of Belfast, Belfast BT7 1NN

Birmingham Dr Mark Simmons (ug-admis-chem-eng@bham.ac.uk), School of Chemical Engineering, University of Birmingham, Edgbaston, Birmingham B15 2TT

Cambridge Cambridge Admissions Office (admissions@cam.ac.uk), University of Cambridge, Fitzwilliam House, 32 Trumpington Street, Cambridge CB2 1QY

Edinburgh Undergraduate Admissions Office (sciengug@ed.ac.uk), College of Science and Engineering, University of Edinburgh, The King's Buildings, West Mains Road, Edinburgh EH9 3JY

Heriot-Watt Admissions Tutor (chemeng@eps.hw.ac.uk), Chemical Engineering, School of Engineering and Physical Sciences, Heriot-Watt University, Riccarton, Edinburgh EH14 4AS

Imperial College London Dr Omar K Matar (o.matar@imperial.ac.uk), Department of Chemical Engineering, Imperial College London, South Kensington Campus, London SW7 2AZ

Leeds Mrs S A Cook (spemeadmissions@leeds.ac.uk), Admissions Secretary, Department of Chemical Engineering, University of Leeds, Leeds LS2 9JT

London South Bank Admissions Office, London South Bank University, 103 Borough Road, London SE1 0AA

Loughborough Dr R B Wilcockson (r.b.wilcockson@lboro.ac.uk), Department of Chemical Engineering, Loughborough University, Loughborough LE11 3TU

Manchester Dr Stuart Holmes (ug.ceas@manchester.ac.uk), Department of Chemical Engineering, University of Manchester, Manchester M60 1QD

Chemical Engineering

71

Newcastle Dr M J Willis (chemeng.ugadmin@ncl.ac.uk), Admissions Tutor, Department of Chemical and Process Engineering, University of Newcastle upon Tyne, Claremont Road, Newcastle upon Tyne NE1 7RU

Nottingham Dr E Lester (edward.lester@nottingham.ac.uk), School of Chemical, Environmental and Mining Engineering, University of Nottingham, University Park, Nottingham NG7 2RD

Oxford Deputy Administrator (Academic), Department of Engineering Science, Oxford University, Parks Road, Oxford OX1 3PJ

Paisley Dr K J Forster (k.forster@paisley.ac.uk), Department of Chemistry and Chemical Engineering, University of Paisley, Paisley PA1 2BE

Sheffield Dr N V Russell, Department of Chemical and Process Engineering, University of Sheffield, Mappin Street, Sheffield S1 3JD

Strathclyde Dr D M Muir (david.m.muir@strath.ac.uk), Department of Chemical Engineering, University of Strathclyde, Glasgow G1 1XJ

Surrey Miss Natasha Baines (eng-admissions@surrey.ac.uk), School of Engineering, University of Surrey, Guildford GU2 7XH

Swansea Dr R Bryant (r.bryant@swansea.ac.uk), School of Engineering, University of Wales Swansea, Singleton Park, Swansea SA2 8PP

Teesside Dr A M Gerard, School of Science and Technology, University of Teesside, Middlesbrough TS1 3BA

UCL Biochemical engineering Dr P A Dalby (p.dalby@ucl.ac.uk), Department of Biochemical Engineering; Chemical engineering Dr P Angeli (p.angeli@ucl.ac.uk), Department of Chemical Engineering; both at University College London, Torrington Place, London WC1E 7JE

Chapter 1: Introduction

See the *Introduction* to this Guide (page 3) for an overview of how to use each chapter in the individual subject parts, including this one, together with general information about engineering and the structure of engineering courses.

The scope of this Guide   This Guide covers first-degree courses in civil engineering, structural engineering, environmental engineering and building, together with a number of courses in closely related fields.

What is civil and environmental engineering?   The term civil engineering goes back to the early days of the profession, and was used to distinguish it from military engineering. At that time it included all aspects of engineering, but as time went on, specialisations such as mechanical and electrical engineering expanded to form distinct branches of engineering in their own right.

Civil and environmental engineering now refers to the body of engineering knowledge, skills and practices required for the design and construction of buildings (from low-rise housing to multi-storey tower blocks using advanced composite materials); transport infrastructure such as roads, railways, bridges, airports and tunnels; water supply and waste-water treatment infrastructure including drainage systems, dams and reservoirs; docks and harbours, and offshore drilling platforms. Civil and environmental engineers are likely to become increasingly involved in the development of the innovative coastal and flood protection measures that may be needed to cope with the effects of climate change.

Civil engineering has advanced considerably from the trial and error methods engineering design and construction originally used, which had little scientific basis. Today's civil engineers must have a firm grasp of engineering science, and be able to undertake investigation, research, planning and design. They have to ask the right questions – often difficult ones – and provide the ideas and creativity needed to find answers.

In creating and improving our built environment, engineers have to take particular care to protect the natural environment. Environmental engineering courses have developed from this background and reflect our increased understanding and awareness of our environment. They seek to combine a basic civil engineering education with a fuller understanding of the environmental impacts of projects and processes such as waste disposal, which has, in fact, always been an intrinsic part of civil engineering. The integration of these issues is an important element of the design decision process: it is a new challenge for engineers, and has led to the development of specialised environmental engineering courses. Some courses also seek to equip graduates to play an environmentally informed role in the development of policy in areas such as transportation, water supply, urban rehabilitation and regeneration, and pollution control.

Courses in environmental technology, as opposed to environmental engineering, often lead to a BSc degree rather than a BEng or MEng. They are usually science-based courses providing the environmental scientist with a basic engineering awareness, and promote the integration of environmental science into construction, remediation, reclamation and resource development projects, where their contribution complements the work of engineers. The emphasis on scientific study in environmental technology courses gives a deeper understanding of a subject such as the chemistry of ground contaminants, but may not allow as detailed a study of the possible remedial engineering measures. For these reasons, they are not covered in this Guide, but appear in the *Environmental Sciences* Degree Course Guide.

Social responsibility

The built environment in which we live inevitably impacts on the natural environment. It is now understood that what we do to the natural environment will, in turn, affect our lives. Society increasingly looks to engineers to mitigate the effects of new development on the natural environment and to remedy past mistakes. All civil and environmental engineers must appreciate the implications of their work, and will be required to articulate these wider issues to the general public and their representatives. Holding the broader view, they will often be required to lead and manage large projects requiring the contribution of other specialists and engineers. Team working is essential.

Civil and environmental engineers at home and abroad

Civil engineering and construction work provides many work opportunities in the UK and abroad. Engineers trained in Britain have for many years played an outstanding role in developing countries. It could be argued that in terms of public health, civil and environmental engineers have saved more lives through the provision of clean drinking water and waste-water management than any other profession.

As a civil or environmental engineer, you will often be on site supervising large projects away from head office. You will need to enjoy working outdoors and will routinely be required to demonstrate self-reliance, initiative and engineering ingenuity, together with an adaptability to local conditions. These attributes will develop rapidly through your professional career and will be every bit as important as the scientific knowledge and technical understanding you acquire during your university course.

Specialisms within civil and environmental engineering and related subjects

The role of civil and environmental engineers has expanded enormously over the last 30 years or so, and many options for career specialisation are now available. In recognition of this, there are a number of professional bodies catering for more specialised branches of the profession, in addition to the Institution of Civil Engineers, which covers all types of civil engineer.

For example, the Institution of Structural Engineers regulates and looks after the interests of structural engineers, who are concerned specifically with the behaviour, design and construction of structural frameworks. The content of structural engineering courses overlaps considerably with that of civil engineering courses, but as you would expect, in the later stages more emphasis is placed on the understanding of structures.

Building services engineering is another major branch of the profession with its own professional body, the Chartered Institution of Building Services Engineers. Professional engineers within this branch design and manage the installation of the services required to keep a building functioning and usable by its occupants. This includes: heating; lighting; ventilation; air conditioning; the supply of power, gas and water; lifts and escalators; telecommunications and data communications; and security and fire alarm systems.

Municipal engineers deal specifically with the requirements and problems associated with areas of population. They need to know about building, water treatment and distribution, and public health engineering. They also have to be sensitive to a wide range of environmental issues.

The responsibilities of environmental engineers are wide-ranging, and may overlap with those described in the preceding paragraphs. They include areas such as planning, transportation, development and construction, all now recognised as essential for the sustainable delivery of the infrastructure solutions to meet society's needs. The Chartered Institution of Water and Environmental Management is the professional body covering many of these activities, along with the Institution of Civil Engineers.

Because of the great variety of work involved in civil and environmental engineering, nearly all courses begin with general coverage, but most provide opportunities for specialisation later. Of course, you may not want to become a specialist, and even when a course has a wide choice of options, many students decide to continue with a fairly broad range of subjects, only specialising after they have graduated and entered the profession.

A job in civil engineering   The large majority of civil engineering and building graduates go straight into employment after their degrees, mostly working in industrial and commercial organisations, with the remainder working in the public sector. Comparatively few go on to further study by, for example, taking Master's courses in specialised areas of engineering, or PhDs.

As you might expect, the demand for civil engineers goes up and down depending on the state of the building and construction industry, which is itself closely linked to the state of the economy. Conditions were difficult during the recession of the early 1990s, but the economic recovery of the mid-1990s brought improved prospects for the building and civil engineering industries, and this has continued into the new millennium. Investment in roads, new hospitals, larger universities, out-of-town shopping complexes, rail links to Kent and extensions of the London Underground system have all provided opportunities, and many civil engineering companies are now developing strong links in continental Europe.

Employers of civil engineers include engineering consultants and contractors, central and local government, transport providers, the water industry, energy and oil companies and large organisations with plant and other facilities to be built and maintained. Although very few graduates start their careers abroad, there are many opportunities overseas for the experienced civil engineer, either working on projects for international firms based in the UK, or employed by organisations based abroad.

Much of the work of civil engineers is project-based. Engineers are often attracted to a particular type of project and will move to organisations that offer the type of work that interests them most. Some specialise in public health engineering, others in such fields as concrete structures, ground engineering or soil mechanics.

Working for consultants or contractors   The large civil engineering contractors offer their clients a one-stop comprehensive service from initial design to project management. Civil engineers working in engineering consultancy usually start their career in the design office, creating designs for roads, dams, railways, bridges, buildings, airports and so on. They progress to more managerial functions, such as site-management evaluation, costing, timing, staffing and supply of materials, liaison with and supervision of contractors, negotiations with clients and the presentation of proposals for new work.

Consultants tend to recruit graduates with good honours degrees or a postgraduate qualification. Personal skills such as leadership qualities and management potential are also high on their list of requirements. UK civil engineering consultants are engaged on work throughout the country and in many parts of the world, so their employees must be prepared to work in many different locations as projects are completed and new contracts come along.

Other private-sector employers   A few civil engineers are recruited directly by oil companies to advise them on the civil and structural engineering problems that arise in the development of pipelines, rigs and submersible plant to operate on the seabed. Water companies also employ some civil engineers to develop and maintain their water supply systems and sewage treatment plants.

Network Rail, which owns and operates the railway infrastructure, needs civil engineers for the maintenance of track, tunnels and bridges, and the building of new facilities.

Public-sector employers   Civil engineering and building graduates are also employed in the public sector, primarily within local or central government. The largest number work for county and district councils, where they take part in the planning process for new bridges, roads, buildings and so on, as well as supervising consultants and contractors involved in detailed design and construction work for their authority. If, for example, a local council decides to build a new swimming pool or an underpass to improve traffic flow, its civil engineers will have some responsibility for the project. The Department for Transport is an important recruiter, with responsibility for all major roads and for approving railway development programmes, though the increasing cost of development has led the government to seek greater private sector involvement in major transport projects.

Docks and harbour boards, development corporations for new towns and industrial sites, and the armed forces are also among those offering opportunities.

For further information on opportunities for civil engineers, visit the website of the Institution of Civil Engineers at www.ice.org.uk.

Environmental engineering careers   Graduates from environmental engineering courses can consider an additional range of opportunities. The work of a graduate environmental engineer will usually involve the investigation, analysis and management of existing environmental problems, or of the problems arising from new development. The projects will reflect the specialism followed and the interests of the employer, ranging from environmental audits of multi-million pound construction projects to the design of individual elements of process plant. Overseas, the work can range from near alternative-technology level to massive water supply or flood protection schemes. During the course of their careers, environmental engineers may also engage in consultancy, design and planning, education and training, environmental and transport planning, environmental management, regulation, research, and surveying, measuring and monitoring.

Employers of environmental engineers include civil engineers, specialist consultancies (environmental, engineering and planning), industry (notably utilities such as water, gas and electricity, and large manufacturers), equipment and plant suppliers, regulatory agencies, education and some international bodies.

Opportunities outside engineering   Civil engineers and building graduates are not, of course, confined to careers related directly to their discipline; after all, around 40% of all graduate vacancies are open to people from any subject. Employers are looking for graduates with well-developed personal skills such as teamwork, organisation and communication skills. Civil engineers and building graduates have a variety of personal skills to offer, and, like other numerate and technically qualified graduates, can compete successfully for a whole range of employment opportunities.

Chapter 2: The courses

TABLE 2a lists the specialised and combined courses at universities and colleges in the UK that lead to the award of an honours degree in civil or environmental engineering, building or a closely related subject. When the table was compiled it was as up to date as possible, but sometimes new courses are announced and existing courses withdrawn, so before you finally fill in your application you should check the UCAS website, www.ucas.com, to make sure the courses you plan to apply for are still on offer.

See Chapter 2 in the *Introduction* (page 6) for advice on how to use TABLE 2a and for an explanation of what the various columns mean.

Table 2a — First-degree courses in **Civil and Environmental Engineering and Building**

Institution / Course title	Degree	Duration (years)	Foundation year	Modes of study	Modular scheme	Course type	No of combined courses
Aberdeen www.abdn.ac.uk							
Civil and environmental engineering ①	BEng/MEng	4, 5	◐	● ▼ ○		● ◐	1
Civil and structural engineering ②	BEng/MEng	4, 5	◐	● ▼ ○		● ◐	1
Civil engineering ③	BEng/MEng	4, 5	◐	● ▼ ○		● ◐	1
Abertay www.abertay.ac.uk							
Civil engineering	BSc	4	◐	● ○ ◐		●	0
Anglia Ruskin www.anglia.ac.uk							
Civil engineering	BSc	3, 4		● ▼ ◐		●	0
Construction management	BSc	3, 4		● ▼ ○ ◐		●	0
Aston www.aston.ac.uk							
Construction management	BSc	3, 4		●	◐	●	0
Bath www.bath.ac.uk							
Civil and architectural engineering	MEng	4, 5		●	◐	●	0
Civil engineering	BEng/MEng	3, 4	●	●	◐	●	0
Belfast www.qub.ac.uk							
Civil engineering ①	BEng/MEng	3, 4, 5	●	●	◐	● ◐	1
Environmental and civil engineering	MEng	4, 5	●	●	◐	●	0
Structural engineering with architecture	MEng	4, 5	●	●	◐	●	0
Birmingham www.bham.ac.uk							
Civil engineering	BEng/MEng	3, 4	●	●	○ ◐	● ◐	1
Bolton www.bolton.ac.uk							
Civil engineering	BSc	3, 4	●	● ▼	◐	●	0
Construction	BSc	3, 4	●	● ▼	◐	●	0
Construction management	BSc	3, 4	●	● ▼	◐	●	0
Bradford www.bradford.ac.uk							
Civil and structural engineering	BEng/MEng	3, 4, 5	●	●	○ ◐	●	0
Brighton www.brighton.ac.uk							
Building studies	BSc	3, 4		●	◐	●	0
Civil engineering	BEng/MEng	3, 4, 5	●	●	◐	●	0
Civil with environmental engineering	BEng	3, 4	●	●	◐	●	0
Construction management	BSc	3, 4	●	●	◐	●	0

(continued) Table 2a

First-degree courses in Civil and Environmental Engineering and Building

Institution / Course title	see combined subject list – Table 2b ① ② ③	Degree	Duration (No. of years)	Foundation year: at this institution	Foundation year: at franchised institution	Foundation year: second-year entry	Modes of study: full time / part time	Modes of study: time abroad	Modes of study: sandwich	Modular scheme	Course type: specialised / combined	No of combined courses
Brighton (continued)												
Project management for construction		BSc	3, 4	●			●		◑		●	0
Bristol www.bris.ac.uk												
Civil engineering		BEng/MEng	3, 4				●	○			●	0
Bristol UWE www.uwe.ac.uk												
Civil engineering		BSc	3, 4	●			● ▬		◑		●	0
Construction management ①		BSc	3, 4	●			● ▬		◑		● ◑	1
Cambridge www.cam.ac.uk												
Civil, structural and environmental engineering		BA/MEng	3, 4				●			🎲	●	0
Cardiff www.cardiff.ac.uk												
Architectural engineering		BEng/MEng	3, 4, 5	●			●	○	◑		●	0
Civil and environmental engineering		BEng/MEng	3, 4, 5	●			●	○	◑		●	0
Civil engineering		BEng/MEng	3, 4, 5	●			●	○	◑		●	0
Central Lancashire www.uclan.ac.uk												
Building services engineering		BEng	3, 4		○		● ▬	○	◑		●	0
Construction project management		BSc	3, 4		○		● ▬	○	◑		●	0
City www.city.ac.uk												
Civil engineering		BEng/MEng	3, 4, 5		○		●		◑		● ◑	2
Colchester I www.colch-inst.ac.uk												
Construction management		BSc	3				●				●	
Coventry www.coventry.ac.uk												
Civil and structural engineering		BEng	3	●			● ▬				●	0
Civil engineering ①		BEng	3, 4	●			● ▬		◑		● ◑	1
Civil engineering construction		BEng	3, 4	●			● ▬		◑		●	0
Civil engineering design		BEng	4, 5				● ▬		◑		●	0
Construction management		BSc	3				● ▬	○			●	0
Derby www.derby.ac.uk												
Construction management		BSc	3	●			●				●	0
Dundee www.dundee.ac.uk												
Civil engineering ①		BEng	3, 4			◐	● ▬	○			● ◑	1
Civil engineering design and management		MEng	4, 5				●				●	0
Durham www.durham.ac.uk												
Civil engineering		MEng	4				●	○			●	0
East London www.uel.ac.uk												
Civil engineering		BSc/BEng	3, 4, 5	●			●		◑		●	0
Edinburgh www.ed.ac.uk												
Civil and environmental engineering		BEng/MEng	4, 5			◐	●				●	0
Civil engineering ①		BEng/MEng	4, 5			◐	●	○			● ◑	1
Structural and fire safety engineering		BEng/MEng	4, 5			◐	●				●	0
Structural engineering with architecture ②		BEng/MEng	4, 5			◐	●				● ◑	1
Exeter www.exeter.ac.uk												
Civil engineering		BEng/MEng	3, 4				●	○		🎲	●	0
Glamorgan www.glam.ac.uk												
Civil engineering		BEng	3, 4	●			● ▬	○	◑		●	0
Project management (construction)		BSc	3, 4	●			●		◑		●	0
Glasgow www.gla.ac.uk												
Civil engineering		BEng/MEng	5			◐	●				●	0
Civil engineering with architecture ①		BEng/MEng	5				●				● ◑	1

First-degree courses in Civil and Environmental Engineering and Building

Institution / Course title	①②③ combined	Degree	Duration (years)	FY: this inst ●	FY: franchised ○	FY: 2nd-yr ◗	MS: full time ●	MS: part time ▼	MS: time abroad ○	MS: sandwich ◗	Modular ❂	CT: specialised ●	CT: combined ◐	No. of combined courses
Glasgow Caledonian www.gcal.ac.uk														
Construction management		BSc	4							◗		●		0
Environmental civil engineering		BSc	4					▼	○	◗		●		0
Greenwich www.gre.ac.uk														
Civil engineering		BSc/BEng	3, 4	●			●	▼		◗		●		0
Civil engineering with project management		BEng	3, 4				●	▼		◗		●		0
Civil engineering with water and environmental management		BEng	3, 4				●	▼		◗		●		0
Design and construction management ①		BSc	3				●						◐	1
Heriot-Watt www.hw.ac.uk														
Architectural engineering		BSc/BEng/MEng	4, 5				●		○	◗		●		0
Civil and environmental engineering		BEng/MEng	4, 5			◗	●		○			●		0
Civil engineering		BEng/MEng	4, 5			◗	●		○			●		0
Construction management		BSc	4			◗	●					●		0
Structural engineering ①		BEng/MEng	4, 5			◗	●		○			●	◐	1
Imperial College London www.imperial.ac.uk														
Civil and environmental engineering		MEng	4				●		○			●		0
Civil engineering		MEng	4				●		○			●		0
Kingston www.kingston.ac.uk														
Civil engineering		BEng	3, 4	●			●	▼		◗		●		0
Construction management		BEng	3, 4	●			●			◗		●		0
Leeds www.leeds.ac.uk														
Architectural engineering		BEng/MEng	3, 4	●			●		○			●		0
Civil and environmental engineering		MEng	4	●			●					●		0
Civil and structural engineering		BEng/MEng	3, 4	●			●		○		❂	●		0
Civil engineering with construction management ①		MEng	4	●			●		○			●	◐	1
Civil engineering with transport		BEng	3	●			●					●		0
Energy and environmental engineering ②		BEng/MEng	3, 4	●									◐	1
Leeds Metropolitan www.lmu.ac.uk														
Civil engineering		BSc	3, 4				●	▼	○	◗		●		0
Construction management		BSc	3, 4				●	▼	○	◗		●		0
Liverpool www.liv.ac.uk														
Civil and environmental engineering		MEng	4	●			●					●		0
Civil and maritime engineering		MEng	4	●			●					●		0
Civil and structural engineering		MEng	4	●			●					●		0
Civil engineering		BEng	3	●			●					●		0
Liverpool John Moores www.ljmu.ac.uk														
Building design technology and management		BSc	3	●			●					●		0
Civil engineering		BSc/BEng	3	●	○		●	▼				●		0
Construction management		BSc	3, 4				●	▼		◗		●		0
London South Bank www.sbu.ac.uk														
Architectural engineering		BEng	3, 4	●			●			◗		●		0
Building services engineering		BSc	3, 4	●			●	▼		◗		●		0
Civil engineering		BSc/BEng	3, 4	●			●			◗		●		0
Construction management		BSc	3, 4	●			●	▼	○	◗		●		0
Loughborough www.lboro.ac.uk														
Architectural engineering and design management		BSc	3, 4	●			●			◗		●		0
Civil engineering		BEng/MEng	3, 4, 5	●			●		○	◗		●		0
Construction engineering management		BSc	4							◗		●		0

First-degree courses in **Civil and Environmental Engineering and Building** — (continued) Table 2a

Institution / Course title	①②③ see combined subject list – Table 2b	Degree	Duration (years)	Foundation year: at this institution	at franchised institution	second-year entry	Modes of study: full/part time	time abroad	sandwich	Modular scheme	Course type (specialised/combined)	No. of combined courses
Manchester www.man.ac.uk												
Civil and structural engineering		MEng	4	●			●				●	0
Civil engineering		BEng/MEng	3, 4	●			●	O			●	0
Napier www.napier.ac.uk												
Civil and timber engineering		BEng	4				●				●	0
Civil and transportation engineering		BEng	4		◐		●		◑		●	0
Civil engineering ①		BSc	4		◐		●		◑		● ◕	1
Construction and project management ②		BSc	4				●				● ◕	1
Construction engineering		BSc	4		◐		●				●	0
Energy and environmental engineering ③		BEng	4		◐		●	O	◑		◕	1
Newcastle www.ncl.ac.uk												
Civil and structural engineering		BEng/MEng	3, 4	●			●	O			●	0
Civil engineering		BEng/MEng	3, 4, 5	●			●	O			●	0
Environmental engineering		BEng/MEng	3, 4	●			●	O			●	0
North East Wales I www.newi.ac.uk												
Construction management		BSc	3	●			●				●	0
Northumbria www.northumbria.ac.uk												
Building design management		BSc	4				◕			✿	●	0
Building project management		BSc	4				◕				●	0
Building services engineering		BEng	3, 4	●			●			✿	●	0
Construction management		BSc	4	●			◕				●	0
Nottingham www.nottingham.ac.uk												
Architectural environment engineering		BEng	3	●			●				●	0
Civil engineering ①		BEng/MEng	3, 4	●			●	O		✿	● ◕	2
Environmental engineering		BEng/MEng	3, 4	●			●		◑	✿	●	0
Nottingham Trent www.ntu.ac.uk												
Civil and structural engineering		BEng	3, 4	● O			●		◑		●	0
Civil engineering		BSc/BEng/MEng	3, 4	● O			●	O	◑		●	0
Construction management		BSc	4				●	O	◑		●	0
Oxford www.ox.ac.uk												
Civil engineering		MEng	4	●			●					0
Oxford Brookes www.brookes.ac.uk												
Building		BSc	4	●			▼		◑		●	0
Construction management		BSc	4	●			▼		◑		●	0
Paisley www.paisley.ac.uk												
Civil engineering (BEng)		BEng	4		◐				◑		●	0
Plymouth www.plymouth.ac.uk												
Civil and coastal engineering		BEng/MEng	3, 4, 5	●			●	O	◑		●	0
Civil and coastal engineering (BSc)		BSc	3, 4	●			●	O	◑		●	0
Civil engineering		BEng/MEng	3, 4, 5	●			●	O	◑		●	0
Civil engineering (BSc)		BSc	3, 4	●			●	O	◑		●	0
Construction management and the environment		BSc	3	● O			●				●	0
Portsmouth www.port.ac.uk												
Civil engineering		BEng/MEng	3, 4, 5	●			● ▼	O	◑		●	0
Construction engineering management		BEng	3, 4	●			● ▼	O	◑		●	0
Reading www.rdg.ac.uk												
Building construction and management		BSc	3	●			●			✿	●	0
Construction management and surveying ①		BSc	3	●			●				◕	1

(continued) Table 2a

First-degree courses in **Civil and Environmental Engineering and Building**

Key to columns:
- Foundation year: ● at this institution; ○ at franchised institution; ◑ second-year entry
- Modes of study: ● full time; ▼ part time; ○ time abroad; ◑ sandwich
- Course type: ● specialised; ◑ combined

Institution / Course title	see combined subject list – Table 2b	Degree	Duration (number of years)	Foundation year	Modes of study	Modular scheme	Course type	No of combined courses
Robert Gordon www.rgu.ac.uk								
Construction, design and management		BSc	4	◑	● ▼ ○	◉	●	0
Mechanical and environmental engineering		BSc	4, 5		● ▼		●	0
Salford www.salford.ac.uk								
Civil engineering ①		BSc/BEng/MEng	3, 4, 5	●	● ○ ◑		● ◑	1
Construction management		BSc	4		● ◑		●	0
Construction project management		BSc	3		●		●	0
Structural engineering with computer science		BEng	3, 4		● ◑		●	0
Sheffield www.sheffield.ac.uk								
Civil and structural engineering		MEng	4	●	● ○		●	0
Civil engineering ①		BEng/MEng	3, 4	●	● ○		● ◑	2
Structural engineering and architecture ②		MEng	4	●	●		◑	1
Structural engineering design		MEng	4	●	●		●	0
Structural engineering with architectural studies ③		MEng	4	●	●		● ◑	1
Sheffield Hallam www.shu.ac.uk								
Construction management		BSc	4	●	▼ ◑		●	0
Southampton www.soton.ac.uk								
Civil engineering ①		BEng/MEng	3, 4	○	● ○ ◑		● ◑	2
Environmental engineering ②		BEng/MEng	3, 4	○	● ○ ◑		● ◑	1
Southampton Solent www.solent.ac.uk								
Construction management		BSc	3	● ○	●		●	0
Strathclyde www.strath.ac.uk								
Architectural engineering		BEng/MEng	4, 5	◑	●		●	0
Civil engineering ①		BEng/MEng	4, 5	◑	● ○		● ◑	2
Surrey www.surrey.ac.uk								
Civil engineering		BEng/MEng	3, 4, 5	●	● ○ ◑		●	0
Swansea www.swan.ac.uk								
Civil engineering		BEng/MEng	3, 4	●	● ○ ◑		● ◑	1
Swansea IHE www.sihe.ac.uk								
Civil engineering and environmental management ①		BSc	3	●	● ▼	◉	◑	1
Construction management		BSc	3	●	● ▼		●	0
Project and construction management		BSc	3	●	● ▼		●	0
Teesside www.tees.ac.uk								
Civil engineering		BEng	3, 4	●	● ○ ◑		●	0
Civil engineering with disaster management		BEng	3, 4	●	● ◑		●	0
UCE Birmingham www.uce.ac.uk								
Construction management and economics		BSc	3, 4		● ▼ ◑		●	0
UCL www.ucl.ac.uk								
Civil engineering		BEng/MEng	3, 4	●	● ○		● ◑	0
Environmental engineering		BEng/MEng	3, 4		● ○		●	0
Ulster www.ulster.ac.uk								
Building services engineering		BEng	3		●		●	0
Civil engineering		BEng	4		▼ ○ ◑		●	0
Construction engineering ①		BSc	3, 4		● ▼ ◑		● ◑	1
Environmental engineering		BEng	4		▼ ◑		●	0
Wales (Newport) www.newport.ac.uk								
Building studies		BSc	3	○	●		●	0
Civil and construction engineering		BSc	3		●		●	0

Institution / Course title	①②③ see combined subject list – Table 2b	Degree	Duration (Number of years)	Foundation year (● at this institution, ○ at franchised institution, ◑ second-year entry)	Modes of study (● full time; ● part time, ○ time abroad, ◑ sandwich)	Modular scheme	Course type (● specialised; ◑ combined)	No of combined courses
Warwick www.warwick.ac.uk								
Civil engineering		BEng/MEng	3, 4, 5		● ○		●	0
Westminster www.wmin.ac.uk								
Architectural engineering		BSc	3	●	● ○		●	0
Building engineering		BSc	3, 4	●	● ●	◑ ♦	●	0
Construction management ①		BSc	3, 4	●	● ●	◑ ♦	● ◑	1
Wolverhampton www.wlv.ac.uk								
Civil engineering		BSc	3, 4	●	● ● ◑		●	0
Construction management		BSc	3		● ●		●	0
Construction project management ①		BSc	3		●		◑	6

Subjects available in combination with civil engineering

TABLE 2b shows those subjects that can be taken in roughly equal proportions with one of the subjects covered by this part of the Guide in the combined degrees listed in TABLE 2a. See Chapter 2 in the *Introduction* (page 6) for general information about combined courses and for an explanation of how to use TABLE 2b.

Subjects to combine with Civil and Environmental Engineering or Building

Architecture   City, Edinburgh②, Glasgow①, Heriot-Watt①, Napier①, Sheffield② ③, Southampton①
Business studies   Sheffield①, Westminster①, Wolverhampton①
Computer science   Salford①
Computing   Napier②, Wolverhampton①
Construction   Edinburgh①, Leeds①
Design   Greenwich①
Energy studies   Leeds②
Energy technology   Napier②
Engineering   Wolverhampton①
Environmental engineering   Belfast①

Environmental management   Strathclyde①, Swansea IHE①, Wolverhampton①
European studies   Aberdeen① ②, Southampton① ②, Strathclyde①
French   Nottingham①
German   Nottingham①
Health studies   Wolverhampton①
Management studies   Aberdeen②, Birmingham, Coventry①, Dundee①, Swansea, Ulster①
Modern languages   Sheffield①
Sports studies   Wolverhampton①
Surveying   City, Reading①

Other courses that may interest you

Heriot-Watt offers a degree in Civil engineering financial management, and West of England one in Rivers and coastal engineering. In addition, you may find courses in the *Integrated Engineering and Mechatronics* part of this Guide of interest. Architecture and architectural technology courses are listed in the *Architecture, Planning and Surveying* Guide.

Chapter 3: The style and content of the courses

Industrial experience and time abroad   Many courses provide a range of opportunities for spending a period of industrial training in the UK or abroad or of study abroad. TABLE 3a gives information about the possibilities for the courses in this Guide. See Chapter 3 in the *Introduction* (page 11) for further information.

Table 3a — Time abroad and sandwich courses

Key:
- Institution / Course title — ①②: see notes after table
- Named 'international' variant of the course
- Location: ● Europe; ○ North America; ◗ industry; ◑ academic institution
- Maximum time abroad (months)
- Time abroad assessed
- Language study: ○ optional; ● compulsory; * contributes to assessment
- Socrates–Erasmus
- Sandwich courses: ● thick; ○ thin
- Arranged by: ● institution; ○ student

Institution / Course title	Named international variant	Location	Maximum time abroad (months)	Time abroad assessed	Language study	Socrates–Erasmus	Sandwich courses	Arranged by
Aberdeen								
Civil and environmental engineering	●	● ○ ◗ ◑	12	●	○*	●		
Civil and structural engineering	●	● ○ ◗ ◑	12	●	○*	●		
Civil engineering	●	● ○ ◗ ◑	12	●	○*	●		
Abertay								
Civil engineering		● ○ ◗	12				●	● ○
Anglia Ruskin								
Civil engineering ③					○*		●	○
Construction management ① ③	●	◗ ◑	12	●	○	●	●	● ○
Bath								
Civil and architectural engineering ③		● ○ ◗ ◑	12		○			○
Civil engineering		● ○ ◗	12		○			○
Belfast								
Civil engineering ③ ④					○*			
Birmingham								
Civil engineering ① ③ ④	●	● ○ ◗ ◑	12	●	○*	●	●	● ○
Bolton								
Civil engineering ③							●	● ○
Construction ③							●	● ○
Construction management ③							●	● ○
Bradford								
Civil and structural engineering ③		◗ ◑	12			●	●	●
Brighton								
Civil engineering ③		● ◗	12		○	●	●	● ○
Civil with environmental engineering ③		● ◗ ◑	12		○	●	●	● ○
Bristol								
Civil engineering ①	●	● ○ ◑	12	●	○*	●		
Bristol UWE								
Civil engineering ③							●	● ○
Construction management ③		● ○ ◗	12	●	○*		●	● ○
Cambridge								
Civil, structural and environmental engineering ④					○*			
Cardiff								
Architectural engineering		● ○ ◗	15				●	○
Civil and environmental engineering ③	●	● ○ ◗ ◑		●	●*		●	● ○
Civil engineering		● ○ ◗	15				●	○

Time abroad and sandwich courses

Ⓘ ②: see notes after table

Institution / Course title	Named 'international' variant of the course	Location: ● Europe; ○ North America; ◗ industry; ◑ academic institution	Maximum time abroad (months)	Time abroad assessed	Language study: ○ optional; ● compulsory; * contributes to assessment	Socrates–Erasmus	Sandwich courses: ● thick; ○ thin	Arranged by: ● institution; ○ student
Central Lancashire								
Building services engineering ③ ④		● ○ ◗ ◑	12	●	○*	●	●	● ○
Construction project management ① ③ ④	●	● ◗ ◑	12	●	○*	●	●	● ○
City								
Civil engineering ③		● ○ ◗ ◑	12	●	○	●	●	●
Coventry								
Civil and structural engineering ③							●	● ○
Civil engineering ③							●	● ○
Civil engineering construction ③							●	● ○
Construction management ③	●	● ◑	12	●	○*	●	●	○
Dundee								
Civil engineering ③		○ ◑	12	●	○*			
Durham								
Civil engineering ③	●	● ○ ◗ ◑	12	●	○			
East London								
Civil engineering ③							● ○	○
Edinburgh								
Civil and environmental engineering		● ○				●		
Civil engineering ③		○ ◑	12	●	○*	●		
Exeter								
Civil engineering ③		● ○ ◑	6	●	○*	●		
Glamorgan								
Civil engineering ①		◗					●	● ○
Project management (construction) ③	●	● ◗ ◑	15	●	○	●	●	○
Glasgow								
Civil engineering						●		
Civil engineering with architecture						●		
Glasgow Caledonian								
Environmental civil engineering						●	●	○
Greenwich								
Civil engineering ③						●	●	●
Civil engineering with project management ③						●	●	●
Civil engineering with water and environmental management ③							●	
Heriot-Watt								
Civil and environmental engineering ③	●	● ○ ◗ ◑	12	●	○*	●		
Civil engineering ③	●	● ○ ◗ ◑	12	●	○*	●		
Construction management	●	● ◗ ◑				●		
Structural engineering ③	●	● ○ ◗ ◑	12	●	○*	●		
Imperial College London								
Civil and environmental engineering	●	● ◑	12	●	●*	●		
Civil engineering	●	● ◑	12	●	●*	●		
Kingston								
Civil engineering ③					○		●	●
Construction management ③							●	●
Leeds								
Architectural engineering ③		○ ◑	12	●				
Civil and environmental engineering ③								
Civil and structural engineering ③	●	● ◑	12	●	○*	●		
Civil engineering with construction management ③		○ ◑	12	●				

(continued) Table 3a — Time abroad and sandwich courses

Institution / Course title ① ②: see notes after table	Named 'international' variant of the course	Location: ● Europe; ○ North America; ➤ industry; ◗ academic institution	Maximum time abroad (months)	Time abroad assessed	Language study: ○ optional; ● compulsory; * contributes to assessment	Socrates–Erasmus	Sandwich courses: ● thick; ○ thin	Arranged by: ● institution; ○ student
Leeds (continued)								
Energy and environmental engineering ③								
Leeds Metropolitan								
Civil engineering	●	➤	12	●			●	● ○
Construction management		○ ◗	12	●			●	● ○
Liverpool								
Civil and environmental engineering ③					○*			
Civil and maritime engineering ③					○*			
Civil and structural engineering ③					○*			
Civil engineering ③					○*			
Liverpool John Moores								
Civil engineering					○*	●	●	
Construction management					○*	●	●	
London South Bank								
Architectural engineering						●	●	
Building services engineering ①						●	●	● ○
Civil engineering ①	●		1	●		●	●	● ○
Construction management ① ③	●	➤ ◗	12		○*	●	●	○
Loughborough								
Architectural engineering and design management ③	●	➤	12			●	●	● ○
Civil engineering ③	●	◗	4	●	○*	●	●	● ○
Construction engineering management	●	○ ➤	4	●	○*		○	
Manchester								
Civil and structural engineering					○*			
Civil engineering	●	● ○	9	●	○*	●		
Napier								
Civil and transportation engineering		● ○ ◗	12		○*		○	● ○
Civil engineering		● ○ ◗	12		○*		○	● ○
Energy and environmental engineering		● ➤	6	●	○	●	○	● ○
Newcastle								
Civil and structural engineering ③		● ○ ◗	12	●	○*	●		● ○
Civil engineering ③		● ○ ◗	12	●	○*	●	●	● ○
Environmental engineering ③		● ○ ◗	12	●	○*	●		
Northumbria								
Building services engineering						●		● ○
Construction management								● ○
Nottingham								
Civil engineering ③	●	● ○ ◗	10	●	○*	●		
Environmental engineering ① ③					○*	●		
Nottingham Trent								
Civil and structural engineering		➤	12		○	●	●	● ○
Civil engineering		➤	12		○	●	●	● ○
Construction management	●	○ ➤ ◗	12	●	○	●	●	● ○
Oxford								
Civil engineering ③					○*			
Oxford Brookes								
Building ④	●	○ ➤ ◗	15	●	○	●	●	
Construction management							●	● ○

Institution / Course title (① ②: see notes after table)	Named 'international' variant of the course	Location: ● Europe; ○ North America; ◗ industry; ◑ academic institution				Maximum time abroad (months)	Time abroad assessed	Language study: ○ optional; ● compulsory; * contributes to assessment	Socrates–Erasmus	Sandwich courses: ● thick; ○ thin	Arranged by: ● institution; ○ student
Plymouth											
Civil and coastal engineering		●	○	◗		12		○*		●	● ○
Civil and coastal engineering (BSc)			○			9	●	○*		●	● ○
Civil engineering ③		●	○	◗		12	●	○*		●	● ○
Civil engineering (BSc)			○			9	●	○*		●	● ○
Construction management and the environment		●	○	◗		12	●	○*		●	○
Portsmouth											
Civil engineering ③		●	○	◗	◑	12	●	○*	●	●	● ○
Construction engineering management ③										●	● ○
Robert Gordon											
Construction, design and management ① ③		●		◗		6	●	○	●	○ ●	● ○
Salford											
Civil engineering	●	●		◗		12	●	○*	●	●	● ○
Sheffield											
Civil and structural engineering	●	●	○		◑	12	●	○*	●		
Civil engineering	●	●	○		◑	12	●	○*	●		
Structural engineering and architecture ③		●					●	○*	●		
Structural engineering design		●				3	●	○*	●		
Structural engineering with architectural studies		●				3	●	○*	●		
Sheffield Hallam											
Construction management ③				◗		9	●	○*	●	●	●
Southampton											
Civil engineering ③ ④	●	●		◗	◑	16	●	●*	●	●	● ○
Environmental engineering ③ ④	●	●		◗	◑	16	●	●*	●	●	● ○
Strathclyde											
Civil engineering ④	●	●	○		◑	8	●	○*	●		
Surrey											
Civil engineering		●	○	◗	◑		●		●	●	●
Swansea											
Civil engineering ③	●	●		◗	◑	12	●	○*	●		
Swansea IHE											
Civil engineering and environmental management ③		●		◗		1		○			
Project and construction management		●		◗		1					
Teesside											
Civil engineering ③		●		◗	◑	12	●		●	●	● ○
UCE Birmingham											
Construction management and economics ① ② ③ ④		●		◗		12	●	○			
UCL											
Civil engineering ③	●	●	○		◑	12	●	○*	●		
Environmental engineering ③	●	●	○		◑	12	●	○*	●		
Ulster											
Civil engineering ④		●	○	◗	◑	12	●		●	●	●
Construction engineering								○	●	●	
Environmental engineering ④		●	○	◗		12	●		●	●	● ○
Warwick											
Civil engineering ③		●	○		◑	12	●	○*	●	●	○
Westminster											
Architectural engineering ③		●				6	●	○*	●		
Building engineering ① ② ③ ④								○*		●	● ○

(continued) **Table 3a** Time abroad and sandwich courses									
Institution Course title ① ②: see notes after table	**Named 'international' variant of the course**	**Location:** ● Europe; ○ North America; ❱ industry; ◑ academic institution	**Maximum time abroad** (months)	**Time abroad assessed**	**Language study:** ○ optional; ● compulsory; * contributes to assessment	**Socrates–Erasmus**	**Sandwich courses:** ● thick; ○ thin	**Arranged by:** ● institution; ○ student	
Westminster (continued) Construction management ① ② ③ ④					○*		●	● ○
Wolverhampton Construction management							●	○

① A year of industrial experience before the course starts is recommended for students on non-sandwich courses
② A year of industrial experience before the course starts is recommended for students on sandwich courses
③ Students on non-sandwich courses can take a year out for industrial experience
④ Minimum period of vacation industrial experience for non-sandwich students (weeks) : Belfast <u>Civil engineering</u> 12 ; Birmingham <u>Civil engineering</u> 8 ; Cambridge <u>Civil, structural and environmental engineering</u> 8 ; Central Lancashire <u>Building services engineering</u> 48 <u>Construction project management</u> 48 ; Oxford Brookes <u>Building</u> 10 ; Southampton <u>Civil engineering</u> 20 <u>Environmental engineering</u> 20 ; Strathclyde <u>Civil engineering</u> 8 ; UCE Birmingham <u>Construction management and economics</u> 8 ; Ulster <u>Civil engineering</u> 46 <u>Environmental engineering</u> 52 ; Westminster <u>Building engineering</u> 4 <u>Construction management</u> 4

Engineering applications   See Chapter 3 in the *Introduction* (page 11) for general information about the inclusion of industrial applications in civil and environmental engineering courses.

A typical course   Many civil engineering courses have similar content and structure, so it is possible to give a reasonable idea of how you can expect to spend your time on a typical course. Remember that any specific course is likely to differ from this typical course in some respects. The numbers given are a rough guide to the hours spent per week in formal teaching (lectures, tutorials and so on) on each topic.

Table 3b	The content of a typical **Civil Engineering** course					
	Year 1	Time (hours per week)	**Year 2**	Time (hours per week)	**Year 3**	Time (hours per week)
	Structural mechanics	2	Structural analysis	2	Structural engineering	2
	Geology	2	Design	2	Geotechnical engineering	2
	Fluid mechanics	2	Mathematics	2	Hydraulics/hydrology	2
	Mathematics	3	Soil mechanics	2	Construction management	2
	Computing	1	Fluid mechanics	2	Specialised options	6
	Surveying	1	Materials technology	1	Design project	5
	Drawing/CAD	3	Computing	1		
	Design	1	Surveying	2		
	Engineering science	1	Laboratory/fieldwork	4		
	Laboratory/fieldwork	3				

Introductory period   The early stages of courses usually concentrate on building a firm basis in mathematics, properties of materials, applied mechanics, fluid mechanics and engineering drawing. Most branches of engineering are built on similar foundations, so a number of institutions begin many of their engineering courses with a common introductory period: see TABLE 3c for where this is the case and at what point specialisation occurs.

Specialised content   You will probably be introduced to some specialised civil engineering subjects at an early stage, and as the course progresses, they will take an increasing proportion of your time. By the final year you will spend virtually all your time studying specialised subjects. It is at this stage that most courses allow you to develop particular interests through a choice of what can in some cases be very specialised options: see TABLE 3c for more details.

The emphasis and specific content of the civil engineering topics vary a little from course to course, but the 'typical course' shown above provides a reasonable illustration of what to expect.

The work covered under surveying also varies, but might typically include electronic distance measurement, the principles of setting-out for construction work and photogrammetry (the use of aerial photographs for surveying), as well as the basic principles and practice of surveying.

On the design side, different courses again give different emphasis to topics, which may include concrete technology, roads and bridges, water and waste-water engineering, and highway and traffic engineering. In some courses these subjects are detailed in the prospectus; in others they appear under the general heading of civil engineering, since many studies in design and construction are common to all civil engineering work.

The building science parts of civil engineering are closely related to architectural studies, and include subjects such as engineering services, acoustics, lighting, building materials and architectural aerodynamics.

Building   Building and associated courses are generally narrower in scope than civil and environmental engineering courses, but they can provide a good preparation for a field of work that is already very large and is continuing to expand. The design, construction, use and maintenance of buildings and the services they contain are becoming ever more complex, requiring well-trained engineers.

Courses in building place a greater emphasis than civil and environmental engineering courses on subjects such as economics, quantity surveying and industrial organisation. If you are interested in building work generally, a course in building engineering would bring you into contact with many engineering aspects of building services including structural engineering and building services engineering, as well as other aspects of design and construction.

Final-year specialisation   The requirements for professional recognition mean that many civil engineering courses offer very little choice until the final year, and even then the choice may be fairly limited. TABLE 3c shows you how much of your time will be spent on options in your final year and where particular areas of specialisation are available. Topics available only on the MEng variant of courses offered as either BEng or MEng are shown as ▾. Remember that the titles given to topics may not correspond exactly to those used in individual prospectuses.

Table 3c — Final-year specialisation

Institution / Course title ①②: see notes after table	T=term; S=semester; Y=year	Time of specialisation after a broad introduction	Time spent on options %	Coastal/offshore engineering	Concrete structures	Contract/project management	Fluid mechanics	Geotechnical engineering	HVAC/building services	Hydraulics	Materials engineering	Numerical analysis	Public health engineering	Steel structures	Structural analysis	Transportation/highways	Waste/water engineering	Environmental impact assessment	Environmental engineering	GIS	Env conservation and management	Energy and the environment	Environmental law
Aberdeen																						
Civil and environmental engineering	Y 3		15	●	●	●	●	●	●					●					●				
Civil and structural engineering	Y 3				●	●	●	●	●					●	●								
Civil engineering ①	Y 3		15	▾	●	●	●	●	●					●	●	●			●				
Abertay																						
Civil engineering ①																						
Anglia Ruskin																						
Civil engineering			20	●		●		●		●			●	●	●	●	●						
Construction management ①			60			●																	
Bath																						
Civil and architectural engineering ①			15		▾		▾	▾						▾	▾								
Civil engineering ②			15																				
Belfast																						
Civil engineering			30	●	●	●	●		●		●	●	●	●	●	●	●						

Final-year specialisation

Institution / Course title / ①②: see notes after table	T=term; S=semester; Y=year / Time of specialisation after a broad introduction	Time spent on options %	Coastal/offshore engineering	Concrete structures	Contract/project management	Fluid mechanics	Geotechnical engineering	HVAC/building services	Hydraulics	Materials engineering	Numerical analysis	Public health engineering	Steel structures	Structural analysis	Transportation/highways	Waste/water engineering	Environmental impact assessment	Environmental engineering	GIS	Env conservation and management	Energy and the environment	Environmental law
Belfast (continued)																						
Structural engineering with architecture		30	●				●	●					●	●								
Birmingham																						
Civil engineering ①		100	●	●	●	●	●		●		●	●	●	●	●	●	●	●				
Bolton																						
Civil engineering ①		15		●			●						●	●	●							
Construction ②		60			●													●				
Construction management ③		15			●																	
Bradford																						
Civil and structural engineering ①	T 2	17		●									●	●		▼	▼					
Brighton																						
Civil engineering ①		55	●	●	●	●			●				●	●	●	●	●	●				
Civil with environmental engineering		55	●	●	●	●			●			●	●	●	●	●	●	●		●	●	●
Bristol																						
Civil engineering ①		33	▼	▼			●		●		●	●		●		●	▼	▼				
Bristol UWE																						
Construction management ①					●																	
Cambridge																						
Civil, structural and environmental engineering ①	Y 3	100	●	●		●	●		●				●	●								
Cardiff																						
Architectural engineering ①		25		●	●		●	●					●					●				
Civil and environmental engineering	S 2	25		●	●	●			●		●		●		●	●	●	●				
Civil engineering ②	S 2	25	●	●	●		●		●		●		●	●	●		●				●	
Central Lancashire																						
Building services engineering ①				●	●			●	●		●		●									
Construction project management ②		20			●		●			●	●											
City																						
Civil engineering ①		40									●	●		●	●	●			●			
Coventry																						
Civil and structural engineering ①		15	●	●	●		●	●					●	●	●							
Civil engineering		15	●	●	●	●				●	●		●									
Civil engineering construction ②		15		●						●	●	●	●									
Construction management		30		●																		
Derby																						
Construction management	Y 2																					
Dundee																						
Civil engineering	Y 2	60	●	●	●	●			●		●		●	●		●	●	●				
Durham																						
Civil engineering	Y 3	100	●	●	●	●			●				●	●								
East London																						
Civil engineering ①		16					●							●	▼	●	▼					
Edinburgh																						
Civil and environmental engineering			●	●	●	●			●						●	●	●			●		
Civil engineering ①	Y 3	15			●		●				●	●	●		●		●					
Exeter																						
Civil engineering	Y 2	25		●	●	●			●	●	●	●	●	●				●				

(continued) Table 3c

Final-year specialisation

Institution / Course title ① ② : see notes after table	T=term; S=semester; Y=year	Time of specialisation after a broad introduction	Time spent on options %	Coastal/offshore engineering	Concrete structures	Contract/project management	Fluid mechanics	Geotechnical engineering	HVAC/building services	Hydraulics	Materials engineering	Numerical analysis	Public health engineering	Steel structures	Structural analysis	Transportation/highways	Waste/water engineering	Environmental impact assessment	Environmental engineering	GIS	Env conservation and management	Energy and the environment	Environmental law
Glamorgan																							
Civil engineering		30			●	●	●			●				●	●	●	●						
Project management (construction) ①		20				●																	
Glasgow																							
Civil engineering		35			●	●		●				◐		●		●	●		◐				
Civil engineering with architecture		35			●	●		●				◐		●		●	●		◐				
Glasgow Caledonian																							
Environmental civil engineering		16				●																	
Greenwich																							
Civil engineering ①		25			●	●		●								●	●				●		
Civil engineering with project management		25			●	●		●		●		●											
Civil engineering with water and environmental management		25			●			●		●							●						
Heriot-Watt																							
Civil and environmental engineering		50		●	●			●				●	●	●	●	●	●		●				
Civil engineering ①		50		●	●			●				●	●	●	●	●	●		●				
Construction management ②		40				●			●														
Structural engineering		50			●						●	●		●	●	●							
Imperial College London																							
Civil and environmental engineering ①		100						●	●	●		●	●	●	●	●	●						
Civil engineering ②		100						●	●	●	●	●	●	●	●	●	●						
Kingston																							
Civil engineering		25						●						●		●							
Construction management	Y 2	10				●													●				
Leeds																							
Architectural engineering		20			●	●	●	●	●	●		●	●	●	●	●	●	●	●			●	
Civil and environmental engineering		30			●	●	●	●	●	●		●	●	●	●	●	●	●	●			●	●
Civil and structural engineering		30			●	●	●	●	●	●		●	●	●	●	●	●	●	●			●	●
Civil engineering with construction management		30			●	●	●	●	●	●		●	●	●	●	●	●	●	●			●	●
Leeds Metropolitan																							
Civil engineering		25				●		●		●						●							
Construction management ①						●			●													
Liverpool																							
Civil and environmental engineering		10		●	●	●	●	●		●				●	●	●	●	●	●				
Civil and maritime engineering		0		●	●	●	●	●		●				●	●	●	●	●	●				
Civil and structural engineering		20		●	●	●	●	●		●				●	●	●	●	●	●				
Civil engineering		50		●	●	●	●	●		●				●	●	●	●	●	●				
Liverpool John Moores																							
Civil engineering		15			●	●	●			●	●	●	●	●	●	●			●				
London South Bank																							
Architectural engineering	Y 3	50			●	●			◐				●										
Building services engineering	Y 2	50							●														
Civil engineering ①	Y 2	70		●		◐		●		●		●			●	●	●						
Construction management	Y 2	12				●																	

Final-year specialisation

Institution / Course title (① ⑩: see notes after table)	Time of specialisation after a broad introduction (T=term; S=semester; Y=year)	Time spent on options %	Coastal/offshore engineering	Concrete structures	Contract/project management	Fluid mechanics	Geotechnical engineering	HVAC/building services	Hydraulics	Materials engineering	Numerical analysis	Public health engineering	Steel structures	Structural analysis	Transportation/highways	Waste/water engineering	Environmental impact assessment	Environmental engineering	GIS	Env conservation and management	Energy and the environment	Environmental law
Loughborough																					
Architectural engineering and design management ①	Y 2	30			●		●	●														
Civil engineering ②		25		●		●	●		●			●	●	●	●	●		●				
Construction engineering management ③		25																				
Manchester																					
Civil and structural engineering ①	Y 3	25		●	●	●	●						●	●	●	●	●					
Civil engineering ②		25		●		◡	●						●		◡	◡						
Napier																					
Civil and transportation engineering		0																				
Civil engineering ①		15					●					●		●	●							
Newcastle																					
Civil and structural engineering ①	Y 3	10	●	●	◡		●				●		●	●		◡		●		◡	◡	
Civil engineering ②		20		●	◡		●		●				●	●	◡	◡		●		◡	◡	
Environmental engineering ③	Y 3	10		●	◡		●						●			◡		●				
Northumbria																					
Building services engineering	Y 3	50			●			●														
Nottingham																					
Architectural environment engineering		25						●														
Civil engineering		80		●	●	●	●		●	●			●		●	●						
Environmental engineering		25			◡		●		●									●		●		●
Nottingham Trent																					
Civil and structural engineering ①		20		●	●	●	●				●	●	●	●		●		●				
Civil engineering ②		30		●	●	●	●			●	●	●	●	●		●		●	●		●	
Construction management ③	Y 4	10																		●		
Oxford																					
Civil engineering ①	Y 3	50	◡	●	●	●	●		●	●			●	●		◡						
Oxford Brookes																					
Building ①		100		●				●														
Paisley																					
Civil engineering (BEng)		25	●	●		●	●				●		●		●	●						
Plymouth																					
Civil and coastal engineering		30	●	●		●	●			●		●	●	●	●							
Civil and coastal engineering (BSc)		20	●												●							
Civil engineering		30	●	●	●	●				●			●	●	◡							
Civil engineering (BSc)		20	●												●							
Portsmouth																					
Civil engineering ①		17	●	●	●				●	●			●	●			●					
Construction engineering management		17	●	●	●	●	●						●	●			●					
Reading																					
Building construction and management ①		25		●				●														
Robert Gordon																					
Construction, design and management	Y 2																					

(continued) Table 3c

Institution / Course title ① ② : see notes after table	T=term; S=semester; Y=year	Time of specialisation after a broad introduction	Time spent on options %	Coastal/offshore engineering	Concrete structures	Contract/project management	Fluid mechanics	Geotechnical engineering	HVAC/building services	Hydraulics	Materials engineering	Numerical analysis	Public health engineering	Steel structures	Structural analysis	Transportation/highways	Waste/water engineering	Environmental impact assessment	Environmental engineering	GIS	Env conservation and management	Energy and the environment	Environmental law
Robert Gordon (continued)																						
Mechanical and environmental engineering	S 2	25				●												●	●				
Salford																						
Civil engineering		30	●	●						●			●	●		●	●	●				
Sheffield																						
Civil and structural engineering ①		20		●	●	●	●		●	●	●		●	●			●	●				
Civil engineering		20		●	●	●	●		●	●	●		●	●			●	●				
Structural engineering and architecture ②		15		●			●	●					●	●			●	●				
Structural engineering design		20		●	●	●	●			●			●	●			●	●				
Structural engineering with architectural studies		20		●	●					●			●	●			●	●				
Sheffield Hallam																						
Construction management ①		18																				
Southampton																						
Civil engineering ①		40	◐												◐							
Environmental engineering ②	Y 2	40	◐												◐							
Strathclyde																						
Civil engineering		15	●	●		◐						●	●				●	●				
Surrey																						
Civil engineering ①	Y 2	25		●	●	●	●		●	●	●		●				●	●				
Swansea																						
Civil engineering ①		10					●			●				●			●	●				
Swansea IHE																						
Civil engineering and environmental management	T 2	10	●														●	●				
Project and construction management	T 2	10															●				●	
Teesside																						
Civil engineering				●			●	●		●			●	●			●	●				
UCE Birmingham																						
Construction management and economics		10																				
UCL																						
Civil engineering			●	●	●	●	●		●	●	●	●	●	●	●	●						
Environmental engineering			●			●						●					●	●		●	●	●
Ulster																						
Civil engineering		17		●									●	●	●			●				
Construction engineering		33			●																	
Environmental engineering ①		16				●																
Wales (Newport)																						
Civil and construction engineering			●	●	●	●		●	●	●			●	●		●		●		●		
Warwick																						
Civil engineering ①	Y 2	38	●	●	●	●		◐	●	●	◐		●	●		◐		◐		◐		◐
Westminster																						
Architectural engineering		25		●		●							●	●		●						
Building engineering ①		37		●	●			●		●				●								
Construction management ②		37		●	●			●		●				●								

(continued) **Table 3c** Institution / Course title ① ② : see notes after table	T=term; S=semester; Y=year	Time of specialisation after a broad introduction	Time spent on options %	Coastal/offshore engineering	Concrete structures	Contract/project management	Fluid mechanics	Geotechnical engineering	HVAC/building services	Hydraulics	Materials engineering	Numerical analysis	Public health engineering	Steel structures	Structural analysis	Transportation/highways	Waste/water engineering	Environmental impact assessment	Environmental engineering	GIS	Env conservation and management	Energy and the environment	Environmental law
Wolverhampton																						
Civil engineering ①			15			●																	
Construction management ②			25			●																	

Aberdeen ① Safety and reliability engineering

Abertay ① Building control; structural appraisals; value management; maintenance

Anglia Ruskin ① Facilities management; commercial management; building management

Bath ① ② Conservation of historic buildings; facade engineering; bridge engineering; lightweight structures

Birmingham ① Advanced railway engineering; transport planning

Bolton ① Project management; structural planning ② Housing studies; value management ③ Commercial management; facilities management; value management

Bradford ① Soil mechanics

Brighton ① Engineering geology

Bristol ① Earthquake engineering; history of civil engineering; timber engineering

Bristol UWE ① Facilities management; production management; residential management; sustainability

Cambridge ① Architectural engineering; earthquake engineering; foundation engineering

Cardiff ① Integrated building design; analysing architecture; issues in modern architecture; construction ② Soil mechanics

Central Lancashire ① ② Fire safety

City ① Remote sensing

Coventry ① Timber structures; masonry structures ② Construction management; finance

East London ① Hydrology; surveying

Edinburgh ① Fire safety engineering

Glamorgan ① Facilities management

Greenwich ① Bridge design and assessment

Heriot-Watt ① Civil engineering systems; engineering geology; finite element analysis ② Facilities management; safety and security; value management; building appraisal

Imperial College London ① Robotics and automated construction ② Robotics and automated construction; systems analysis

Leeds Metropolitan ① Construction technology; documenting and estimating; surveying

London South Bank ① Foundation engineering; soil mechanics; masonry and timber engineering; open channel hydraulics; advanced mathematics

Loughborough ① Building services; construction economics ② Construction technology; maintenance and repair; photogrammetry ③ Civil engineering measurement; pre-construction information technology

Manchester ① ② Earthquake engineering; fire engineering; conservation of structures

Napier ① Surveying

Newcastle ① Finite element analysis; seismic-resistant design; surveying; engineering geology ② Surveying ③ Environmental engineering for developing countries; management of waste and contaminated land; surveying; engineering geology

Nottingham Trent ① Engineering surveying; structural defects; masonry design; timber design; environmental issues; plate and shell structures; rock mechanics ② Engineering surveying; structural defects; masonry design; timber design; environmental issues; plate and shell structures; air pollution ③ Building or residential development

Oxford ① Wind engineering; engineering geology; soils

Oxford Brookes ① Development appraisal; maintenance and facilities management; building analysis

Plymouth ① Diving and underwater technology

Portsmouth ① Engineering for natural disasters

Reading ① Wide range of options

Sheffield ① Solar and wind energy; geology; bridge engineering; robot technology; structural dynamics ② Science and technology; structural glazing; fire engineering

Sheffield Hallam ① Construction engineering; land surveying; facilities management; refurbishment and development studies

Southampton ① ② Engineering for developing countries

Surrey ① Bridge engineering; applied mathematics and mechanics; wind engineering

Swansea ① Engineering systems

Ulster ① Environmental issues

Warwick ① Sustainability; fluid dynamics

Westminster ① ② Construction design and safety; refurbishment; dispute resolution; facilities management; fire safety; advanced technology

Wolverhampton ① Construction; construction finance ② Advanced construction IT; facilities management; plant management

MEng courses TABLE 2a lists many courses that can lead to the award of either a BEng or an MEng degree. For these courses, most of the tables in this part of the Guide give information specifically for the BEng stream (the tables show information for MEng courses that are MEng only). Where both degrees are available, much of the information will also apply to the MEng streams, but TABLE 3d shows you where there are differences for the MEng stream, and at what point the MEng course separates from the BEng.

Table 3d — MEng course differences						
Institution	**Course title**	**MEng separates from BEng**	**More engineering**	**More management**	**More languages**	**Other differences**
Aberdeen	*All courses*	Year 3	●	●		More extensive group project; engineering analysis and methods; engineering and project management
Bath	Civil and architectural engineering	Sem 4	●	●		More optional; more project work
	Civil engineering	Sem 4	●	●		
Belfast	Civil engineering	Year 3				
Birmingham	Civil engineering	Year 3	●	●	●	More project work; students taught to greater depth
Bradford	Civil and structural engineering	Year 3	●	●		Technical and managerial aspects studied in greater depth
Brighton	Civil engineering	Year 3		●		
Bristol	Civil engineering	Year 3	●	●		Large design project
Cambridge	Civil, structural and environmental engineering	Year 3	●			
Cardiff	Architectural engineering	Year 3				More integrated design
	Civil and environmental engineering	Year 3	●	●		More project work and design
	Civil engineering	Year 3				More integrated design
City	Civil engineering	Year 3				
Edinburgh	Civil and environmental engineering	Year 3	●	●	●	
	Civil engineering	Year 4				Final year: examination + 50% of time on project
	Structural and fire safety engineering	Year 4	●			
	Structural engineering with architecture		●			
Exeter	Civil engineering	Year 3	●			Group design project; more independent learning
Glasgow	Civil engineering	Year 3		●		Final year devoted largely to real case studies
	Civil engineering with architecture	Year 3		●		Final year devoted largely to real case studies
Heriot-Watt	Civil and environmental engineering	Year 3				Personal and professional development; study and industrial experience abroad
	Civil engineering	Year 3				Personal and professional development; study and industrial experience abroad (C/E Europe)
	Structural engineering	Year 3				Personal and professional development; study and industrial experience abroad (C/E Europe)
Leeds	Architectural engineering	Level 3	●	●		Extended project work
	Civil and structural engineering	Level 3	●	●		Extended project work
Liverpool John Moores	Civil engineering	Year 3				
Loughborough	Civil engineering	Year 3	●	●		Outdoor management course; optional sponsorship scheme
Manchester	Civil engineering	Year 3	●	●		
Newcastle	*All courses*	Year 3	●	●	●	Multidisciplinary design project
Northumbria	Building services engineering		●			More innovative; on deeper, more concentrated level

(continued) **Table 3d** Institution	MEng course differences						
	Course title	MEng separates from BEng	More engineering	More management	More languages	Other differences
Nottingham	Civil engineering	Year 3				
	Environmental engineering	Year 3	●	●		
Plymouth	Civil and coastal engineering	Stage 3	●	●		Multidisciplinary group project; additional management and technical subjects
	Civil engineering	Stage 3	●	●		Multidisciplinary group project; additional management and technical subjects
Portsmouth	Civil engineering	Year 3	●			Greater breadth and depth in engineering topics
Robert Gordon	Mechanical and environmental engineering	Year 3	●	●		Major multidisciplinary project in year 5
Salford	Civil engineering	Year 1				
Sheffield	Civil and structural engineering	Year 3				
	Civil engineering	Year 3				Study abroad; 90% project work in final year
Southampton	*Both courses*	Year 3		●		Multidisciplinary design/project work
Strathclyde	Civil engineering	Year 3	●	●		
Surrey	Civil engineering	Year 3	●	●		Greater breadth and depth
Swansea	Civil engineering	Level 3	●	●		
UCL	Civil engineering	Year 3	●			Final year specialisation in integrated design and communication skills
	Environmental engineering	Year 3	●			Final-year specialisation in integrated design and communication skills
Warwick	Civil engineering	Year 3	●			Optional year abroad; year 4 multidisciplinary group project

General studies and additional subjects   Specialised engineering courses run the risk of focusing the student's attention entirely on technical subjects. To prevent this, some institutions include topics on liberal, complementary or general studies in their specialised courses.

Many degree courses include industrial relations, economics and other social science subjects as separate topics, while others include coverage of them in the teaching of engineering subjects. In some cases these topics are compulsory. The reason for including them in civil engineering courses is to stress the concern engineers should have for the economic and social aspects of their work. Indeed, inclusion of these aspects is a requirement for accreditation by the professional bodies. Several institutions offer a compulsory topic entitled 'the engineer in society' to meet this requirement; it usually covers subjects such as law, architecture, politics, accountancy, management techniques and the contribution of the civil engineer to the environment. In some courses it is included in all years: in others it is a final-year subject only.

TABLE 3e gives information about where particular supporting topics are available as compulsory (●) or optional components (○), or where there are both compulsory and optional components of a topic (◑). Remember that the titles given to topics may not be exactly the same as those used in individual prospectuses. TABLE 3a contains information about language teaching; you will need to look at prospectuses or ask institutions direct for information about other subjects.

Where a course is part of a modular degree scheme, there will probably be more opportunities to take a wide range of other subjects, though the requirements for professional qualification may mean that the choice is more restricted than it would be if you were following a course in a non-engineering subject. TABLE 2a shows where a course is part of a modular degree scheme allowing a relatively free choice of modules from a large number of subjects.

Table 3e

Subsidiary and supporting subjects

● compulsory; ○ optional; ◐ compulsory + options

Institution	Course title	Architecture	Communication skills	Drawing and sketching	Economics	Environment	Health and safety	History of engineering/building	Law	Leadership skills	Management	Planning	Others
Aberdeen	Civil and environmental engineering		●	●	●		●	●	○	○	●	●	
	Civil and structural engineering		●	●	●		●		○	○	●	●	
	Civil engineering		●	●	●	○	●		○	○	●	●	
Abertay	Civil engineering		●	●		●	●				●	●	
Anglia Ruskin	Civil engineering	○	●	○	●	○	●	●	◐	◐	◐	◐	
	Construction management	○	●	○	●	●	●	●	●	●	●	○	
Bath	Both courses	●	●	●		●	●	●	●	●			
Belfast	Civil engineering	○	●	●	●		●	●	●	○	●	○	
	Structural engineering with architecture	◐	●	●			●	●	○		●	○	
Birmingham	Civil engineering		◐	◐	◐	◐	●		○		●		
Bolton	Civil engineering		●	●	●	●	●	●	●	●	●	●	Surveying
	Construction	○	●	○	○	○	●	○	○	○	○	○	Construction economics; development studies
	Construction management		●	●	●	●	●		●	●	●	◐	
Brighton	Civil engineering		●	●		◐	○	○		●	◐		
	Civil with environmental engineering		●	●		◐	●				○		
Bristol	Civil engineering	○	●	●	●		○	●	○	●	●		Surveying
Bristol UWE	Construction management	○	●	●	●	●	●		○	●		●	●
Cambridge	Civil, structural and environmental engineering				○					○			
Cardiff	Architectural engineering	●	●	●	●	●				●	●	●	
	Civil and environmental engineering		●	●	●	●	●		●		●		
	Civil engineering		●	●	●		●		●	●	●	●	
Central Lancashire	Building services engineering		◐	○	○	◐	○	○	◐	◐	◐	○	
	Construction project management		◐	○	◐	○	◐	◐	◐	◐	◐	◐	
City	Civil engineering		●	●		●	●	●	●	●	●	●	
Coventry	Civil and structural engineering		●	●	○					●	●		
	Civil engineering		●	●	○	●	●		●		●		
	Civil engineering construction		●	●	●		●		●		●		
	Construction management		●	●	●	●	●		●	○	●	○	
Derby	Construction management	○	●	○	○	○	●	○	●	○	●	○	
Dundee	Civil engineering	○	●	●	○	●	●	●	○	●	○	○	
Durham	Civil engineering		●	●		●	●		●		●		
East London	Civil engineering		●	●									Ethical, legal and social issues

Subsidiary and supporting subjects

● compulsory; ○ optional; ◑ compulsory + options

Institution	Course title	Architecture	Communication skills	Drawing and sketching	Economics	Environment	Health and safety	History of engineering/building	Law	Leadership skills	Management	Planning	Others
Edinburgh	Civil and environmental engineering		●	●	●	●	●			●	●	●	
	Civil engineering	◑	●	●	○	◑	●			●	◑		
Exeter	Civil engineering		●	●	●	●	●		●		●		Geotechnical field course
Glamorgan	Civil engineering		●	●			●						
	Project management (construction)		●	●	●	●	●		●		●	●	Mathematics
Glasgow	*Both courses*		●	●	◑	●	●		●		◑	●	
Glasgow Caledonian	Environmental civil engineering		●	●	●	●	●		●	●	◑	●	
Greenwich	Civil engineering		●	○		●	●		●		●		
	Civil engineering with project management		●	○		●	●		●		●		
	Civil engineering with water and environmental management		●	○		●	●		●		●		
Heriot-Watt	Civil and environmental engineering		●	●	●	◑	●		●	○	●		
	Civil engineering		●	●	●	◑	●		●	○	●		
	Construction management	●	●	●	●	◑	◑	●	●	●	◑	◑	Negotiation and marketing; corporate strategy
	Structural engineering	○	●	●	●	◑	●		●	○	●		
Imperial College London	*Both courses*		●	●	●	◑			○		●	◑	Civil engineering in context; humanities; risk analysis
Kingston	Civil engineering		●	●	●	○	●		●	●	●	●	
	Construction management		●	●	●	●			●	●	●	●	
Leeds	Architectural engineering	●	●	●	◑	●	●	●	○	○	●	●	
	Civil and environmental engineering		●	●	○	●	●	○	●	○	●	○	
	Civil and structural engineering	○	●	●	◑	●	●	◑	◑	○	●	○	
	Civil engineering with construction management	○	●	●	◑	●	●	○	●	○	●	○	
Leeds Metropolitan	Civil engineering											●	Mathematics and statistics
	Construction management					●	●	●	●				
Liverpool	*All courses*		●	●	○	●	○				●		
Liverpool John Moores	Civil engineering		●	●	●	●	●				●		
	Construction management		●	●	●	●		○	●		●	●	
London South Bank	Architectural engineering	●	●	●	○	○	●	●	○	●	●	○	
	Building services engineering		●		●				●	●	●		
	Civil engineering		●	●	●							●	
	Construction management		●		●				●		●		
Loughborough	Architectural engineering and design management								○		○	○	Management information systems
	Civil engineering		●	●	●	●	●		●	●	●	●	Human resource management; financial management
	Construction engineering management		●	●		●	●	●	●	●	●	●	Development economics; language; management information systems
Manchester	Civil and structural engineering	○	●	●		○	●	○			●	●	
	Civil engineering	○	●	●		●	●	○			●	●	
Napier	Civil and transportation engineering		●	●	●		●	●			●	●	
	Civil engineering		●	●			●	●			●	●	

Civil and Environmental Engineering and Building

(continued) Table 3e

Institution	Course title	Architecture	Communication skills	Drawing and sketching	Economics	Environment	Health and safety	History of engineering/building	Law	Leadership skills	Management	Planning	Others
Newcastle	Civil and structural engineering	●	●	●	●	●	●	●	○		○	●	Engineering ethics
	Civil engineering		●	●	●	●	●		◑		◑	◑	Engineering ethics
	Environmental engineering		●	●	●	●	●		◑	○	◑	◑	Engineering ethics
Northumbria	Building design management		○		○		○				○		
	Building project management		○		○		○		○		○	○	
	Building services engineering	○	●	●			●				●		
	Construction management		●		●	○	●		○		●		
Nottingham	Architectural environment engineering	○	●	●	●	●	●	●	●	●	●	○	
	Civil engineering		●	●	◑	◑	○	○	○	○	◑	◑	Geology
	Environmental engineering		●	●	●	●	●		●		●	○	
Nottingham Trent	Civil and structural engineering		●	●	●	●	●	●	●	●	●	●	
	Civil engineering		●	●	●	●	●	●	●	●	●	●	
	Construction management	○	●	●	●	●	●	○	●	●	●	○	
Oxford	Civil engineering		●	●		●	●				○		Computing
Oxford Brookes	Building	○	●	○	○	○	●		○	●	●	○	Surveying; developing countries; EU
Paisley	Civil engineering (BEng)	●	●	●	●	●	●	●	●	●	●	●	
Plymouth	Civil and coastal engineering		●	●		●	●		○	●	●	●	Diving and underwater technology
	Civil and coastal engineering (BSc)		●	●		●	●			●	●	●	IT; diving technology; virtual reality
	Civil engineering		●	●		●	●		○	●	●	●	Diving and underwater technology
	Civil engineering (BSc)		●	●		●	●			●	●	●	IT; diving technology; virtual reality
Portsmouth	Civil engineering		●	●	●	●	●	○	●	●	●		Surveying; diving
	Construction engineering management		●	●	●	●	●	○	○		●		Surveying; diving
Reading	Building construction and management		●	●			●	●	●	●	●	●	
Robert Gordon	Construction, design and management	●	●	●	●		●		●	●	●	●	
	Mechanical and environmental engineering		●			●	●				●		
Salford	Civil engineering		●	●	●		●			●	●	●	
Sheffield	Civil and structural engineering	○	●	●		●	●	●		●	●	●	
	Civil engineering	○	●	●		●	●	●		●	●	●	
	Structural engineering and architecture	●	●	●		●	●	●		●	●	●	
	Structural engineering design	●	●	●		●	●	●		●	●	●	
	Structural engineering with architectural studies	●	●	●		●	●	●		●	●	●	
Sheffield Hallam	Construction management		●	◑	●	○	◑	●	●	○	●	●	
Southampton	Civil engineering	○	●	●		●	●		○	●	●	○	
	Environmental engineering	○	●	●		●	●		○	●	◑	○	
Strathclyde	Civil engineering	○	●	●	○		●	○	●	●	●	●	
Surrey	Civil engineering	○	●	●		○	●	●	○	●	●	○	Computing systems
Swansea	Civil engineering		●	●	●	○	●	●			●		
Swansea IHE	Civil engineering and environmental management		●		●	●	○		●		●	○	
	Project and construction management		●			●	●	○	◑	◑	●		
Teesside	Civil engineering					●	●	●			●		

Subsidiary and supporting subjects

● compulsory; ○ optional; ◑ compulsory + options

Subsidiary and supporting subjects

● compulsory; ○ optional; ◐ compulsory + options

Institution	Course title	Architecture	Communication skills	Drawing and sketching	Economics	Environment	Health and safety	History of engineering/building	Law	Leadership skills	Management	Planning	Others
UCE Birmingham	Construction management and economics		●	●	●	●	●	●	●	●	●	○	
UCL	Civil engineering		●	◐	○	○	●		○	○	○	○	
	Environmental engineering		●	○	○	●	●			○	○	○	○
Ulster	Civil engineering		●	●	●	●	●			●	●	●	●
	Construction engineering	●	●	●	●	●	●	●	●	●	●	●	
	Environmental engineering		●			●	●				●		
Wales (Newport)	Building studies				●	●	●		●		●		
	Civil and construction engineering					●	●		●		●		
Warwick	Civil engineering		●	●	●	○	●			●		●	●
Westminster	Architectural engineering	●	●	●	○	●	○	●	○		○	○	
	Building engineering	○	●	●	●	●	○	○	○		●	●	○
	Construction management	○	●	●	●	●	○	○	●		●	●	○
Wolverhampton	Civil engineering		●	●			●		○	○	●	●	
	Construction management		●	●	●	●	●	○	●	●	●	●	

Chapter 4: Teaching and assessment methods

See Chapter 4 in the *Introduction* (page 16) for general information about teaching and assessment methods used in all types of engineering course, including those covered in this part of the Guide. It also explains how to interpret TABLE 4, which gives information about projects and assessment methods used on individual courses. An additional feature of most civil engineering courses is the inclusion of compulsory field work in surveying, in addition to class field exercises. They may also include geology field work, which often has a compulsory residential element.

Table 4 — Assessment methods

Institution	Course title	Frequency of assessment (Key: ● term; ◐ semester; ○ year)	Years of exams contributing to final degree (years of exams not contributing to final degree)	Coursework: minimum/maximum %	Project/dissertation: minimum/maximum %	Time spent on projects in: first/intermediate/final years %			Group projects: ● compulsory; ○ optional	Orals: ◐ if borderline; ● for projects; ● everyone
Aberdeen	Civil and environmental engineering	◐	(1),(2),**3,4,5**	15/**20**	39/**42**	20	20	50	●	○
	Civil and structural engineering	◐	(1),(2),**3,4,5**	15/**20**	39/**42**	20	20	50	●	○
	Civil engineering	◐	(1),(2),**3,4,5**	15/**20**	40/**40**	20	20	50	●	○
Abertay	Civil engineering	◐	(1),(2),(3),**4**	30/**30**	20/**20**	30	40	50	●	○
Anglia Ruskin	Civil engineering	◐	(1),**2,3,4**	60/**70**	10/**20**	5	10	25		○
	Construction management	◐	(1),**2,3,4**	40/**40**	30/**30**	30	50	60	○	◐○
Bath	Civil and architectural engineering	◐	(1),(2),**3,4,5**	10/**10**	40/**40**	25	25	35	●	●
	Civil engineering	◐	(1),(2),**3,4**	10/**10**	30/**30**	25	25	25	●	○●
Belfast	Civil engineering	◐	(1),**2,3,4**	**15**	**12**	0	0	35		○
	Structural engineering with architecture	◐	(1),**2,3,4**	10/**15**	20/**35**	0	35	45		○
Birmingham	Civil engineering	◐	(1),**2,3,4**	15/**20**	30/**35**	10	15	30	●	○
Bolton	Civil engineering	◐	(1),**2,3**	38/**38**	25/**25**	10	20	30	●	○
	Construction	◐	(1),**2,3**	25/**50**	33/**33**	20	20	40	●	◐○
	Construction management	◐	(1),**2,3**	25/**45**	33/**33**	20	20	30	●	◐○
Bradford	Civil and structural engineering	◐	(1),**2,3,4,5**			0	0	25	●	
Brighton	Civil engineering	◐	(1),**2,3,4**	35/**45**	20/**20**	20	25	40	●	
	Civil with environmental engineering	◐	(1),**2,3**	35/**45**	20/**20**	20	25	40	●	○
Bristol	Civil engineering	○	(1),**2,3,4**	10/**15**	25/**25**	10	25	35	●	◐○
Bristol UWE	Civil engineering	◐	(1),**2,3**							
	Construction management	◐	(1),**2,3**	40/**60**	17/**17**	20	25	25	●	◐
Cambridge	Civil, structural and environmental engineering	○	(1),(2),**3,4**	0/**20**	50/**50**	5	20	50	●	○
Cardiff	Architectural engineering	◐	(1),**2,3**	15	15	25	25	33	●	○●
	Civil and environmental engineering	◐	(1),**2,3,4,5**	15/**25**	15/**25**	2	10	20	●	
	Civil engineering	◐	(1),**2,3,4**	15	15	25	25	30	●	◐○●
Central Lancashire	Building services engineering	◐	(1),(2),**3,4**	20/**30**	20/**30**	20	20	30		◐
	Construction project management	◐	(1),**2,3,4**	30/**50**	20/**20**	20	25	40	○	◐
City	Civil engineering		**1,2,3,4**	20/**30**	10/**15**	15	20	25	●	○
Coventry	Civil and structural engineering	○	(1),**2,3**	25/**25**	25/**25**	40	40	40	●	
	Civil engineering	○	(1),**2,3**	25/**25**	25/**25**	40	40	40	●	◐
	Civil engineering construction	○	(1),**2,3**	25/**25**	25/**25**	40	40	40	●	◐
	Construction management	○	(1),**2,3**	30/**44**	20/**20**	10	10	25	●	◐

Key for frequency of assessment column: ◑ term; ◐ semester; ○ year

Group projects: ● compulsory; ○ optional

Orals: ◑ if borderline; ● everyone; ○ for projects

Institution	Course title	Frequency of assessment	Years of exams contributing to final degree (years of exams not contributing to final degree)	Coursework: minimum/maximum %	Project/dissertation: minimum/maximum %	Time spent on projects in: first/intermediate/final years %			Group projects: compulsory; optional	Orals
Derby	Construction management	○	(1),**2,3**	12/**12**	50/**50**	20	30	40	●	◑
Dundee	Civil engineering	◑	(1),**2,3,4,5**	30/**40**	25/**40**	20	20	50	●	
Durham	Civil engineering	○	(1),**2,3,4**	5/**10**	25/**25**	5	10	50	●	○○
East London	Civil engineering	◑	(1),**2,3**							●
Edinburgh	Civil and environmental engineering	◑	(1),(2),**3,4,5**	20	25	20	40	50	●	
	Civil engineering	◑	(1),(2),**3,4,5**	20/**20**	20/**20**	26	35	50	●	●
	Structural and fire safety engineering	◑	(1),(2),**3,4,5**			20	40	40	●	
	Structural engineering with architecture	◑	(1),(2),**3,4,5**	5/**15**	35/**50**	20	40	40	●	○
Exeter	Civil engineering	◑	(1),**2,3,4**	15/**20**	25/**35**		25	25	●	○●
Glamorgan	Civil engineering	◑	(1),**2,3**	30/**30**	20/**20**	10	15	30	●	○●
	Project management (construction)	◑	(1),(2),**3**	34/**47**	20/**20**	10	50	20	●	○●
Glasgow	Civil engineering	◑	(1),(2),**3,4,5**		40/**55**	10	20	60	●	○●
	Civil engineering with architecture	◑	(1),(2),**3,4,5**		50/**65**	25	30	65	●	○●
Glasgow Caledonian	Environmental civil engineering	◑	(1),(2),**3,4**	**60**	33/**33**	10	40	40	●	
Greenwich	Civil engineering	◑◐	(1),**2,3**	20/**20**	25/**25**	35	35	35	●	○
	Civil engineering with project management	◑◐	(1),**2,3**	20/**20**	25/**25**	35	35	35		○
	Civil engineering with water and environmental management	◑◐	(1),**2,3**	20/**20**	25/**25**	35	35	35		○
Heriot-Watt	Civil and environmental engineering	◑◐	(1),(2),**3,4,5**	10/**35**	12/**12**	5	10	20	●	○○
	Civil engineering	◑◐	(1),(2),**3,4,5**	10/**35**	12/**12**	5	10	20	●	○○
	Construction management	◑◐	(1),(2),(3),**4**		50/**50**	12	16	50		○●
	Structural engineering	◑◐	(1),(2),**3,4**	10/**35**	12/**12**	5	10	20		○○
Imperial College London	Civil and environmental engineering	○	**1,2,3,4**	17/**20**	25/**30**	10	25	40		○
	Civil engineering	○	**1,2,3,4**	17/**20**	25/**30**	10	25	40	●	○
Kingston	Civil engineering	◑	(1),**2,3**	30/**35**	10/**15**	10	20	20	●	○○
	Construction management	◑	(1),**2,3**	30/**40**	30/**40**	10	10	30	●	○○
Leeds	Architectural engineering	◑	(1),**2,3,4**	10/**20**	15/**30**	20	20	40	●	○
	Civil and environmental engineering	◑	(1),(2),**3,4**	20/**20**	30/**30**	20	20	40	●	○
	Civil and structural engineering	◑	(1),**2,3,4**	10/**20**	15/**30**	20	20	40	●	○
	Civil engineering with construction management	◑	(1),(2),**3,4**	20/**20**	30/**30**	20	20	40	●	○
	Energy and environmental engineering	◑	(1),**2,3,4**	25/**30**	20/**25**		15	30	●	○○
Liverpool	Civil and environmental engineering	◑	(1),**2,3,4**	30/**30**	10/**15**	5	5	15	●	○○
	Civil and maritime engineering	◑	(1),**2,3,4**	30/**30**	10/**15**	5	5	15	●	○○
	Civil and structural engineering	◑	(1),**2,3,4**	30/**30**	10/**15**	5	5	15	●	○○
	Civil engineering	◑	(1),**2,3**	30/**30**	10/**15**	5	5	15	○	○○
Liverpool John Moores	Civil engineering	◑	(1),**2,3,4**	**10**	**30**	0	0	30	●	○
	Construction management	◑	(1),**2,3,4**	15	20	30	40	50	●	○○
London South Bank	Architectural engineering	◑	(1),**2,3**	30	20	15	20	30		○○
	Building services engineering	◑	(1),**2,3**	20/**40**	10/**25**	10	15	20	●	◑
	Civil engineering	◑	(1),**2,3**	20/**40**	10/**25**	10	20	30	●	◑
	Construction management	◑	(1),**2,3**	20/**40**	10/**25**	10	20	30	●	◑
Loughborough	Architectural engineering and design management	◑	(1),**2,3,4**	25/**75**	25/**75**	50	50	75	○	
	Civil engineering	◑	(1),**2,3,4**	25/**30**	20/**35**	8	20	50	●	○
	Construction engineering management	◑	(1),**2,3,4**	35/**35**	15/**15**			25	●	○○

(continued) Table 4 — Assessment methods

Key for frequency of assessment column: ◑ term; ◐ semester; ○ year

Frequency of assessment · Years of exams contributing to final degree (years of exams not contributing to final degree) · Coursework: minimum/maximum % · Project/dissertation: minimum/maximum % · Time spent on projects in: first/intermediate/final years % · Group projects: ● compulsory; ○ optional · Orals: ◐ if borderline; ● for projects; ○ everyone

Institution	Course title	Freq.	Years of exams	Coursework	Project/diss.	First	Int.	Final	Group	Orals
Manchester	Civil and structural engineering	◑	(1),(2),3,4	10/20	20/20	10	12	20	●	
	Civil engineering	◑	(1),2,3,4	10/20	20/20	10	12	20	●	
Napier	Civil and transportation engineering	◑	(1),(2),3,4	16/16	31/31	0	0	30	●	○
	Civil engineering	◑	(1),(2),3,4	20/20	30/30	0	20	30	●	○
	Energy and environmental engineering	◑	(1),(2),3,4	30/30	25/25	10	15	25	●	○●
Newcastle	Civil and structural engineering	◑	(1),2,3,4	10/20	20/20	15	15	20	●	◑
	Civil engineering	◑	(1),2,3,4	10/20	12/20	15	15	20	●	◑
	Environmental engineering	◑	(1),2,3,4	10/20	12/20	15	15	20	●	◑
Northumbria	Building design management	◑	(1),2,(3),4							
	Building project management	◐	2,3,4							
	Building services engineering	◑		30/50	50/50	20	30	50	○	◑
	Construction management	◐	2,4							
Nottingham	Civil engineering	◑	(1),2,3,4	10/20	13/40	10	10	20	●	◐○●
	Environmental engineering	◑	(1),2,3	20/50	10/15	5	20	25	○	◑
Nottingham Trent	Civil and structural engineering	◑	(1),2,3,4	50/60	17/20	20	20	30	●	○
	Civil engineering	◑	(1),2,3,4	50/60	17/20	20	20	30	●	○
	Construction management	◑	(1),2,4	30/40	30/40	20	30	40	●	○
Oxford	Civil engineering		(1),3,4	13/13	25/25	10	15	50	●	◑
Oxford Brookes	Building	◑	(1),2,4	40/60	10/12	30	40	50	●	◐○
	Construction management	◑	(1),2,4							
Paisley	Civil engineering (BEng)	◑	(1),(2),3,4,5	30/40	20/20	5	15	25		○
Plymouth	Civil and coastal engineering	◑	(1),2,3,4	25/30	25/40	10	20	30		○
	Civil and coastal engineering (BSc)	◑	(1),2,3	25/30	25/40	10	20	30		○
	Civil engineering	◑	(1),2,3,4	25/30	25/40	10	20	30	●	○
	Civil engineering (BSc)	◑	(1),2,3	25/30	25/40	10	20	30	●	○
	Construction management and the environment	◑	(1),2,3	60	40	20	50	70	●	
Portsmouth	Civil engineering	◑	(1),2,3,4	10/30	10/20	10	10	50	●	○
	Construction engineering management	◑	(1),2,3	10/30	10/20	10	10	30	●	○
Reading	Building construction and management	○	(1),(2),3	0/10	33/33	5	15	20	●	◑
Robert Gordon	Construction, design and management	◑	(1),(2),(3),4	20/20	50/50	30	35	50	●	◐○
	Mechanical and environmental engineering	◑	(1),(2),3,4,5	25	25	33	20	25	●	◐○●
Salford	Civil engineering	◑	(1),2,3	20	12	10	10	20	●	○
Sheffield	Civil and structural engineering	◑	(1),2,3,4	10/20	15/20	15	20	80	●	○
	Civil engineering	◑	(1),2,3,4	5/15	20/60	10	20	80	●	○
	Structural engineering and architecture	◑	(1),2,3,4	5/10	40/60	15	30	90	●	○
	Structural engineering design	◑	(1),2,3,4	5/15	20/60	10	20	80	●	○
	Structural engineering with architectural studies	◑	(1),2,3,4	5/15	20/60	10	20	80	●	○
Sheffield Hallam	Construction management	◑	(1),(2),4	20/50	50/50	10	20	33	●	◑
Southampton	Civil engineering	◑	(1),2,3,4	15/20	20/30	5	15	40	●	◐○
	Environmental engineering	◑	(1),2,3,4	15/20	20/30	5	15	40	●	◑
Strathclyde	Civil engineering	◑	(1),2,3,4,5	0/15	40/40	15	15	40	●	◑
Surrey	Civil engineering	◑	(1),2,3,4	10/20	10/15	10	10	25	●	○
Swansea	Civil engineering	◑	(1),2,3,4	10	15	10	10	15	●	○

Assessment methods

Institution	Course title	Frequency of assessment (Key for frequency of assessment column: ◐ term; ◑ semester; ○ year)	Years of exams contributing to final degree (years of exams not contributing to final degree)	Coursework: minimum/**maximum** %	Project/dissertation: minimum/**maximum** %	Time spent on projects in: first/intermediate/final years %			Group projects: ● compulsory; ○ optional	Orals: ◐ if borderline; ○ for projects; ● everyone
Swansea IHE	Civil engineering and environmental management	◑	(1),**2,3**	30/**30**	30/**30**	10	20	40	●	◑
	Project and construction management	◑	(1),**2,3**	30/**30**	30/**30**	10	20	40	●	◑
Teesside	Civil engineering	◑	(1),**2,3**		40/**50**	25	35	45	○	◑○
UCE Birmingham	Construction management and economics	○	(1),**2,3**	50/**60**	20/**25**	50	50	65	●	◑
UCL	*Both courses*	○	**1,2,3,4**	20/**30**	8/**12**	20	20	30	●	○
Ulster	Civil engineering	◑	(1),**2,4**	32/**32**	17/**17**			17		◑○
	Construction engineering	◑	(1),(2),(3),**4**	37/**40**	30/**30**	20	30	40		◑○
	Environmental engineering	◑	(1),**2,4**	25/**35**	30/**35**	15	20	30	●	◑○
Wales (Newport)	Building studies	○	(1),**2,3**		**18**			33		
	Civil and construction engineering	◑	(1),**2,3**			10	25	50		
Warwick	Civil engineering	○	(1),**2,3,4**	18/**24**	18/**18**	12	12	25		○●
Westminster	Architectural engineering	◑	(1),**2,3**	10/**10**	55/**65**	40	45	50		○
	Building engineering	◑	(1),**2,3,4**	43/**44**	12/**14**	25	50	75		◑
	Construction management	◑	(1),**2,3,4**	43/**44**	12/**14**	25	50	75		◑
Wolverhampton	Civil engineering	◑	(1),**2,3,4**	50/**60**	20/**30**	5	15	30		○
	Construction management	◑	(1),**2,3,4**	40/**40**	17/**17**	10	25	50	●	◑○

Chapter 5: Entrance requirements

See Chapter 5 in the *Introduction* (page 18) for general information about entrance requirements that applies to all types of engineering course, including those covered in this part of the Guide. It also explains how to interpret TABLE 5, which gives information about entrance requirements for individual courses.

Table 5 — Entrance requirements

Institution	Course title	Number of students (includes other courses)	Typical offers (BSc/BEng)			Typical offers (MEng)			● compulsory; ○ preferred	
			UCAS tariff points	A-levels	SCQF Highers	UCAS tariff points	A-levels	SCQF Highers	A-level Mathematics	A-level Physics
Aberdeen	All courses	(170)		CCD	ABBBC		BCC	ABBBC	●	●
Abertay	Civil engineering	25	200						○	○
Anglia Ruskin	Civil engineering	20	240		BBBC					
	Construction management	30	240		BBCC					
Aston	Construction management		240–280		BBBBC					
Bath	Civil and architectural engineering	30				340	AAB		●	○
	Civil engineering	20	260			340			●	○
Belfast	Civil engineering	100		BCC	BBBC		BBB		●	○
	Environmental and civil engineering						BBB			
	Structural engineering with architecture	20					BBB		●	○
Birmingham	Civil engineering	(75)		ABB	BBBBB		ABB	BBBBB	●	○
Bolton	Civil engineering	20	200						○	○
	Construction	10	220							
	Construction management	15	220							
Bradford	Civil and structural engineering	(50)	240			300			○	○
Brighton	Building studies		180							
	Civil engineering	44	240			300			●	○
	Civil with environmental engineering	15	220							○
	Construction management		180							
	Project management for construction		180							
Bristol	Civil engineering	65		ABB	AAAAB		ABB	AAAAB	●	●
Bristol UWE	Civil engineering	25	200–260						○	○
	Construction management	30	180–220							
Cambridge	Civil, structural and environmental engineering	(300)		AAA			AAA		●	●
Cardiff	Architectural engineering	20	240–280			300			●	○
	Civil and environmental engineering	15	240			240			●	○
	Civil engineering	45	240–280			300			●	○
Central Lancashire	Building services engineering	10	240		BBBB				○	○
	Construction project management	20	240		BBBC					
City	Civil engineering	(46)	230	BBB	ABBB	300			●	○
Colchester I	Construction management		40							
Coventry	Civil and structural engineering	25	260						○	○
	Civil engineering	25	300						○	○
	Civil engineering construction	18	200						○	○
	Civil engineering design		200						○	○

Entrance requirements

Institution	Course title	Number of students (includes other courses)	Typical offers (BSc/BEng) UCAS tariff points	A-levels	SCQF Highers	Typical offers (MEng) UCAS tariff points	A-levels	SCQF Highers	● compulsory, O preferred	A-level Mathematics	A-level Physics
Coventry (continued)	Construction management	20	220								
Derby	Construction management	5	140–180							●	
Dundee	Civil engineering	40	240–300							●	O
	Civil engineering design and management		260–320								
Durham	Civil engineering	(140)					ABB	AAAA		●	O
East London	Civil engineering	45	120–225							●	●
Edinburgh	Civil and environmental engineering	10		BBB	BBBB		BBB	BBBB		●	O
	Civil engineering	50		BBB	BBBB		BBB	BBBB		●	O
	Structural and fire safety engineering	25		BBB	BBBB		BBB	BBBB		●	O
	Structural engineering with architecture	10		BBB	BBBB		BBB	BBBB		●	O
Exeter	Civil engineering	30	240			300				●	O
Glamorgan	Civil engineering	20	140–180							●	O
	Project management (construction)	10	180–240								
Glasgow	Civil engineering	40		CCC	ABBB					●	O
	Civil engineering with architecture	18		CCC	ABBB					●	O
Glasgow Caledonian	Construction management			BBC	BBBC						
	Environmental civil engineering	20	220	BCC	BBCC					●	O
Greenwich	Civil engineering	35	180–240	CCD	BBCC					●	O
	Civil engineering with project management	10	240	CCC	BBB					●	O
	Civil engineering with water and environmental management	10	240	CCC	BBB					●	O
	Design and construction management		180		CCC						
Heriot-Watt	Architectural engineering			CCD	BBBC		CCD	BBBC		●	
	Civil and environmental engineering			CDD	BBBC		CCC	AABB		●	
	Civil engineering	50		CDD	BBBC		CCC	AABB		●	●
	Construction management	12	220		BBBC					O	O
	Structural engineering	20		CDD	BBBC		CCC	AABB		●	
Imperial College London	Civil and environmental engineering	25					ABB			●	●
	Civil engineering	(65)					ABB			●	●
Kingston	Civil engineering	30	120–220							●	O
	Construction management	10	180		BCC						
Leeds	Architectural engineering	(150)		CCC			BBC			●	
	Civil and environmental engineering	(150)					BBB	AABB		●	
	Civil and structural engineering	(150)		CCC	BBBC		BBB	AABB		●	
	Civil engineering with construction management	(150)					BBB	AABB		●	
	Civil engineering with transport			CCC							
	Energy and environmental engineering	15		BCC			BCC			O	O
Leeds Metropolitan	Civil engineering	50	160								
	Construction management	25	140								
Liverpool	Civil and environmental engineering	10				300	BBB	AAAA		●	O
	Civil and maritime engineering	10				300	BBB	AAAA		●	O
	Civil and structural engineering	20				300	BBB	AAAA		●	O
	Civil engineering	40	260	BCC	AABB					●	O

Civil and Environmental Engineering and Building

(continued) Table 5

Entrance requirements

Institution	Course title	Number of students (includes other courses)	Typical offers (BSc/BEng) UCAS tariff points	A-levels	SCQF Highers	Typical offers (MEng) UCAS tariff points	A-levels	SCQF Highers	● compulsory; ○ preferred A-level Mathematics	A-level Physics
Liverpool John Moores	Building design technology and management		180		BBC					
	Civil engineering	15	260						●	○
	Construction management	35	180						○	○
London South Bank	Architectural engineering	15		CC	BBB				●	○
	Building services engineering	25		CC	BBB				○	○
	Civil engineering	30		CC	BBB				●	○
	Construction management	40		CC	BBB				○	○
Loughborough	Architectural engineering and design management	20	260	BCC					○	○
	Civil engineering	90	260			300			●	○
	Construction engineering management	30	260							
Manchester	Civil and structural engineering	(75)				320	AAB		●	○
	Civil engineering	(75)	300	ABB		320	AAB		●	○
Napier	Civil and timber engineering		200	CDD	CCCC					
	Civil and transportation engineering	20	220	CCD	BBCC				●	○
	Civil engineering	35	200	CDD	CCCC					
	Construction and project management		220	CDD	CCCC					
	Construction engineering		200	CDD	CCCC					
	Energy and environmental engineering	20	220						●	○
Newcastle	Civil and structural engineering	25		BBB			AAB		●	○
	Civil engineering	60		BBB			AAB		●	○
	Environmental engineering			BBB			AAB		●	○
North East Wales I	Construction management		120							
Northumbria	Building design management	16	240							
	Building project management		240							
	Building services engineering	30	240						●	○
	Construction management		240		CCCCC					
Nottingham	Architectural environment engineering	20		CCC					●	●
	Civil engineering	90		ABB			ABB		●	○
	Environmental engineering	30		BCC			BBC		●	○
Nottingham Trent	Civil and structural engineering	(40)	260						●	○
	Civil engineering	(60)	200–240						○	○
	Construction management	45	220		CCCCC				○	○
Oxford	Civil engineering	(170)					AAA	AAAAB	●	●
Oxford Brookes	Building	20		BCC						
	Construction management			BCC						
Paisley	Civil engineering (BEng)	50		CD	BBC				●	○
Plymouth	Civil and coastal engineering	35	240			300			●	○
	Civil and coastal engineering (BSc)	20	160						○	○
	Civil engineering	35	240			300			●	○
	Civil engineering (BSc)	20	160						○	○
	Construction management and the environment	40	220–230							
Portsmouth	Civil engineering	50	240			300			●	
	Construction engineering management	50	180–240						○	

(continued) Table 5 — Entrance requirements

Institution	Course title	Number of students (includes other courses)	Typical offers (BSc/BEng) UCAS tariff points	A-levels	SCQF Highers	Typical offers (MEng) UCAS tariff points	A-levels	SCQF Highers	A-level Mathematics ● compulsory; ○ preferred	A-level Physics
Reading	Building construction and management	15	260–280		BBBB				○	○
	Construction management and surveying		260–280							
Robert Gordon	Construction, design and management	10	160		BCC				○	○
	Mechanical and environmental engineering	30	220–240	CD	BBCC				●	○
Salford	Civil engineering	50	240		BBBBC	300		AABBB	●	
	Construction management		240							
	Construction project management		240							
Sheffield	Civil and structural engineering	80					ABB		●	
	Civil engineering	80		BBB			ABB		●	
	Structural engineering and architecture	20					ABB		●	
	Structural engineering design						ABB		●	
	Structural engineering with architectural studies	20					ABB		●	
Sheffield Hallam	Construction management	40	230		BBBCC					
Southampton	Civil engineering	50	350	BBB		370	ABB		●	○
	Environmental engineering	15	350	BBB		370	ABB		●	
Southampton Solent	Construction management		100							
Strathclyde	Architectural engineering			BCC	BBBBC		BBB	BBBBB		
	Civil engineering	70		BCC	BBBB		BBB	AAAB	●	●
Surrey	Civil engineering	70	260	BCC	BBBBB	300	ABC	AABBB	●	○
Swansea	Civil engineering	50	260			320			●	
Swansea IHE	Civil engineering and environmental management	10	80–340						○	○
	Construction management		80–340							
	Project and construction management	10	80–340						○	○
Teesside	Civil engineering		180–240						○	○
	Civil engineering with disaster management		160–240							
UCE Birmingham	Construction management and economics	20	180	CDD	CCCC					
UCL	*Both courses*	(60)		AAB			AAB		○	○
Ulster	Building services engineering		260							
	Civil engineering	15	260						●	○
	Construction engineering	30	280							
	Environmental engineering	20	260						●	○
Wales (Newport)	Building studies	20	160							
	Civil and construction engineering	10	160							
Warwick	Civil engineering	60		BBB			AAB		○	○
Westminster	Architectural engineering	40	300		BBBB				○	○
	Building engineering	40	220		BBBB				○	○
	Construction management	40	220		BBBB				○	
Wolverhampton	Civil engineering	10	160–220		BBBB				○	
	Construction management	25	160–220						○	
	Construction project management		160–220							

Civil and Environmental Engineering and Building

109

See Chapter 6 in the *Introduction* (page 20) for general information about professional qualification that applies to all types of engineering course, including those covered in this part of the Guide.

Civil, structural and building services engineering   The Institution of Civil Engineers (ICE), the Institution of Structural Engineers (IStructE) and the Chartered Institution of Building Services Engineers (CIBSE) are jointly responsible, through the Joint Board of Moderators, for accreditation of courses in the area of civil, environmental, structural and building services engineering. Successful completion of an accredited course can satisfy the Engineering Council's academic requirements for achieving Chartered or Incorporated Engineer status. Corporate membership of one of these three bodies is also sufficient to satisfy the Engineering Council's requirement that you belong to a nominated professional body or one of their affiliates. See Chapters 6 (page 20) and 7 (page 24) in the *Introduction* for more information about professional qualification and the status of Chartered Engineer and Incorporated Engineer, and for the addresses of the professional institutions.

Building   To obtain full membership of the Chartered Institute of Building (MCIOB), you must either hold a recognised degree or successfully complete the member examination. Following that, you are required to pass the professional interview, which is used to assess your professional competence and personal qualities, and provide evidence of the nature, duration and level of experience you have obtained in building practice. Those who are academically qualified for the member class, but lack the necessary practical experience, may be admitted to the graduate class.

The Chartered Institute of Building is not one of the Engineering Council's nominated professional bodies or one of their affiliates, so membership does not contribute to satisfying the requirements for achieving Chartered or Incorporated Engineer status.

Accredited courses   Some courses are accredited by several professional institutions, but accreditation by an individual institution may be conditional on your having taken certain options, which may not coincide with those required by another institution. TABLE 6 shows which institutions have accredited individual courses. In the case of the ICE, I StructE and CIBSE, where accreditation is available at either Chartered or Incorporated level, ● indicates that Chartered level applies, ○, Incorporated. In some instances, either level of accreditation may be awarded, depending on whether you take a BEng or MEng degree; this is shown as ◐ in the table.

Remember that the process of accreditation goes on continuously, so you should check with the professional institutions for the current position on specific courses. Information has not been supplied for courses that are not included in the table.

Table 6 — Accreditation by professional institutions

Legend:
- ● accreditation at Chartered Engineer level
- ○ accreditation at Incorporated Engineer level
- ◑ accreditation level depends on degree taken

Institution	Course title	Institution of Civil Engineers	Institution of Structural Engineers	Chartered Institution of Building Services Engineers	Chartered Institute of Building	Energy Institute	Chartered Institution of Water and Environmental Management	Royal Institution of Chartered Surveyors	Institute of Mechanical Engineers	Royal Institute of British Architects
Aberdeen	All courses	●	●							
Abertay	Civil engineering		○							
Anglia Ruskin	Civil engineering		○							
	Construction management				●					
Bath	Civil and architectural engineering	●	●							
	Civil engineering	●	●							
Belfast	Civil engineering	●	●							
	Structural engineering with architecture		●							
Birmingham	Civil engineering	●	●							
Bolton	Civil engineering		○							
	Construction									
	Construction management				●					
Bradford	Civil and structural engineering	◑	◑							
Brighton	Civil engineering	●	●							
	Civil with environmental engineering									
Bristol	Civil engineering	●	●							
Bristol UWE	Civil engineering		○							
	Construction management				●					
Cambridge	Civil, structural and environmental engineering	●	●							
Cardiff	Architectural engineering		●							
	Civil and environmental engineering									
	Civil engineering	●	●							
Central Lancashire	Building services engineering			●						
	Construction project management				●					
City	Civil engineering	●	●							
Coventry	Civil and structural engineering	●	●							
	Civil engineering	●	●							
	Civil engineering construction		○							
	Construction management				●					
Derby	Construction management									
Dundee	Civil engineering	●	●							
Durham	Civil engineering	●	●							
East London	Civil engineering	●	●							
Edinburgh	Civil and environmental engineering	●								
	Civil engineering	●	●							
	Structural and fire safety engineering									
	Structural engineering with architecture									
Exeter	Civil engineering	●	●							
Glamorgan	Civil engineering	●	●							
	Project management (construction)									
Glasgow	Both courses	●	●							
Glasgow Caledonian	Environmental civil engineering		○							
Greenwich	Civil engineering	◑	◑							
	Civil engineering with project management	●	●							
	Civil engineering with water and environmental management	●	●							

(continued) Table 6

Accreditation by professional institutions

Key:
● accreditation at Chartered Engineer level
○ accreditation at Incorporated Engineer level
◑ accreditation level depends on degree taken

Institution	Course title	Institution of Civil Engineers	Institution of Structural Engineers	Chartered Institution of Building Services Engineers	Chartered Institute of Building	Energy Institute	Chartered Institution of Water and Environmental Management	Royal Institution of Chartered Surveyors	Institute of Mechanical Engineers	Royal Institute of British Architects
Heriot-Watt	Civil and environmental engineering	●								
	Civil engineering	●	●							
	Construction management				●			●		
	Structural engineering	●	●							
Imperial College London	Both courses	●	●	●						
Kingston	Civil engineering	●	●							
	Construction management									
Leeds	Architectural engineering	●	●							
	Civil and environmental engineering	●	●							
	Civil and structural engineering	●	●							
	Civil engineering with construction management	●	●							
Leeds Metropolitan	Civil engineering	●	●							
	Construction management				●					
Liverpool	All courses	●	●							
Liverpool John Moores	Civil engineering	●	●							
	Construction management				●					
London South Bank	Architectural engineering									
	Building services engineering			●		●				
	Civil engineering	●	●							
	Construction management				●					
Loughborough	Architectural engineering and design management				●					
	Civil engineering	●	●							
	Construction engineering management				●					
Manchester	Both courses	●	●							
Napier	Civil and transportation engineering	●	●							
	Civil engineering		○							
Newcastle	Civil and structural engineering	●	●							
	Civil engineering	●								
	Environmental engineering	●								
Northumbria	Building design management									
	Building project management				●					
	Building services engineering				●					
	Construction management									
Nottingham	Architectural environment engineering				●					
	Civil engineering	●	●					●		
	Environmental engineering									
Nottingham Trent	Civil and structural engineering	●	●							
	Civil engineering	◑	◑							
	Construction management				●			●		
Oxford	Civil engineering	●	●	●					●	
Oxford Brookes	Both courses									
Paisley	Civil engineering (BEng)	●	●							
Plymouth	Civil and coastal engineering							●		
	Civil and coastal engineering (BSc)	●	●					●		
	Civil engineering	●	●					●		
	Civil engineering (BSc)	●	●							

Accreditation by professional institutions

● accreditation at Chartered Engineer level
○ accreditation at Incorporated Engineer level;
◑ accreditation level depends on degree taken

Institution	Course title	Institution of Civil Engineers	Institution of Structural Engineers	Chartered Institution of Building Services Engineers	Chartered Institute of Building	Energy Institute	Chartered Institution of Water and Environmental Management	Royal Institution of Chartered Surveyors	Institute of Mechanical Engineers	Royal Institute of British Architects
Portsmouth	Civil engineering	●	●							
	Construction engineering management		○							
Reading	Building construction and management				●					
Robert Gordon	Construction, design and management									
	Mechanical and environmental engineering								●	
Salford	Civil engineering	●	●							
Sheffield	Civil and structural engineering	●	●							
	Civil engineering	●	●							
	Structural engineering and architecture	●	●							●
	Structural engineering design	●	●							
	Structural engineering with architectural studies	●	●							
Sheffield Hallam	Construction management				●			●		
Southampton	Civil engineering	●	●							
	Environmental engineering	●								
Strathclyde	Civil engineering	●	●							
Surrey	Civil engineering	●								
Swansea	Civil engineering	●	●							
Swansea IHE	Civil engineering and environmental management									
	Project and construction management									
Teesside	Civil engineering									
UCE Birmingham	Construction management and economics				●					
UCL	Civil engineering	●	●							
	Environmental engineering	●	●				●			
Ulster	Civil engineering	●	●							
	Construction engineering				●					
	Environmental engineering			●			●			
Wales (Newport)	Building studies									
	Civil and construction engineering									
Warwick	Civil engineering	●	●							
Westminster	Architectural engineering									●
	Building engineering				●			●		
	Construction management				●			●		
Wolverhampton	Civil engineering									
	Construction management				●					

See Chapter 7 in the *Introduction* (page 24) for a list of sources of information that apply to all types of engineering course, including those covered in this part of the Guide. It also lists a number of books that can give some general background to engineering and the work of engineers. Visit the websites of the professional institutions for information on careers, qualifications and course accreditation, as well as information about publications, many of which you can download. Chapter 7 in each of the other individual subject parts of the Guide contains further suggestions for background reading in particular engineering disciplines.

The following books and websites give some background information specifically for civil engineering and building.

Structures – Or Why Things Don't Fall Down J E Gordon. Penguin, 1991, £10.99

www.nceplus.co.uk The website of *New Civil Engineer* magazine

www.careersinconstruction.com A site listing current job vacancies, which will give you an idea of opportunities and salaries

The courses This Guide gives you information to help you narrow down your choice of courses. Your next step is to find out more about the courses that particularly interest you. Prospectuses cover many of the aspects you are most likely to want to know about, but some departments produce their own publications giving more specific details of their courses. University and college websites are shown in TABLE 2a.

You can also write to the contacts listed below.

Aberdeen Student Recruitment and Admissions Service (sras@abdn.ac.uk), University of Aberdeen, Regent Walk, Aberdeen AB24 3FX

Abertay Student Recruitment Office (sro@abertay.ac.uk), University of Abertay Dundee, Bell Street, Dundee DD1 1HG

Anglia Ruskin Contact Centre (answers@anglia.ac.uk), Anglia Ruskin University, Bishop Hall Lane, Chelmsford CM1 1SQ

Aston Schools Liaison and Careers (j.r.seymour@aston.ac.uk), Aston University, Aston Triangle, Birmingham B4 7ET

Bath Dr P J Walker (abs.adm@bath.ac.uk), School of Architecture and Civil Engineering, University of Bath, Claverton Down, Bath BA2 7AY

Belfast Dr H T Johnston, Advisor of Studies, Department of Civil Engineering, The Queen's University of Belfast, Belfast BT7 1NN

Birmingham Dr C Carliell-Marquet (uga-civeng@bham.ac.uk), School of Civil Engineering, University of Birmingham, Edgbaston, Birmingham B15 2TT

Bolton Civil engineering Peter Bullman (sam1@bolton.ac.uk); Construction Construction management Alan Cornthwaite (ac5@bolton.ac.uk); both at Department of the Built Environment, Bolton University, Deane Road, Bolton BL3 5AB

Bradford Mr Jack Bradley (ug-eng-enquiries@bradford.ac.uk), Admissions Tutor, School of Engineering, Design and Technology, University of Bradford, Bradford BD7 1DP

Brighton Construction management Project management for construction David Rutter (d.k.rutter@brighton.ac.uk); All other courses W J Williams (w.j.williams@brighton.ac.uk); both at School of the Environment, University of Brighton, Brighton BN2 4GJ

Bristol Admissions Tutor (ceng-ugadmissions@bristol.ac.uk), Department of Civil Engineering, University of Bristol, Queen's Building, University Walk, Bristol BS8 1TR

Bristol UWE Admissions office (fbe.entry@uwe.ac.uk), Faculty of the Built Environment, University of the West of England Bristol, Coldharbour Lane, Frenchay, Bristol BS16 1QY

Cambridge Cambridge Admissions Office (admissions@cam.ac.uk), University of Cambridge, Fitzwilliam House, 32 Trumpington Street, Cambridge CB2 1QY

Cardiff Dr S Bentley (bentleysp@cardiff.ac.uk), School of Engineering, Cardiff University, PO Box 917, Cardiff CF24 1XH

Central Lancashire Building services engineering Mr Mike Murray; Construction project management Mr Ian Wardle (iwardle@uclan.ac.uk), Course Leader; both at Department of Built Environment, University of Central Lancashire, Preston PR1 2HE

City Undergraduate Admissions Office (ugadmissions@city.ac.uk), City University, Northampton Square, London EC1V 0HB

Colchester I Course Enquiry Line (info@colchester.ac.uk), Colchester Institute, Sheepen Road, Colchester, Essex CO3 3LL

Coventry Administrative Assistant (s.buckley@coventry.ac.uk), School of Science and the Environment, Coventry University, Priory Street, Coventry CV1 5FB

Derby R E Parker (r.e.parker@derby.ac.uk), School of Arts, Design and Technology, University of Derby, Kedleston Road, Derby DE22 1GB

Dundee Dr M R Jones (m.r.jones@dundee.ac.uk), Admissions Tutor, Department of Civil Engineering, University of Dundee, Dundee DD1 4HN

Durham Dr Tim Short (engineering.admissions@durham.ac.uk), School of Engineering, University of Durham, South Road, Durham DH1 3LE

East London Student Admissions Office (admiss@uel.ac.uk), University of East London, Docklands Campus, 4–6 University Way, London E16 2RD

Edinburgh Undergraduate Admissions Office (sciengug@ed.ac.uk), College of Science and Engineering, University of Edinburgh, The King's Buildings, West Mains Road, Edinburgh EH9 3JY

Exeter Admissions Secretary (eng-admissions@exeter.ac.uk), Department of Engineering, School of Engineering, Computer Science and Mathematics, University of Exeter, North Park Road, Exeter EX4 4QF

Glamorgan Civil engineering D K Pugh, School of the Built Environment; Project management (construction) Roy Garwood, School of Design and Advanced Technology; both at University of Glamorgan, Pontypridd, Mid Glamorgan CF37 1DL

Glasgow Dr I McConnochie (mcconnochie@civil.gla.ac.uk), Department of Civil Engineering, Glasgow University, Glasgow G12 8QQ

Glasgow Caledonian School of the Built and Natural Environment, Glasgow Caledonian University, Cowcaddens Road, Glasgow G4 0BA

Greenwich Admissions Co-ordinator (eng-courseinfo@gre.ac.uk), Medway School of Engineering, University of Greenwich, Central Avenue, Chatham Maritime ME4 4TD

Heriot-Watt Construction management Admissions Tutor (b.s.robertson@hw.ac.uk), Department of Building Engineering and Surveying; All other courses Dr W J Ferguson (w.j.ferguson@hw.ac.uk), Admissions Officer, Department of Civil Engineering; both at Heriot-Watt University, Edinburgh EH14 4AS

Imperial College London Mrs L C Green (l.green@imperial.ac.uk), Academic Administrator, Department of Civil and Environmental Engineering, Imperial College London, South Kensington Campus, London SW7 2AZ

Kingston Student Information and Advice Centre, Cooper House, Kingston University, 40–46 Surbiton Road, Kingston upon Thames KT1 2HX

Leeds Energy and environmental engineering Admissions Tutor (spemeadmissions@leeds.ac.uk), Department of Fuel and Energy; All other courses Dr Terry Cousens (ugcivil@leeds.ac.uk), School of Civil Engineering; both at University of Leeds, Leeds LS2 9JT

Leeds Metropolitan Course Enquiries (course-enquiries@leedsmet.ac.uk), Leeds Metropolitan University, Civic Quarter, Leeds LS1 3HE

Liverpool Admissions Tutor (ugeng@liv.ac.uk), Department of Engineering, University of Liverpool, Brownlow Hill, Liverpool L69 3GH

Liverpool John Moores Student Recruitment Team (recruitment@ljmu.ac.uk), Liverpool John Moores University, Roscoe Court, 4 Rodney Street, Liverpool L1 2TZ

London South Bank Admissions Office, London South Bank University, 103 Borough Road, London SE1 0AA

Loughborough Admissions Tutor (civ.eng@lboro.ac.uk), Department of Civil and Building Engineering, Loughborough University, Loughborough LE11 3TU

Manchester Dr A J Bell (ug.mace@manchester.ac.uk), School of Mechanical, Aerospace and Civil Engineering, University of Manchester, PO Box 88, Manchester M60 1QD

Napier Civil and timber engineering Civil and transportation engineering David Read; Energy and environmental engineering Larry Sutherland, School of Engineering; All other courses John McNeill; all at Napier University, 10 Colinton Road, Edinburgh EH10 5DT

Newcastle Enquiries Service (enquiries@ncl.ac.uk), University of Newcastle upon Tyne, Newcastle upon Tyne NE1 7RU

North East Wales I Admissions Office (enquiries@newi.ac.uk), North East Wales Institute, NEWI Plas Coch, Mold Road, Wrexham LL11 2AW

Northumbria Admissions (er.educationliaison@northumbria.ac.uk), University of Northumbria, Trinity Building, Northumberland Road, Newcastle upon Tyne NE1 8ST

Nottingham Architectural environment engineering Miss Kim O'Reilly (kim.o'reilly@nottingham.ac.uk), School of the Built Environment; Civil engineering W H Askew (w.askew@nottingham.ac.uk), School of Civil Engineering; Environmental engineering Martin Waller (martin.waller@nottingham.ac.uk), School of Chemical, Environmental and Mining Engineering; all at University of Nottingham, University Park, Nottingham NG7 2RD

Nottingham Trent School of the Built Environment (sbe.ugqueries@ntu.ac.uk), Nottingham Trent University, Burton Street, Nottingham NG1 4BU

Oxford Deputy Administrator (Academic), Department of Engineering Science, Oxford University, Parks Road, Oxford OX1 3PJ

Oxford Brookes Albert McIlveen, School of Architecture, Oxford Brookes University, Headington, Oxford OX3 8TA

Paisley Professor P G Smith, Department of Civil Engineering, University of Paisley, High Street, Paisley PA1 2BE

Plymouth Construction management and the environment Steve Goodhew (sgoodhew@plymouth.ac.uk), School of Civil and Structural Engineering, University of Plymouth, Drake Circus, Plymouth PL4 8AA; All other courses Dr Les Hamill (technology@plymouth.ac.uk), School of Civil and Structural Engineering, University of Plymouth, Palace Court, Palace Street, Plymouth PL1 2DE

Portsmouth Admissions Tutor (civilweb@port.ac.uk), Department of Civil Engineering, University of Portsmouth, Lion Gate Building, Lion Terrace, Portsmouth PO1 3HF

Reading Mr Keith Hutchinson (k.hutchinson@reading.ac.uk), Faculty of Urban and Regional Studies, University of Reading, Whiteknights, PO Box 219, Reading RG6 2AW

Robert Gordon Paul Begg (p.begg@rgu.ac.uk), The Robert Gordon University, Schoolhill, Aberdeen AB9 1FR

Salford Ray Baker (r.baker@salford.ac.uk), School of Environment and Life Sciences, University of Salford, Salford M5 4WT

Sheffield Undergraduate Admissions Tutor, Department of Civil and Structural Engineering, University of Sheffield, Mappin Street, Sheffield S1 3JD

Sheffield Hallam Marketing Manager, Department of Environment and Development, Faculty of Development and Society, Sheffield Hallam University, Howard Street, Sheffield S1 1WB

Southampton Undergraduate Admissions Co-ordinator, School of Civil Engineering and the Environment, University of Southampton, Southampton SO17 1BJ

Southampton Solent Steve Bralee, Built Environment Division, Southampton Solent University, East Park Terrace, Southampton SO14 0YN

Strathclyde Academic Selector (n.s.ferguson@strath.ac.uk), Department of Civil Engineering, University of Strathclyde, Glasgow G4 0NG

Surrey Miss Natasha Baines (eng-admissions@surrey.ac.uk), School of Engineering, University of Surrey, Guildford GU2 7XH

Swansea Professor D Peric, Department of Civil Engineering, University of Wales Swansea, Singleton Park, Swansea SA2 8PP

Swansea IHE Civil engineering and environmental management Construction management John Greville (j.greville@sihe.ac.uk); Project and construction management Margaret Grills (margaret.grills@sihe.ac.uk); both at Swansea Institute of Higher Education, Townhill Road, Swansea SA2 0UT

Teesside Admissions Officer (registry@tees.ac.uk), University of Teesside, Middlesbrough TS1 3BA

UCE Birmingham Admissions Tutor (built.environment@uce.ac.uk), School of Property and Construction, University of Central England in Birmingham, Perry Barr, Birmingham B42 2SU

UCL Admissions Tutor (civeng-admissions@ucl.ac.uk), Department of Civil and Environmental Engineering, University College London, Gower Street, London WC1E 6BT

Ulster Building services engineering Civil engineering Dr W Cousins (w.cousins@ulst.ac.uk); Construction engineering S E McCaughey; both at School of the Built Environment; Environmental engineering Faculty Administration Officer, Faculty of Engineering; all at University of Ulster, Shore Road, Newtownabbey BT37 0QB

Wales (Newport) Admissions (admissions@newport.ac.uk), University of Wales College, Newport, PO Box 101, Newport NP18 3YH

Warwick Director of Undergraduate Admissions, Department of Engineering, University of Warwick, Coventry CV4 7AL

Westminster Admissions and Marketing Office (admissions@wmin.ac.uk), University of Westminster, 35 Marylebone Road, London NW1 5LS

Wolverhampton Civil engineering Paul Havell; Construction management Construction project management Mr J Billingham (in5410@wlv.ac.uk); both at SEBE, University of Wolverhampton, Wulfruna Street, Wolverhampton WV1 1SB

See the *Introduction* to this Guide (page 3) for an overview of how to use each chapter in the individual subject parts, including this one, together with general information about engineering and the structure of engineering courses.

The scope of this Guide   The courses in this Guide are concerned with a number of reasonably distinct engineering fields: electrical engineering, electronic engineering and information engineering. Having said that, the boundaries between them are often ill-defined and most courses include aspects of these fields.

Electrical engineering   At one time, the term electrical engineering embraced all engineering fields involving the use of electricity, but it is now largely taken to refer to heavy current or power engineering. It is the longest established of the engineering branches covered in this part of the Guide and deals with the generation, distribution and commercial use of electrical power. You will learn about the theory and practice underlying the use of devices such as generators, transformers, motors and transmission lines. Such devices can be found everywhere in industry, commercial premises and the home, so there is continuing demand for innovation to achieve greater efficiency, to exploit new materials and to address new markets. However, it is true to say that electrical engineering has not experienced the same rapid growth in recent years that electronic and information engineering have seen.

Electronic engineering   Electronics and electronic engineering are concerned largely with the properties and use of semiconductor devices such as transistors and integrated circuits, though courses may also include some work on other electronic devices with specialised applications. (There is no very clear distinction between courses with the title 'electronics' and 'electronic engineering', though the former tend to be more concerned with the underlying science and less with the design of electronic systems.)

Much of our everyday life at home, school or work has been revolutionised over the past generation by the development of electronics. Electronic control is increasingly finding its way into devices as simple as electric drills and washing machines, and as complex as entire robotically operated factories. The progress in developing electronic devices has been phenomenal. In computer electronics, for instance, there has been a consistent doubling of power and halving of cost every few years. A large part of this has been made possible by the integration of more and more functionality on a single chip. This has meant that electronic engineers can now concentrate on design at the system level, using standard chips to carry out standard functions, but connecting them together to carry out specific system functions. The use of programmable chips also allows the engineer to get customised behaviour from a standard, and therefore cheap, component. It would be wrong, however, to think that this has reduced the task of the electronic engineer to just fitting building blocks together. In particular, electronic engineers are required to design the individual integrated circuits themselves. Even at the system level, the high performance required by many applications means that the time taken for a signal to cross a printed circuit board becomes significant, so great care has to be taken to ensure that operations that are supposed to be synchronised do in fact take place at the same time. The explosion in mobile radio communications has created a great demand for designers of radio circuits that operate at very high frequencies (above 1GHz), and of very complex digital integrated circuits with low power consumption for battery operation. Another field of rapid growth is optoelectronics, for transmission, and photonics, for processing of information using modulated light signals.

Information engineering   The third field of engineering is the most recent and perhaps least clearly defined. It is concerned with the transmission, processing and interpretation of information, usually using digital techniques. Well established in telecommunications, these techniques are crucial to, for example, CD and DVD players, and they are now being applied to computer recognition of speech, and computer vision for robots. There is no universally agreed term for this branch of engineering, though 'information engineering' is the least ambiguous; specialised courses in this field that appear in this Guide have titles such as 'telecommunications engineering', 'network engineering' and 'communications technology'. Information technology is a close relation, but tends more to computer science, and is used in some cases to describe general computing courses (this Guide contains hardware-oriented computer courses; see the *Mathematics, Statistics and Computer Science* Guide for information technology, as well as general computer science courses). Note, incidentally, that 'information science' has almost no connection with information engineering: it deals with the organisation of information at a higher level, and in the past was largely concerned with paper-based systems and librarianship, though it is now much more computer-oriented.

Information engineering draws on and overlaps considerably with electronic engineering, but tends to involve higher-level or more specialised systems. The ever-increasing use of computers and the rapid development of technology means that increasing numbers of information engineers are going to be required for the foreseeable future. For example, the general trend away from large centralised mainframe computers to distributed systems where most of the information processing is carried out locally at the departmental or desktop level has placed increasing demands on communication systems. The expected growth in multimedia applications will multiply these demands many-fold. Similar developments are occurring in the home, where a revolution in entertainment and information services is coming about through services such as cable TV and broadband internet access.

What do electrical and electronic engineering graduates do?

Electrical and electronic equipment is found in nearly all work, domestic and leisure contexts. This means that electronic and electrical engineers are employed by an ever-widening range of organisations, including manufacturers, retailers, banks and information providers of all kinds.

Most graduates in electrical or electronic engineering enter careers directly related to their degree, mostly working within industry and commerce. Significant numbers work in the computing industry for employers such as computer manufacturers and software firms. However, the computing industries have tended to go through periods of boom and recession, making their demand for graduates change rapidly from year to year.

Although most electrical and electronic engineers want to use their degree in their future work, many graduate vacancies are open to people from any subject background. In addition to their technical skills, electrical and electronic engineers develop a number of transferable skills through their course activities and sandwich placements. These include analytical problem-solving, project management and teamwork. Students who take the opportunity to develop personal skills as well as technical skills can find employment opportunities with a wide range of employers, including areas such as finance, management, buying and selling.

Electrical engineering   One major area employing electrical engineers is electricity generation and distribution, which has become more competitive, with more companies entering the business and many large firms with big industrial complexes producing their own electricity. The railway companies are another group of employers using electrical engineers to supply, install and maintain equipment.

Large electric motors are used extensively in industry for transport, such as cranes and conveyors, and for driving machinery. They are also used outside factories in applications such as lifts and escalators. The manufacturers of this machinery employ electrical engineers for design, development, installation, repair and maintenance. Many larger users also employ their own engineers to monitor the installation and maintenance of equipment. Designers and suppliers of heating, ventilation, refrigeration and air-conditioning equipment also employ electrical engineers.

Electronic engineering   Electronic engineers occupy a number of different roles in industry. Some are employed in research, design and development teams to think up novel and marketable ideas, then develop these ideas into attractive and cost-effective products, and finally turn them into successful manufactured goods. There are, however, a wide variety of other jobs requiring specialist engineering skills and knowledge. These include technical writing and patent work, customer services, technical marketing and engineering consultancy. Larger systems may also need specialist installation, repair and maintenance.

The electronics industries include telecommunications businesses and the manufacturers of instrumentation and control systems, and computer and peripheral equipment. They also include companies involved in the design and manufacture of electronic components – integrated circuits, semiconductors and the other building blocks of our modern electronically based society.

Information technology   Information technology is a vast and varied business area, employing many electronic engineers. Employers include the providers of telephone and data communication services, including fixed-line and mobile phones, broadband services and cable TV, as well as computer manufacturers and the producers of peripheral equipment such as printers and modems.

The rapid growth in the use of the internet has highlighted the important economic and strategic role of this sector. Opportunities in IT cover the whole spectrum from research through production engineering to training customers and after-sales service, and may include working on software in addition to electronics.

The information technology and entertainment industries are becoming increasingly interrelated through developments such as the use of cable TV lines for telephones or broadband. In addition to these, broadcasting organisations, video production companies, recording companies and theatres all have a few openings.

Industrial electronics   Many electronic engineers are employed by manufacturing companies to install, maintain and optimise sensing and automatic control equipment. They develop control processes that use electronic instrumentation, provide and support communications equipment, and are involved in research to develop novel techniques. Some engineers work for the producers and suppliers of the instrumentation and process control equipment; a few are employed by contractors who design major plant for heavy industries such as chemicals; and others become consultants advising on ways to improve manufacturing efficiency.

Oil exploration is another specialised area of employment. Seismic surveys are carried out on land or at sea with sensitive electronic equipment, which must be calibrated, maintained and operated by electronic engineers. Knowledge of what is happening down oil wells is gained by wire-line logging, scanning the inside of wells with electronic devices.

Medical electronics   Rapid progress is being made in the development of ever more sophisticated medical equipment such as brain and body scanners, ultrasonic imaging equipment and other diagnostic tools. This has meant that medical electronics has become an expanding field of employment. Some engineers are employed by the producers of such equipment; others work in hospitals on its calibration and use.

Defence and the armed forces   The military relies increasingly on sophisticated electronics in areas such as satellite surveillance, laser-guided bombs and secure communications systems. Although vacancies in the field of defence electronics have been reduced, there are still some opportunities to work for defence contractors and aerospace companies.

The armed forces are also important recruiters. The Army, Royal Navy and RAF all need engineers to maintain their equipment and liaise with suppliers.

Further education or training   Some graduates in electrical and electronic engineering stay on in higher education, mainly to study for higher degrees. Often they take Master's courses to increase their knowledge in areas such as robotics, computer-aided design, digital electronics or integrated circuits. Some take more general courses, such as integrated manufacturing systems, which introduce them to all the other technologies involved in production processes. Others take research degrees to develop their research skills. However, funding for postgraduate study is not automatic, and there can be competition for places on courses.

Chapter 2: The courses

The Guide includes a wide variety of courses, ranging from those covering the whole of electrical and electronic engineering to specialised courses dealing with a particular application. However, the specialised courses often start with a broad introduction, and the more wide-ranging courses often allow specialisation in later years, so, in practice, the material covered by courses with very different titles may be very similar.

You should also note that although TABLE 2a lists a wide range of courses at many institutions, there may be even more on offer that we do not list here, mainly because they are very specialised. A few of these are mentioned at the end of this chapter, but you should also check the institutions' websites and prospectuses to make sure you are considering the full range of possibilities.

TABLE 2a lists the specialised and combined courses at universities and colleges in the UK that lead to the award of an honours degree in electrical or electronic engineering. When the table was compiled it was as up to date as possible, but sometimes new courses are announced and existing courses withdrawn, so before you finally fill in your application you should check the UCAS website, www.ucas.com, to make sure the courses you plan to apply for are still on offer.

See Chapter 2 in the *Introduction* (page 6) for advice on how to use TABLE 2a and for an explanation of what the various columns mean.

Table 2a — First-degree courses in Electrical and Electronic Engineering

Institution / Course title	see combined subject list – Table 2b	Degree	Duration (No of years)	Foundation year (● at this institution / ○ at franchised institution / ◐ second-year entry)			Modes of study (● full time; ⊸ part time / ○ time abroad / ◑ sandwich)			Modular scheme	Course type (● specialised; ◑ combined)	No of combined courses
Aberdeen www.abdn.ac.uk												
Electrical and electronic engineering ①		BEng/MEng	4, 5			◐	● ⊸	○			● ◑	4
Electronic and computer engineering		BEng/MEng	5			◐	●				●	0
Electronic engineering with communications		BEng/MEng	5			◐	●				●	0
Electronics and computer/software engineering ③		BEng	4			◐	● ⊸	○			● ◑	2
Anglia Ruskin www.anglia.ac.uk												
Electronics		BSc	3	●			● ⊸	○		✿	● ◑	3
Aston www.aston.ac.uk												
Communications engineering		BEng	3, 4	●	○		●		◑		●	0
Electrical and electronic engineering		BEng	3, 4	●	○		●		◑		●	0
Electronic engineering ①		BEng	3, 4				●		◑		● ◑	2
Electronic systems engineering ②		MEng	4, 5				●		◑		● ◑	1
Internet engineering		BEng	3, 4	●			●				●	0
Bangor www.bangor.ac.uk												
Communications and computer systems		BEng/MEng	3, 4	●			●				●	0
Electronic engineering		BEng/MEng	3, 4	●			●				●	0
Bath www.bath.ac.uk												
Computers, electronics and communications		BEng/MEng	3, 4, 5	●			●		○ ◑		●	0
Electrical and electronic engineering		BEng/MEng	3, 4, 5	●			●		○ ◑		●	0

First-degree courses in **Electrical and Electronic Engineering**

Institution / Course title	①②③ see combined subject list – Table 2b	Degree	Duration (Number of years)	Foundation year: ● at this institution	○ at franchised institution	◑ second-year entry	Modes of study: ● full time, ▼ part time	○ time abroad	◑ sandwich	✦ Modular scheme	Course type: ● specialised, ◑ combined	No of combined courses
Bath (continued)												
Electrical engineering and applied electronics		BEng/MEng	3, 4, 5	●			●	○	◑		●	0
Electrical power engineering		BEng/MEng	3, 4, 5	●			●	○	◑		●	0
Electronic and communication engineering		BEng/MEng	3, 4, 5	●			●	○	◑		●	0
Electronics with space science and technology		BEng/MEng	3, 4, 5	●			●	○	◑		●	0
Bedfordshire www.beds.ac.uk												
Computer networking		BSc	3	●			●				●	0
Belfast www.qub.ac.uk												
Electrical and electronic engineering		BEng/MEng	3, 4, 5	●			●		◑		●	0
Electronic and software engineering		BEng/MEng	3, 4, 5	●			●		◑		●	0
Birmingham www.bham.ac.uk												
Communication systems engineering ①		BEng/MEng	3, 4	●			●	○	◑		● ◑	2
Computer systems engineering ②		BEng/MEng	3, 4	●			●	○	◑		● ◑	2
Electronic and communications engineering		BEng/MEng	3, 4	●			●	○	◑		●	0
Electronic and computer engineering		BEng/MEng	3, 4	●			●	○	◑		●	0
Electronic and electrical engineering ③		BEng/MEng	3, 4	●			●	○	◑		● ◑	1
Electronic engineering ④		BEng/MEng	3, 4	●			●	○	◑		● ◑	2
Bolton www.bolton.ac.uk												
Electronic and computer engineering		BEng	3, 4	●			●				●	0
Internet communications and networks		BSc	3, 4	●			●				●	0
Bournemouth www.bournemouth.ac.uk												
Electronics		BEng	3, 4		○		●		◑		●	0
Bradford www.bradford.ac.uk												
Electrical and electronic engineering		BEng/MEng	3, 4, 5	●			●		◑		●	0
Electronic, telecommunications and internet engineering		BEng/MEng	3, 4, 5	●			●		◑		●	0
Brighton www.brighton.ac.uk												
Audio electronics		BEng	3, 4	●			● ▼	○	◑		●	0
Digital electronics, computing and communication		BEng	3, 4	●			●	○	◑		●	0
Electrical and electronic engineering		BEng	3, 4	●			● ▼	○	◑		●	0
Bristol www.bris.ac.uk												
Communications and multimedia engineering		MEng	4	●			●				●	0
Computer systems engineering		MEng	4	●			●	○			●	0
Electrical and electronic engineering ①		BEng/MEng	3, 4	●			●	○			● ◑	1
Electronic and communications engineering		BEng/MEng	3, 4	●			●				●	0
Electronic engineering		BEng	3	●			●				●	0
Bristol UWE www.uwe.ac.uk												
Computer systems engineering		BSc	3, 4	●			●		◑		●	0
Computing and telecommunications		BSc	3, 4	●	○		●	○	◑		●	0
Digital systems engineering		BEng/MEng	3, 4, 5	●	○		● ▼	○	◑	✦	●	0
Electrical and electronic engineering		BEng/MEng	3, 4, 5	●	○		● ▼	○	◑	✦	●	0
Electronic engineering		BEng/MEng	3, 4, 5	●	○		● ▼	○	◑	✦	●	0
Music systems engineering		BSc	3, 4	●	○		●	○	◑		●	0
Brunel www.brunel.ac.uk												
Computer systems engineering		BEng	3, 4	●			●	○	◑		●	0
Electronic and electrical engineering		BEng	3, 4	●			●	○	◑		●	0
Electronic and microelectronic engineering		BEng	3, 4	●			●	○	◑		●	0
Electronic/electrical engineering (communication systems)		BEng	3, 4	●			●	○	◑		●	0
Electronic/electrical engineering (control systems)		BEng	3, 4	●			●	○	◑		●	0
Electronic/electrical engineering (power electronics systems)		BEng	3, 4	●			●	○	◑		●	0

First-degree courses in **Electrical and Electronic Engineering**

Sidebar: **Electrical and Electronic Engineering**

Institution / Course title	① ② ③ see combined subject list – Table 2b	Degree	Duration (Number of years)	Foundation year	● at this institution	○ at franchised institution	◑ second-year entry	Modes of study ● full time; ► part time	○ time abroad	◑ sandwich	Modular scheme	Course type ● specialised; ◑ combined	No of combined courses
Brunel (continued)													
Internet engineering		BEng	3, 4		●			●		◑		●	0
Buckinghamshire Chilterns UC www.bcuc.ac.uk													
Internet technology		BSc	3		●			●				●	0
Cambridge www.cam.ac.uk													
Electrical and electronic engineering		BA/MEng	3, 4		●			●			✿	●	0
Electrical and information sciences		BA/MEng	3, 4		●			●			✿	●	0
Information and computer engineering		BA/MEng	3, 4		●			●			✿	●	0
Cardiff www.cardiff.ac.uk													
Computer systems engineering		BEng/MEng	3, 4, 5		●			●	○	◑		●	0
Electrical and electronic engineering		BEng/MEng	3, 4, 5		●			●	○	◑		●	0
Electronic engineering		BEng/MEng	3, 4, 5		●			●	○	◑		●	0
Central Lancashire www.uclan.ac.uk													
Computer engineering		BEng	3		●			●	○			●	0
Digital communications		BEng	3, 4		●			●	○	◑		●	0
Digital signal and image processing		BEng	3, 4		●			●	○	◑		●	0
Electronic engineering		BEng	3, 4	●		○		● ►	○	◑		●	0
City www.city.ac.uk													
Communications		BEng	3, 4	●		○		●	○	◑		●	0
Computer systems engineering		BSc/BEng	3, 4	●		○		●	○	◑		●	0
Electrical and electronic engineering		BSc/Eng/MEng	3, 4, 5	●		○	○	●	○	◑		●	0
Systems and control engineering		BEng/MEng	3, 4, 5	●		○		●	○	◑		●	0
Coventry www.coventry.ac.uk													
Communications engineering		BEng	3, 4		●			●		◑		●	0
Computer hardware and software engineering		BEng	3, 4		●			●		◑		●	0
Computers, networking and communications technology		BEng	3, 4		●			●		◑		●	0
Electrical systems engineering		BEng	3, 4		●			●		◑		●	0
Electronics engineering		BEng	3, 4		●			● ►		◑		●	0
De Montfort www.dmu.ac.uk													
Broadcast technology		BSc	3		●			●				●	0
Electronic engineering		BEng	3, 4			○		●	○	◑		●	0
Derby www.derby.ac.uk													
Computer networks ①		BSc	4					►		◑	●	● ◑	15
Electrical and electronic engineering		BSc	3	●	●			● ►				●	0
Electronics		BSc	3	●	●			●				●	0
Dundee www.dundee.ac.uk													
Electronic and electrical engineering		BEng/MEng	4, 5				◑	● ►	○			●	0
Electronic engineering ①		BEng/MEng	4, 5				◑	● ►				● ◑	2
Electronic engineering and microcomputer systems		BEng/MEng	4, 5				◑	●				●	0
Electronics and computing ②		BEng	4				◑	●				◑	1
Microelectronics and photonics		BEng/MEng	4, 5				◑	●				●	0
Physics and microelectronics ③		BSc	4				◑	●				◑	1
Durham www.durham.ac.uk													
Communications engineering		MEng	4		●			●	○			●	0
Computer engineering		MEng	4		●			●	○			●	0
Electronic engineering		MEng	4		●			●	○			●	0
East Anglia www.uea.ac.uk													
Computer systems engineering		BSc	3		●			●				●	0

First-degree courses in **Electrical and Electronic Engineering**

Institution / Course title	①②③ see combined subject list – Table 2b	Degree	Duration (Number of years)	Foundation year (● at this institution)	(○ at franchised institution)	(◑ second-year entry)	Modes of study (● full time)	(▼ part time)	(○ time abroad)	(◑ sandwich)	Modular scheme	Course type (● specialised)	(◑ combined)	No of combined courses
East London www.uel.ac.uk														
Computing and electronics ①		BEng	3				●					●	◑	4
Electrical and electronic engineering ②		BEng	3, 4	●	○		●	▼	○	◑		●	◑	4
Edinburgh www.ed.ac.uk														
Electrical engineering ①		MEng	4, 5			◑	●				✿	●	◑	1
Electronics ②		BEng/MEng	4, 5			◑	●		○	◑	✿	●	◑	2
Electronics and electrical engineering ③		BEng/MEng	4, 5			◑	●					●	◑	1
Electronics and electrical engineering (communications)		BEng	4			◑	●					●		0
Essex www.essex.ac.uk														
Audio engineering		BEng	3	●			●					●		0
Computer engineering		BEng	3	●			●					●		0
Computers and networks		BEng	3	●			●					●		0
Computers and telecommunications		BEng	3	●			●					●		0
Electronic engineering		BEng	3	●			●					●		0
Electronics and computers ①		BSc	4				●					●	◑	1
Network and internet technology		BEng	3				●					●		0
Optoelectronics and communications systems		BEng	3				●					●		0
Telecommunications engineering		BEng	3	●			●					●		0
Exeter www.exeter.ac.uk														
Electronic engineering		BEng/MEng	3, 4				●		○		✿	●		0
Internet engineering		BSc	3				●					●		0
Glamorgan www.glam.ac.uk														
Computer systems engineering		BEng/MEng	3, 4, 5	●			●			◑		●		0
Electrical and electronic engineering		BEng/MEng	3, 4, 5	●			●			◑		●		0
Electronic and communication engineering		BEng/MEng	3, 4, 5	●			●			◑		●		0
Electronic engineering		BEng/MEng	3, 4, 5	●			●			◑		●		0
Electronics ①		BSc	3, 4	●			●			◑		●	◑	1
Mobile telecommunications		BEng/MEng	3, 4	●			●			◑		●		0
Glasgow www.gla.ac.uk														
Audio and video engineering		BEng/MEng	4, 5			◑	●		○			●		0
Electronic and software engineering ①		BEng/MEng	4, 5			◑	●		○			●	◑	1
Electronics and electrical engineering		BEng/MEng	4, 5			◑	●		○			●		0
Electronics with music ②		BEng	4			◑	●		○			●	◑	1
Microcomputer systems engineering		BEng	4			◑	●		○			●		0
Glasgow Caledonian www.gcal.ac.uk														
Computer engineering		BSc	4			◑	●					●		0
Electrical power engineering		BEng	4			◑	●	▼				●		0
Electronic engineering (BEng)		BEng	4			◑	●	▼		◑		●		0
Electronic engineering (BSc)		BSc	4			◑	●					●		0
Telecommunications engineering		BSc	4			◑	●					●		0
Greenwich www.gre.ac.uk														
Communications systems and software engineering ①		BEng	3, 4	●			●	▼		◑			◑	1
Computer networking		BEng	3, 4	●	○		●	▼		◑		●		0
Computer systems and software engineering ②		BSc/BEng	3, 4	●	○		●	▼		◑		●	◑	1
Control and instrumentation engineering		BEng	3	●			●	▼		◑		●		0
Electrical and electronic engineering		BSc/BEng	3, 4	●	○		●	▼		◑		●		0
Electrical and electronic engineering technology		BEng	3, 4	●			●	▼		◑		●		0
Electrical engineering		BEng	3, 4	●			●	▼		◑		●		0
Electronic engineering		BEng	3, 4	●	○		●	▼		◑		●		0

(continued) Table 2a

First-degree courses in **Electrical and Electronic Engineering**

Institution / Course title	combined (see Table 2b)	Degree	Duration (years)	Foundation: at this institution	Foundation: franchised	Foundation: 2nd-year entry	Mode: full time	Mode: part time	Mode: time abroad	Mode: sandwich	Modular scheme	Course type	No of combined courses
Greenwich (continued)													
Electronics and computer systems		BEng	3, 4	●			●			◐		●	0
Internet technologies		BEng	3, 4	●			●	✈		◐		●	0
Heriot-Watt www.hw.ac.uk													
Computing and electronics ①		BEng/MEng	4, 5	●	◐		●			◐		◐ ●	1
Electrical and electronic engineering		BEng/MEng	4, 5			◐	●			◐		●	0
Electronic and photonic engineering		BEng/MEng	4, 5			◐	●			◐		●	0
Robotics and cybertronics		BEng/MEng	4, 5			◐	●			◐		●	0
Hertfordshire www.herts.ac.uk													
Digital communications and electronics		BEng/MEng	3, 4, 5		○		●		○	◐		●	0
Digital systems and computer engineering		BEng/MEng	3, 4, 5		○		●		○	◐		●	0
Electrical and electronic engineering		BEng	3, 4		○		●	✈		◐		●	0
Huddersfield www.hud.ac.uk													
Computer control systems		BEng/MEng	4, 5	●			●			◐		●	0
Electronic and communication engineering		BEng	3, 4	●			●			◐		●	0
Electronic and electrical engineering		BEng	3	●			●			◐		●	0
Electronic design		BSc	4	●			●			◐		●	0
Electronic engineering		BEng/MEng	4, 5	●			●			◐		●	0
Electronic engineering and computer systems		BEng	3, 4	●			●			◐		●	0
Hull www.hull.ac.uk													
Computer systems engineering		BEng	3		○		●					●	0
Electronic engineering		BEng/MEng	3, 4		○		●	✈		◐		●	0
Information and computer control technology		BSc	3				●			◐		●	0
Mobile telecommunications technology		BEng/MEng	3, 4				●			◐		●	0
Imperial College London www.imperial.ac.uk													
Electrical and electronic engineering ①		BEng/MEng	3, 4	●			●		○			● ◐	1
Information systems engineering		BEng/MEng	3, 4	●			●		○			●	0
Kent www.ukc.ac.uk													
Computer systems engineering		BEng/MEng	3, 4	●			●			◐		●	0
Electronic and communications engineering		BEng/MEng	3, 4	●			●			◐		●	0
King's College London www.kcl.ac.uk													
Computer systems and electronics		BEng/MEng	3, 4, 5				●			◐		●	0
Electronic engineering		BEng/MEng	3, 4, 5				●			◐		●	0
Telecommunication engineering		BEng/MEng	3, 4, 5				●			◐		●	0
Kingston www.kingston.ac.uk													
Communication systems		BSc	3				●					●	0
Lancaster www.lancs.ac.uk													
Computer systems engineering		BEng/MEng	3, 4				●			◐		●	0
Electronic communication systems		BEng/MEng	3, 4				●			◐		●	0
Electronic systems engineering		BEng/MEng	3, 4				●			◐		●	0
Leeds www.leeds.ac.uk													
Electronic and communications engineering		BEng/MEng	3, 4	●			●		○			● ◐	0
Electronic and electrical engineering		BEng/MEng	3, 4	●			●		○			● ◐	0
Electronic engineering		BEng/MEng	3, 4	●			●		○			●	0
Leicester www.le.ac.uk													
Communications and electronic engineering		BEng/MEng	3, 4, 5	●			●		○	◐		●	0
Electrical and electronic engineering		BEng/MEng	3, 4, 5	●			●		○	◐		●	0
Embedded systems engineering		BEng/MEng	3, 4, 5	●			●		○	◐		●	0

First-degree courses in **Electrical and Electronic Engineering**

Institution / Course title	combined / subject list – Table 2b	Degree	Duration (Number of years)	Foundation year: ● at this institution / ○ at franchised institution / ◐ second-year entry			Modes of study: ● full time, ➤ part time / ○ time abroad / ◐ sandwich			Modular scheme	Course type: ● specialised; ◐ combined	No of combined courses
Liverpool www.liv.ac.uk												
Computer science and electronic engineering ①		BEng/MEng	3, 4	●	○		●				◐	1
Electrical and electronic engineering		BEng	4	●			●				●	0
Electrical engineering		BEng	3	●	○		●				●	0
Electrical engineering and electronics		BEng/MEng	3, 4	●	○		●				●	0
Electronic and communication engineering		BEng	3	●	○		●				●	0
Electronics ②		BEng/MEng	3, 4	●	○		●				● ◐	1
Wireless communications and 3G technology		BEng/MEng	3, 4	●	○		●	○			●	0
Liverpool John Moores www.ljmu.ac.uk												
Broadcast engineering		BEng	3, 4	●			●		◐		●	0
Broadcast technology		BSc	3, 4	●			●		◐		●	0
Computer engineering		BEng	3, 4	●			●		◐		●	0
Design electronics		BSc	3, 4	●			●		◐		●	0
Electrical and electronic engineering		BEng	3, 4	●			● ➤		◐		●	0
Networks and telecommunications engineering		BEng	3, 4	●			●		◐		●	0
London Metropolitan www.londonmet.ac.uk												
Audio electronics		BSc	3, 4	●			● ➤	○	◐	◐	●	0
Communications systems		BSc	3, 4	●	○		● ➤	○	◐	◐	●	0
Computer networking ①		BSc	3, 4	●	○		● ➤	○	◐	◐	● ◐	1
Electronic and communications engineering		BEng	3, 4	●			● ➤	○	◐		●	0
Electronics		BSc	3, 4	●			● ➤		◐	◐	●	0
London South Bank www.sbu.ac.uk												
Electrical and electronic engineering		BEng	3, 4	●			● ➤	○	◐		●	0
Internet and multimedia engineering		BEng	3, 4	●			●		◐		●	0
Telecommunications and computer networks engineering		BEng	3, 4	●			●		◐		●	0
Loughborough www.lboro.ac.uk												
Computer network and internet engineering		MEng	4, 5	●			●		◐		●	0
Computer systems engineering		BEng/MEng	3, 4, 5	●			●	○	◐		●	0
Electronic and electrical engineering		BEng/MEng	3, 4, 5	●			●	○	◐		●	0
Electronics and software engineering		MEng	4, 5	●			●	○	◐		●	0
Wireless communication engineering		MEng	4, 5	●			●	○	◐		●	0
Manchester www.man.ac.uk												
Computer engineering ①		BSc/MEng	3, 4	●			●	○	◐		● ◐	2
Computer systems engineering		BEng/MEng	3, 4, 5	●			●		◐		●	0
Computing and communications systems engineering		BEng/MEng	3, 4, 5	●			●		◐		●	0
Electrical and electronic engineering		BEng/MEng	3, 4, 5	●			●		◐		●	0
Electronic systems engineering		BEng/MEng	3, 4, 5	●			●		◐		●	0
Manchester Metropolitan www.mmu.ac.uk												
Communication and electronic engineering		BEng	3, 4	●	○		●	○	◐		●	0
Computer and electronic engineering		BEng	3, 4	●	○		●	○	◐		●	0
Computer and network technology		BSc	3	●	○		●		◐		●	0
Electrical and electronic engineering		BEng	3, 4	●	○		● ➤	○	◐		●	0
Middlesex www.mdx.ac.uk												
Computer networks		BSc	3	●			●				●	0
Napier www.napier.ac.uk												
Computer networks and distributed systems		BEng	4					●	◐		●	0
Electronic and communication engineering		BEng	4		●						●	0
Electronic and computer engineering	◐	BEng	4					◐			●	0
Electronic and electrical engineering	◐	BEng	4				➤	◐			●	0

(continued) Table 2a

First-degree courses in **Electrical and Electronic Engineering**

Legend: ① ② ③ see combined subject list – Table 2b. Foundation year: ● at this institution, ○ at franchised institution, ◑ second-year entry. Modes of study: ● full time, ► part time, ○ time abroad, ◑ sandwich. ✇ Modular scheme. Course type: ● specialised; ◑ combined.

Institution / Course title	Degree	Duration (years)	Foundation year	Modes of study	Modular scheme	Course type	No of combined courses
Newcastle www.ncl.ac.uk							
Computer systems engineering	BEng/MEng	3, 4	●	● ○		●	0
Electrical and electronic engineering	BEng/MEng	3, 4, 5	●	● ○ ○		●	0
Electronic communications	BEng/MEng	3, 4	●	● ○ ○		●	0
Electronic engineering	BEng/MEng	3, 4	●	● ○		●	0
North East Wales I www.newi.ac.uk							
Computer networks ①	BSc	3	●	●		● ◑	1
Electrical and electronic engineering	BEng	3	●	●		●	0
Northumbria www.northumbria.ac.uk							
Communication and electronic engineering	BEng	3, 4	●	●	◑ ✇	●	0
Computer and network technology	BSc	3, 4	●	●	◑ ✇	●	0
Electrical and electronic engineering	BEng	3, 4	●	●	◑ ✇	●	0
Nottingham www.nottingham.ac.uk							
Electrical and electronic engineering ①	BEng/MEng	3, 4	●	● ○		● ◑	3
Electrical engineering ②	BEng/MEng	3, 4	●	● ○		● ◑	2
Electronic and communications engineering	BEng/MEng	3, 4	●	● ○		●	0
Electronic and computer engineering	BEng/MEng	3, 4	●	● ○		●	0
Electronic engineering ③	BEng/MEng	3, 4	●	● ○		● ◑	3
Oxford www.ox.ac.uk							
Electrical engineering	MEng	4	●	●		●	0
Information engineering	MEng	4	●	●		●	0
Oxford Brookes www.brookes.ac.uk							
Electronic systems design	BSc	3	●	●		●	0
Telecommunications	BSc	3	● ►	●		●	0
Paisley www.paisley.ac.uk							
Internet technologies	BSc	4	●	●		●	0
Plymouth www.plymouth.ac.uk							
Communication engineering	BEng	3, 4	● ○	●	○ ◑	●	0
Computer engineering	BEng	3, 4	● ○	●	○ ◑	●	0
Computer systems and networks	BSc	4			◑	●	0
Computer systems engineering	BSc	3	● ○	●		●	0
Electrical and electronic engineering	BEng	3, 4	● ○	● ► ○ ◑		●	0
Electrical and electronic systems	BSc	3, 4	● ○	●	○ ◑	●	0
Electronic communication systems	BSc	3		●	◑	●	0
Electronic engineering	BEng	3, 4	● ○	● ◡ ○ ◑		●	0
Internet technologies and applications	BSc	3		●		●	0
Portsmouth www.port.ac.uk							
Communication systems engineering	BEng/MEng	3, 4, 5	●	●	○ ◑	●	0
Communications engineering	BEng	3, 4	●	●	○ ◑	●	0
Computer engineering	BEng/MEng	3, 4, 5	●	●	○ ◑	●	0
Computer technology	BSc	3, 4	●	●	○ ◑	●	0
Electronic and electrical engineering	BEng/MEng	3, 4, 5	●	●	○ ◑	●	0
Electronic engineering	BEng	3, 4	●	●	○ ◑	●	0
Internet technology	BSc	3, 4	●	●	○ ◑	●	0
Queen Mary www.qmul.ac.uk							
Audio systems engineering	BEng	3	●	●		●	0
Communication engineering	MEng	4	●	●	○	●	0
Computer engineering	BEng/MEng	3, 4	●	●	○	●	0
Electrical and electronic engineering	BEng	3	●	●	○	●	0

First-degree courses in **Electrical and Electronic Engineering**

Institution / Course title	①②③ see combined subject list – Table 2b	Degree	Duration (Number of years)	Foundation year: ● at this institution	○ at franchised institution	◑ second-year entry	Modes of study: ● full time; ▶ part time	○ time abroad	◑ sandwich	Modular scheme: ◐ combined	Course type: ● specialised; ◑ combined	No of combined courses
Queen Mary (continued)												
Electronic engineering		BEng/MEng	3, 4	●			●	○			●	0
Internet engineering		BEng	3	●			●				●	0
Physics and electronics ①		BSc/MSci	3, 4	●			●	○			●◑	1
Telecommunications		BEng	3	●			●	○			●	0
Reading www.rdg.ac.uk												
Computer engineering ①		BSc	3	●			●				●◑	3
Cybernetics ③		MEng	3, 4	●			●				●◑	4
Electronic engineering ③		BEng/MEng	3, 4	●			●				●◑	1
Robert Gordon www.rgu.ac.uk												
Communications and computer engineering		BEng	4				●				●	0
Communications and computer network engineering		BEng	4				●				●	0
Electrical and energy engineering		BEng	4			◑	●		▶		●	0
Electronic and communications engineering ①		BEng	4			◑	●	○	▶		●◑	1
Electronic and computer engineering		BEng	4			◑	●	○	▶		●	0
Electronic and electrical engineering		BSc/BEng/MEng	4, 5			◑	●	○	▶		●	0
St Andrews www.st-and.ac.uk												
Microelectronics and photonics		BEng/MEng	4, 5			◑	●				●	0
Sheffield www.sheffield.ac.uk												
Computer systems engineering ①		BEng/MEng	3, 4	●			●				●◑	2
Data communications engineering ②		BEng/MEng	3, 4	●			●				●◑	1
Digital electronics		MEng	4	●			●				●	0
Electrical engineering ③		BEng/MEng	3, 4	●			●	○			●◑	3
Electronic and communications engineering ④		MEng	4	●			●				●◑	1
Electronic engineering ⑤		BEng/MEng	3, 4	●			●	○			●◑	3
Electronic, control and systems engineering ⑥		BEng/MEng	3, 4	●			●				●◑	2
Engineering (computing, electronics, systems and control)		BEng/MEng	3, 4	●			●				●	0
Microelectronics		MEng	4	●			●				●	0
Systems and control engineering ⑦		BEng/MEng	3, 4	●			●				●◑	2
Sheffield Hallam www.shu.ac.uk												
Computer and network engineering		BSc	3, 4	●	○		●			◐	●	0
Computer networks		BSc	3, 4	●			●			◐	●	0
Electrical and electronic engineering		BEng/MEng	3	●			●			◐	●	0
Electronic engineering ①		BEng	4	●			●			◐	●◑	5
Southampton www.soton.ac.uk												
Computer engineering		BEng/MEng	3, 4	●	○		●	○			●	0
Electrical engineering		BEng/MEng	3, 4	●	○		●	○	◑		●	0
Electronic engineering ①		BEng/MEng	3, 4	●	○		●	○			●◑	4
Southampton Solent www.solent.ac.uk												
Computer network communications		BSc	3	●			●				●	0
Computer network management		BSc	3	●			●				●	0
Computer systems and networks		BSc	3	●			●				●	0
Electronic engineering		BEng	3	●			●		▶		●	0
Staffordshire www.staffs.ac.uk												
Broadcasting technology		BSc	3, 4				●			◐	●	0
Computer systems		BSc/BEng/MEng	3, 4, 5		○		●	○	▶	◐	●	0
Electronic engineering		BEng/MEng	3, 4, 5	●	○		●	○	▶	◐	●	0
Electronic systems design		BEng/MEng	3, 4, 5	●	○		●	○	▶	◐	●	0
Internet technology		BEng/MEng	3, 4, 5	●			●	○		◐	●	0

(continued) Table 2a

First-degree courses in **Electrical and Electronic Engineering**

Institution / Course title	①②③ see combined subject list – Table 2b	Degree	Duration (Number of years)	Foundation year (● at this institution, ○ at franchised institution, ◑ second-year entry)	Modes of study (● full time; ⌐ part time, ○ time abroad, ◑ sandwich)	Modular scheme	Course type (● specialised; ◑ combined)	No of combined courses
Staffordshire (continued)								
Mobile and wireless business systems		BEng/MEng	3, 4		● ○ ◑		●	0
Mobile device technology		BEng/MEng	3, 4		● ○ ◑		●	0
Wireless networking technology		BEng/MEng	3, 4		● ○ ◑		●	0
Strathclyde www.strath.ac.uk								
Computer and electronic systems		BEng/MEng	4, 5	◑	●		●	0
Digital communication and multimedia systems		BEng/MEng	4, 5		●		●	0
Electrical energy systems		BEng/MEng	4, 5	◑	● ○		●	0
Electronic and digital systems		BEng/MEng	4, 5	◑	● ◑ ○		●	0
Electronic and electrical engineering ①		BEng/MEng	4, 5	◑	● ◑ ○		● ◑	2
Surrey www.surrey.ac.uk								
Audio media engineering		BEng/MEng	3, 4, 5		● ◑		●	0
Digital media engineering		BEng/MEng	3, 4, 5		● ◑		●	0
Electronic engineering		BEng/MEng	3, 4, 5	●	● ○ ◑		●	0
Electronics and computer engineering		BEng/MEng	3, 4, 5	●	● ○ ◑		●	0
Electronics with satellite engineering		MEng	4, 5		● ○ ◑		●	0
Telecommunication systems		BEng/MEng	3, 4, 5		● ◑		●	0
Sussex www.sussex.ac.uk								
Computer systems engineering		BEng/MEng	3, 4	●	●		●	0
Electrical and electronic engineering		BEng/MEng	3, 4	●	● ○		●	0
Electronic and communication engineering ①		MEng	4	●	● ○		● ◑	1
Electronic engineering		BEng	3	●	●		●	0
Electronics and digital technology		MEng	4		●		●	0
Swansea www.swan.ac.uk								
Communication systems		BEng/MEng	3, 4, 5	●	● ○ ◑		●	0
Electronic and electrical engineering		BEng/MEng	3, 4, 5	●	● ○ ◑		●	0
Electronics with computing science/management ①		BEng/MEng	3, 4, 5	●	● ○ ◑		● ◑	2
Internet technology ②		BSc	3	●	●		● ◑	1
Mobile communications and internet technology ③		BSc	3	●	●		● ◑	1
Swansea IHE www.sihe.ac.uk								
Computer networks		BSc	3		● ⌐		●	0
Computer systems and electronics		BEng	3	●	● ⌐		●	0
Internet technology and networks		BEng	3	●	● ⌐		●	0
Teesside www.tees.ac.uk								
Electrical and electronic engineering		BEng	3, 4	● ○	● ⌐ ◑		●	0
Instrumentation and control engineering		BEng	3, 4	● ○	● ⌐ ◑		●	0
UCE Birmingham www.uce.ac.uk								
Computing and electronics ①		BSc	3, 4	●	● ⌐ ◑		● ◑	1
Electronic engineering ②		BEng	3, 4	●	● ⌐ ◑		● ◑	1
Telecommunications and networks		BEng	3, 4	●	● ⌐ ◑		●	0
UCL www.ucl.ac.uk								
Electronic and electrical engineering		BEng/MEng	3, 4		● ○ ◑		●	0
Electronic engineering with communications engineering		MEng	4		●		●	0
Electronic engineering with computer science/ nanotechnology		MEng	4		● ○ ◑		●	0
Ulster www.ulster.ac.uk								
Electronics and computer systems		BEng/MEng	4	●	⌐ ◑		●	0
Electronics, communications and software		BEng	4	●	◑		●	0
Engineering (electrical/electronic)		BEng/MEng	4, 5		● ◑		●	0

First-degree courses in **Electrical and Electronic Engineering**

Institution / Course title	①②③ see combined subject list – Table 2b	Degree	Duration (years)	Foundation year (● at this institution)	Modes of study (● full time; ➤ part time)	○ time abroad	◐ sandwich	✿ Modular scheme	Course type (● specialised; ◑ combined)	No of combined courses
Wales (Newport) www.newport.ac.uk										
Electrical engineering		BEng	3		●				●	0
Electronic engineering		BEng	3		●				●	0
Wales (UWIC) www.uwic.ac.uk										
Electrical systems engineering		BSc	3	●	●				●	0
Electronic communication systems		BSc	3	●	● ➤				●	0
Electronic control systems		BSc	3	●	● ➤				●	0
Electronic microcomputer systems		BSc	3	●	● ➤				●	0
Music and audioelectronic systems ①		BSc	3		● ➤				◑	1
Warwick www.warwick.ac.uk										
Electronic engineering		BEng/MEng	3, 4, 5		●	○			●	0
Westminster www.wmin.ac.uk										
Computer systems engineering		BEng	3	●	●				●	0
Computer systems technology		BSc	3	●	●		◑		●	0
Digital communications engineering		BEng	3	●	●				●	0
Digital signal processing		BSc	3	●	● ➤				●	0
Electronic engineering		BEng	3	●	●				●	0
Electronics ①		BSc	3	●	● ➤				● ◑	1
Mobile communications		BSc	3	●	● ➤				●	0
Networks and communications engineering		BSc	3	●	●				●	0
Wolverhampton www.wlv.ac.uk										
Electronics and communications engineering		BEng	3	●	●				●	0
York www.york.ac.uk										
Computer systems and software engineering		MEng	4	●	●				●	0
Electronic and communication engineering		BEng/MEng	3, 4, 5	●	●	○	◐		● ◑	0
Electronic and computer engineering		BEng/MEng	3, 4, 5	●	●	○ ○	◐		● ◑	0
Electronic engineering ①		BEng/MEng	3, 4, 5	●	●	○	◐		● ◑	3
Radio frequency engineering		MEng	4, 5		●		◐		●	0

Subjects available in combination with electrical or electronic engineering

TABLE 2b shows those subjects that can be taken in roughly equal proportions with engineering in the combined degrees listed in TABLE 2a. See Chapter 2 in the *Introduction* (page 6) for general information about combined courses and for an explanation of how to use TABLE 2b.

Electrical and Electronic Engineering

Table 2b Subjects to combine with **Electrical or Electronic Engineering**

Accountancy/accounting Derby①
American studies Derby①
Artificial intelligence Reading②, Robert Gordon①, Southampton①
Audiotechnology Anglia Ruskin, UCE Birmingham②
Biology Derby①
Biomedical engineering Reading②
Business studies Derby①, North East Wales I①, Sheffield① ② ③ ④ ②, Sheffield Hallam①, Strathclyde①, Sussex①, York①
Chinese studies Nottingham①
Communication engineering Aberdeen①, Birmingham②, Sheffield Hallam①
Communication studies East London②
Computer engineering Aberdeen②, Birmingham①
Computer science Anglia Ruskin, Aston①, Edinburgh②, Liverpool① ②, Reading② ③, Swansea① ② ③
Computing Derby①, Dundee②, Essex①, Heriot-Watt①, London Metropolitan①, UCE Birmingham①, Westminster①
Control engineering Aberdeen①, Sheffield Hallam①
Creative writing Derby①
Criminology Derby①
Cultural studies East London②
Dance Derby①
Education Derby①
Electrical engineering Sheffield Hallam①
Electronic engineering Reading① ②
Electronics Manchester①
Engineering Reading①
English Derby①

English literature East London①
Environmental management Derby①
Environmental science Derby①
European studies Aberdeen①, Strathclyde①
Forensic science Derby①
Health studies East London②
Information engineering Sheffield Hallam①
Information technology Glamorgan①, Reading①
International business studies East London②
Internet technology Anglia Ruskin
Management studies Aberdeen①, Aston① ②, Birmingham① ② ③ ④, Bristol①, Dundee①, Edinburgh③, Imperial College London①, Nottingham① ② ③, Sheffield① ② ③ ④ ⑤ ⑥ ⑦, Swansea①
Marketing East London①
Mathematics Manchester①, Nottingham③
Mechanical engineering Edinburgh①
Media technology York①
Microelectronics Southampton①
Modern languages Nottingham① ② ③, Sheffield③ ⑤
Music Glasgow②, Wales (UWIC)①
Music technology York①
Physics Dundee① ②, Queen Mary①
Psychology Derby①, East London①
Software engineering Aberdeen②, Birmingham②, Edinburgh②, Glasgow①, Greenwich① ②
Telecommunications engineering Southampton①

Other courses that may interest you

The following courses are closely related to the courses in this part of the Guide but do not fit neatly into any of the Guides in the series. They include courses where you spend less than half your time studying electrical or electronic engineering, as well as courses providing a more intensive study of a specialised aspect of the subjects.

- Audio/music/video technology (Derby, Glamorgan, Leeds, Queen Mary, Southampton Solent and York)
- Entertainment technology (Portsmouth)
- Sound/broadcast engineering (North East Wales I)
- Medical electronics/Medical instrumentation/Biomedical electronics/Medical systems engineering (Liverpool, Plymouth, Sheffield and Southampton)
- Computer and network security (Essex)
- E-commerce engineering: (Queen Mary).

You may also find courses of interest in the following parts of this Guide:

- *Integrated Engineering and Mechatronics:* for courses combining electrical or electronic engineering with other engineering subjects, and courses that begin with general coverage but allow specialisation in later years
- *Mechanical and Manufacturing Engineering:* for courses in avionics that are not included in this part of the Guide.

Finally, the following Guides also contain courses with links to electrical and electronic engineering:
- *Mathematics, Statistics and Computer Science*
- *Physics and Chemistry:* for applied physics courses.

Chapter 3: The style and content of the courses

Industrial experience and time abroad Many courses provide a range of opportunities for spending a period of industrial training in the UK or abroad or of study abroad. TABLE 3a gives information about the possibilities for the courses in this Guide. See Chapter 3 in the *Introduction* (page 11) for further information.

Table 3a — Time abroad and sandwich courses	Named 'international' variant of the course	Location: ● Europe; ○ North America; ◗ industry; ◑ academic institution	Maximum time abroad (months)	Time abroad assessed	Language study: ○ optional; ● compulsory; * contributes to assessment	Socrates–Erasmus	Sandwich courses: ● thick; ○ thin	Arranged by: ● institution; ○ student
Aberdeen								
Electrical and electronic engineering	●	● ○ ◗ ◑	12	●	○*	●		
Electronics and computer/software engineering	●	◗ ◑	12	●	○*	●		
Anglia Ruskin								
Electronics ③	●	● ○ ◑	6	●	○*	●		
Aston								
Communications engineering							●	● ○
Electrical and electronic engineering							●	● ○
Electronic engineering							●	● ○
Electronic systems engineering	●		12	●	○	●	●	● ○
Internet engineering							●	● ○
Bangor								
Electronic engineering ③						●		
Bath								
Computers, electronics and communications ① ③	●	◗ ◑	9	●	○*	●	●	○
Electrical and electronic engineering ① ③	●	◗ ◑	9	●	○*	●	●	○
Electrical engineering and applied electronics ① ③	●	◗ ◑	9	●	○*	●	●	○
Electronic and communication engineering ① ③	●	◗ ◑	9	●	○*	●	●	○
Electronics with space science and technology ① ③	●	◗ ◑	9	●	○*	●	●	○
Belfast								
Electrical and electronic engineering ① ③	●	○ ◗ ◑	12		○*	●		
Birmingham								
Communication systems engineering ③	●	○ ◗ ◑	12	●	○*	●		
Computer systems engineering ③	●	○ ◗ ◑	12	●	○*	●		
Electronic and electrical engineering ③	●	○ ◗ ◑	12	●	○	●		
Electronic engineering ③	●	○ ◗ ◑	12	●	○*	●		
Bradford								
Electrical and electronic engineering	●	○ ◗	11		○		●	● ○
Electronic, telecommunications and internet engineering	●	○ ◗	11		○		●	● ○
Brighton								
Audio electronics ③ ⑤	●	◗	12		○		●	● ○
Digital electronics, computing and communication ③ ⑤	●	◗			○		●	● ○
Electrical and electronic engineering ③ ⑤	●	◗	12				●	● ○
Bristol								
Communications and multimedia engineering ③							●	○
Computer systems engineering ③		○ ◑	12	●	○			○
Electrical and electronic engineering ③	●	● ◑	12	●	○*	●	●	○

Institution / Course title (① ②: see notes after table)	Named 'international' variant of the course	Location: ● Europe; ○ North America; ◖ industry; ◗ academic institution	Maximum time abroad (months)	Time abroad assessed	Language study: ○ optional; ● compulsory; * contributes to assessment	Socrates–Erasmus	Sandwich courses: ● thick; ○ thin	Arranged by: ● institution; ○ student
Bristol (continued)								
Electronic and communications engineering ③							●	○
Electronic engineering ③							●	○
Bristol UWE								
Computing and telecommunications ③							●	● ○
Digital systems engineering ③		● ○ ◖ ◗	12	●	○*	●	● ●	● ○
Electrical and electronic engineering ③		● ○ ◖ ◗	12	●	○*	●	● ●	● ○
Electronic engineering ③		● ○ ◖ ◗	12	●	○*	●	● ●	● ○
Brunel								
Computer systems engineering ③		● ○ ◖	18	●		●	● ●	● ○
Electronic and electrical engineering		● ○ ◖	18	●	○	●	● ●	● ○
Electronic and microelectronic engineering		● ○ ◖	18	●		●	● ●	● ○
Electronic/electrical engineering (communication systems)		● ○ ◖	18	●		●	● ●	● ○
Electronic/electrical engineering (control systems)		● ○ ◖	18	●		●	● ●	● ○
Electronic/electrical engineering (power electronics systems)		● ○ ◖	18	●		●	● ●	● ○
Internet engineering						●	●	● ○
Cambridge								
Electrical and electronic engineering ③					○*			
Electrical and information sciences ③					○*			
Information and computer engineering ③					○*			
Cardiff								
Computer systems engineering							●	●
Electrical and electronic engineering							●	●
Electronic engineering							●	●
Central Lancashire								
Computer engineering ③		● ○ ◖ ◗	12	●	○*	●	● ●	○
Digital communications ③		● ○ ◖ ◗	12	●	○*	●	● ●	○
Digital signal and image processing ③		● ○ ◖ ◗	12	●	○*	●		○ ○
Electronic engineering ③		● ○ ◖ ◗	12	●	○*	●		○
City								
Computer systems engineering ①		● ○ ◖ ◗	12	●	○	●	●	
Electrical and electronic engineering ③		● ○ ◖ ◗	12	●	○	●	●	
Coventry								
Computers, networking and communications technology						●	● ●	● ○
Electronics engineering ③		● ◖ ◗	12		○	●	● ●	● ○
De Montfort								
Electronic engineering		● ◖	12		○*	●	● ●	● ○
Derby								
Computer networks							●	○
Electrical and electronic engineering ③					○	●	● ●	● ○
Electronics ③		◖	12		○	●	● ●	● ○
Dundee								
Electronic and electrical engineering		● ○ ◗	12		○			
Electronic engineering		○ ◗	12	●	○	●		
Durham								
Communications engineering	●	● ○ ◖ ◗	12	●	○			
Computer engineering ③	●	● ○ ◖ ◗	12	●	○			
Electronic engineering ③	●	● ○ ◖ ◗	12	●	○			

(continued) Table 3a — Time abroad and sandwich courses

① ② : see notes after table

Institution / Course title	Named 'international' variant of the course	Location: ● Europe; ○ North America; ◗ industry; ◐ academic institution	Maximum time abroad (months)	Time abroad assessed	Language study: ○ optional; ● compulsory; * contributes to assessment	Socrates–Erasmus	Sandwich courses: ● thick; ○ thin	Arranged by: ● institution; ○ student
East London								
Electrical and electronic engineering ③	●	◗	12	●			●	● ○
Edinburgh								
Electronics		● ◗ ◐	10	●	○*	● ●		
Electronics and electrical engineering					○*			
Electronics and electrical engineering (communications)					○*			
Essex								
Computer engineering ③								
Computers and networks ③								
Electronic engineering ③								
Telecommunications engineering ③								
Exeter								
Electronic engineering ③	●	● ○ ◐	6	●	○*			
Glamorgan								
Computer systems engineering ③		◗	12					●
Electrical and electronic engineering ③		● ◗ ◐	12	●	○	● ●	●	●
Electronic and communication engineering ③		● ◗	12	●	○	● ●	●	●
Electronic engineering ③		● ◗ ◐	12	●	○	● ●	●	●
Electronics ② ③	●	● ◗ ◐	12	●	●	● ●	●	●
Glasgow								
Electronic and software engineering ③	●	● ◗	6	●	○*	●		
Electronics and electrical engineering ③	●	● ○ ◗ ◐	6	●	○*	●		
Electronics with music ③	●	● ◗	6	●	○*	●		
Microcomputer systems engineering ③	●	● ○ ◗ ◐	6	●	○*	●		
Glasgow Caledonian								
Electrical power engineering ③		● ◗ ◐	12	●	○	● ●		○
Electronic engineering (BEng) ③		● ◗ ◐	12	●	○	● ●		○
Telecommunications engineering	●		9	●		●		
Greenwich								
Communications systems and software engineering						●		
Computer networking ③						●	●	
Computer systems and software engineering ③						●	●	
Control and instrumentation engineering ③						●	●	
Electrical and electronic engineering ③						●	●	
Electrical engineering						●	●	
Electronic engineering ③						●	●	
Electronics and computer systems						●	●	
Internet technologies						●	●	
Heriot-Watt								
Computing and electronics ③	●	◐	3	●	○*	●	●	
Electrical and electronic engineering ③	●	◐	3	●	○*	●	●	
Electronic and photonic engineering ③	●	◐	3	●	○*	●	●	
Robotics and cybertronics ③	●	◐	3	●	○*	●	●	
Hertfordshire								
Digital communications and electronics	● ○	◗ ◐	18	●	○*	● ●	●	○ ○
Digital systems and computer engineering	● ○	◗ ◐	18	●	○*	● ●	●	
Electrical and electronic engineering ③	● ○	◗ ◐	18	●	○*	● ●	●	○ ○
Huddersfield								
Computer control systems	●	◗	12			● ●	●

Time abroad and sandwich courses

Institution — Course title — ① ③: see notes after table	Named 'international' variant of the course	Location: ● Europe; ○ North America; ➤ industry; ◑ academic institution	Maximum time abroad (months)	Time abroad assessed	Language study: ○ optional; ● compulsory; * contributes to assessment	Socrates–Erasmus	Sandwich courses: ● thick; ○ thin	Arranged by: ● institution; ○ student
Huddersfield (continued)								
Electronic and communication engineering						●	●	●
Electronic and electrical engineering						●	●	●
Electronic design						●	●	●
Electronic engineering						●	● ●	●
Electronic engineering and computer systems						●	●	●
Hull								
Computer systems engineering ③	●	➤ ◑	12	●	○*	●	● ●	○
Electronic engineering ③		➤	3	●	○*	●	● ●	○
Information and computer control technology ③		➤	3	●	○*	●		○
Mobile telecommunications technology ③		➤	3	●	○*	●		○
Imperial College London								
Electrical and electronic engineering ① ③	●	● ◑	12	●	○	●		
Information systems engineering	●	◑	12		○	●		
Kent								
Computer systems engineering ① ② ③							●	○
Electronic and communications engineering ① ② ③	●	● ◑	9	●	●*		●	
Leeds								
Electronic and communications engineering ③		● ○ ◑	12	●	○	●		
Electronic and electrical engineering ③		● ○ ◑	12	●	○	●		
Electronic engineering ③		● ○ ◑	12	●	○	●		
Leicester								
Communications and electronic engineering ③	●	● ○ ➤ ◑	12	●	○	●		○
Electrical and electronic engineering ③	●	● ○ ◑	12	●	○	●		○
Liverpool								
Computer science and electronic engineering ③					○*	●		
Electrical and electronic engineering ③	●	➤	12	●	○ ○	●		
Electrical engineering ③					○*	●		
Electrical engineering and electronics ③					○*	●		
Electronic and communication engineering ③					○*	●		
Electronics ③					○*	●		
Wireless communications and 3G technology ③					○*	●		
Liverpool John Moores								
Broadcast engineering ③							●	○
Broadcast technology							●	○ ○
Computer engineering ③					○		●	○ ○
Design electronics					○		●	○ ○
Electrical and electronic engineering ③					○		●	○ ○
Networks and telecommunications engineering ③							●	● ○
London Metropolitan								
Audio electronics ③		● ➤ ◑	12	●	○*	●	● ○	● ○
Communications systems ③		● ➤ ◑	12	●	○*	●	● ○	● ○
Computer networking ③		● ➤ ◑	12	●	○*	●	● ○	● ○
Electronic and communications engineering ③	●	● ➤ ◑	12	●	○*	●	● ○	● ○
Electronics		● ➤ ◑	12	●	○*	●	● ○	● ○
London South Bank								
Electrical and electronic engineering ①		➤ ◑	12	●	○	●	●	● ○
Loughborough								
Computer network and internet engineering ③		● ○ ➤ ◑	12	●		●	●	○

Electrical and Electronic Engineering

(continued) Table 3a — Time abroad and sandwich courses

Location: ● Europe; ○ North America; ▼ industry; ◐ academic institution
Language study: ○ optional; ● compulsory; * contributes to assessment
Sandwich courses: ● thick; ○ thin
Arranged by: ● institution; ○ student
Course title: ① ③ : see notes after table

Institution / Course title	Named 'international' variant of the course	Location	Maximum time abroad (months)	Time abroad assessed	Language study	Socrates–Erasmus	Sandwich courses	Arranged by
Loughborough (continued)								
Computer systems engineering ① ③						●	●	○
Electronic and electrical engineering ③		● ○ ▼ ◐	12	●	○*	●	●	○
Electronics and software engineering ③		● ○ ▼ ◐	12	●		●	●	○
Manchester								
Computer engineering ③		● ◐	12			●	●	○
Computer systems engineering ③							●	● ○
Computing and communications systems engineering ③							●	● ○
Electrical and electronic engineering ③							●	● ○
Electronic systems engineering ③							●	● ○
Manchester Metropolitan								
Communication and electronic engineering	●	● ▼ ◐	12		○*	●	●	● ○
Computer and electronic engineering	●	● ▼ ◐	12		○*	●	●	● ○
Electrical and electronic engineering	●	● ▼ ◐	12		○*	●	●	● ○
Middlesex								
Computer networks ③		● ○ ▼ ◐	12		○*	●	●	
Napier								
Electronic and computer engineering ③		● ▼	6	●	○	●		● ○
Electronic and electrical engineering ③		● ▼	6	●	○	●		● ○
Newcastle								
Computer systems engineering ③		● ○ ▼ ◐	6	●	○*	●		
Electrical and electronic engineering ③		● ○ ▼ ◐	6	●	○*	●		
Electronic communications		● ○ ▼ ◐	6	●	○*	●		
Electronic engineering ③		● ○ ▼ ◐	6	●	○*	●		
Northumbria								
Communication and electronic engineering ③							●	●
Electrical and electronic engineering ③							●	●
Nottingham								
Electrical and electronic engineering ③	●	● ○ ▼ ◐	12	●		●	●	
Electrical engineering ③	●	● ○ ▼ ◐	12	●		●		
Electronic and communications engineering ③	●	● ○ ▼ ◐	12	●		●		
Electronic and computer engineering ③	●	● ○ ▼ ◐	12	●		●		
Electronic engineering ③	●	● ○ ▼ ◐	12	●		●	●	
Oxford								
Electrical engineering ③					○*			
Information engineering ③					○*			
Plymouth								
Communication engineering		● ○ ▼	12				●	● ○
Computer engineering							●	● ○
Computer systems engineering		● ▼					●	● ○
Electrical and electronic engineering ③		● ○ ▼	12		○*	●	●	● ○
Electrical and electronic systems		● ▼					●	○
Electronic communication systems		● ▼					●	● ○
Electronic engineering		● ○ ▼	12				●	● ○
Portsmouth								
Communication systems engineering ③	●	● ○ ▼	12	●	○*	●	●	● ○
Communications engineering ③	●	● ○ ▼	12	●	○*	●	●	○ ○
Computer engineering ③	●	● ○ ▼	12	●	○*		●	● ○
Computer technology ③	●	● ○ ▼	12	●	○*		●	● ○

Legend:
- Named 'international' variant of the course
- Location: ● Europe; ○ North America; ⊸ industry; ◑ academic institution
- Maximum time abroad (months)
- Time abroad assessed
- Language study: ○ optional; ● compulsory; * contributes to assessment
- Socrates–Erasmus
- Sandwich courses: ● thick; ○ thin
- Arranged by: ● institution; ○ student

Institution / Course title	Named int'l variant	Location	Max time abroad (months)	Time abroad assessed	Language study	Socrates–Erasmus	Sandwich courses	Arranged by
Portsmouth (continued)								
Electronic and electrical engineering ③	●	● ○ ⊸	12	●	O*	●	● ●	● ○
Electronic engineering ③	●	● ○ ⊸	12	●	O*	●	● ●	● ○ ○
Internet technology ③	●	● ○ ⊸	12	●	O*	●	● ●	● ○
Queen Mary								
Communication engineering ③		● ○ ◑	12	●	O*	●		
Computer engineering ③		● ○ ◑	6	●	O*	●		
Electrical and electronic engineering ③		● ○ ◑	6	●	O*	●		
Electronic engineering ③		● ○ ◑	6	●	O*	●		
Internet engineering ③		● ○ ◑	6	●	O*	●		
Physics and electronics ③		●	3	●	O*	●		
Telecommunications ③		● ○ ◑	6	●	O*	●		
Reading								
Electronic engineering					O*	●		
Robert Gordon								
Electronic and communications engineering		● ○ ◑	12	●	○	●		
Electronic and computer engineering		● ○ ◑	12	●	○	●		
Electronic and electrical engineering		● ○ ◑	12	●	○	●		
Sheffield								
Computer systems engineering ③					O*			
Electrical engineering ③		● ○ ◑	10	●	●*	●		○
Electronic and communications engineering ③		● ○ ◑	10	●	O*	●		
Electronic engineering ③		● ○ ◑	10	●	●*	●		○ ○
Electronic, control and systems engineering					O*			
Systems and control engineering ③					O*			
Sheffield Hallam								
Computer and network engineering ③		● ⊸ ◑		●	O*	●	●	●
Southampton								
Computer engineering ②		●	12	●	●	●	● ●	○
Electrical engineering ③		● ◑	8	●	O*	●		○
Electronic engineering ②		●	12	●	●	●	● ●	○
Staffordshire								
Computer systems		● ○ ⊸ ◑	12	●	○	●	● ●	● ○
Electronic engineering ③		● ⊸ ◑	18	●	O*	●	● ●	● ○
Electronic systems design ③		● ⊸ ◑	12	●	O*	●	● ●	● ○
Strathclyde								
Computer and electronic systems ③		● ○ ◑	5		O*	●		
Electrical energy systems		● ○ ⊸ ◑	12	●	O*	●		
Electronic and digital systems		● ○ ◑			O*	●		
Electronic and electrical engineering	●	● ○ ⊸ ◑	12	●	O*	●		
Surrey								
Digital media engineering ③							● ●	● ○
Electronic engineering ③	●	● ○ ⊸	13	●	O*	●	● ●	● ○
Electronics and computer engineering ③							● ●	● ○
Telecommunication systems ③							●	
Sussex								
Computer systems engineering ③								
Electrical and electronic engineering ③	●	● ○ ⊸ ◑	10	●	O*	●		
Electronic and communication engineering ③		● ○ ⊸ ◑	10	●	O*	●		

(continued) Table 3a

Time abroad and sandwich courses

ⓘ ⓔ : see notes after table

Location: Europe ●; North America ○; industry ◗; academic institution ◑

Language study: optional ○; compulsory ●; * contributes to assessment

Sandwich courses: thick ●; thin ○

Arranged by: institution ●; student ○

Institution / Course title	Named 'international' variant of the course	Location	Maximum time abroad (months)	Time abroad assessed	Language study	Socrates–Erasmus	Sandwich courses	Arranged by
Swansea								
Communication systems ③	●	● ○ ◗ ◑	12	●	●*	●	●	● ○
Electronic and electrical engineering ③	●	● ○ ◗ ◑	12	●	●*	●	●	● ○
Electronics with computing science/management ③	●	● ○ ◗ ◑	12	●	●*	●	●	○
Internet technology ③								
Mobile communications and internet technology ③								
Teesside								
Both courses ③		● ◗ ◑	12	●			●	● ○
UCE Birmingham								
Computing and electronics ③							●	●
Electronic engineering ③		● ◗ ◑					●	●
Telecommunications and networks ③							●	●
UCL								
Electronic and electrical engineering ③ ④	●	● ○ ◑	12	●	●	●		○
Electronic engineering with computer science/nanotechnology ③ ④	●	● ○ ◑	12	●	●*	●		○
Ulster								
Electronics and computer systems		● ○ ◗ ◑	12	●	●	●	●	●
Warwick								
Electronic engineering ③		● ○ ◗ ◑	12	●	○*	●	●	○
Westminster								
Computer systems engineering ③					●*	●		
Digital communications engineering ③					●*	●		
Electronic engineering ③					●*	●		
York								
Electronic and communication engineering ③		● ○ ◗ ◑	12	●	○	●	● ○	● ○
Electronic and computer engineering ③		● ○ ◗ ◑	12	●	○	●	● ○	● ○
Electronic engineering ③		● ○ ◗ ◑	12	●	○	●	● ○	● ○

① A year of industrial experience before the course starts is recommended for students on non-sandwich courses
② A year of industrial experience before the course starts is recommended for students on sandwich courses
③ Students on non-sandwich courses can take a year out for industrial experience
④ Other patterns of sandwich course : UCL Electronic and electrical engineering Student/firm's choice Electronic engineering with computer science/nanotechnology Student/firm's choice
⑤ Minimum period of vacation industrial experience for non-sandwich students (weeks) : Belfast Electrical and electronic engineering 12 ; Brighton Audio electronics 38 Digital electronics, computing and communication 38 Electrical and electronic engineering 38 ; Cambridge Electrical and electronic engineering 8 Electrical and information sciences 8 Information and computer engineering 8 ; Napier Electronic and computer engineering 20 Electronic and electrical engineering 20

Course content   Courses in electrical and electronic engineering build on a broad range of fundamental principles in mathematics and the physical sciences. This foundation in basic science is essential, as the rate of progress in the field, most particularly in electronics, means that specific knowledge about current implementations can rapidly become out of date. With a thorough understanding of the basic principles underlying the subject, you will be in a much better position to master new developments as they occur later in your career. Although the basic features of all electrical and electronic engineering courses are very similar, individual lecturers in any topic treat their subjects in their own characteristic ways, and often differ more from each other than the material covered might suggest.

Theory All courses give both theoretical and practical education (see Chapter 4 in the *Introduction* (page 16) for more information on practical work). The role of theoretical work is to establish mental models and abstract concepts, which you can then use with confidence when tackling complex or unusual problems. In the early stages of the course you will be taught material from mathematics, electricity, physics, mechanics and the science of engineering materials, with other contributions from chemistry, engineering drawing and thermodynamics. However, the emphasis placed on each subject varies widely between courses.

These subjects are fundamental to many other branches of engineering, and some or all of them may be taught, at least in the early stages of the course, in common with students on other courses. TABLE 3b shows where the course is taught initially in common with other courses and at what point you specialise in electrical and electronic engineering.

The basic subjects are soon joined by more specifically electrical subjects such as systems analysis, control, communications, logic design and many others (see the headings in TABLE 3b for further examples).

Specialisation Most courses offer a very wide range of options in the later years. These often reflect the research interests of the department or individual members of staff, so tend to change more rapidly than other parts of the course to keep track of new developments, or because staff move on to other institutions. This has the advantage of introducing you to the latest developments in the field, and in some cases you may even be taught by the people who were responsible for them.

TABLE 3b shows you where specific subjects are available as compulsory or optional components of the final year, or where there are both compulsory and optional components of a topic. Following the table there are some footnotes showing other topics offered on individual courses, and following that a brief glossary describing what the various topic headings may involve.

Table 3b — Final-year course content

Legend: ○ ⊘ : see notes after table ○ optional; ● compulsory; ⊘ compulsory + options; ▶ available for MEng only. Time of specialisation: S = semester; Y = year.

Institution / Course title	Time of specialisation	Power distribution and machines	RF circuits/antennae	Microwave circuits/radar	Power electronics	Optoelectronics/photonics	Advanced analogue circuits	Device fabrication	VLSI design/CAD	Circuit simulation	Production/manufacture/CIM	High voltage engineering	Software engineering	Computer architectures	Telecommunications	Networks/datacommunications	Control engineering	Automation/robotics	DSP	Image processing/vision	Speech processing	AI/KBS/Neural nets	Management/economics/law
Aberdeen Electrical and electronic engineering	Y 3			○	○		○						○	○	○		○		○				○
Electronics and computer/software engineering	Y 3				○		○				●	●	○				○		●				○

(continued) Table 3b — Final-year course content

Key: O optional; ● compulsory; ◑ compulsory + options; ◐ available for MEng only

S = semester; Y = year

Institution / Course title	Time of specialisation	Power distribution and machines	RF circuits/antennae	Microwave circuits/radar	Power electronics	Optoelectronics/photonics	Advanced analogue circuits	Device fabrication	VLSI design/CAD	Circuit simulation	Production/manufacture/CIM	High voltage engineering	Software engineering	Computer architectures	Telecommunications	Networks/datacommunications	Control engineering	Automation/robotics	DSP	Image processing/vision	Speech processing	AI/IKBS/Neural nets	Management/economics/law
Anglia Ruskin																							
Electronics									O	◑			O	O		◑	◑		O	O	O	O	O
Aston																							
Communications engineering ①			●			O	O			●				●	●	O			O				
Electrical and electronic engineering ②				O						●			●		O	●	O					O	O
Electronic engineering			O	O		O	O	O		●			●	O	●	O			O			O	O
Electronic systems engineering ③			O	O		●	●	O		●			●	O	●	O			●			O	O
Internet engineering ④			O				●			●			●	●	●	O			O			●	●
Bath																							
Computers, electronics and communications						O							●	O	O				●	O		◐	●
Electrical and electronic engineering ①		O	O	O	O	O		◐	●	●			●		O				●		◐	◐	●
Electrical engineering and applied electronics ②		O			O								●			O			●			◐	●
Electronic and communication engineering			O	O		O		◐	◐				●	O	O				●			◐	●
Belfast																							
Electrical and electronic engineering ①		O	O	O	O		O	O	O		O	●	O	O	O	O	O	O					●
Birmingham																							
Communication systems engineering		O	●	◐	O	O	O	O	O		O	O	O	O	●	●	O		●	O	O	O	O
Computer systems engineering			O	◐		O	O	O	O		O		●	●	O				●	O	O	O	O
Electronic and electrical engineering		O	O	◐	O	O	O		O		O		O	O	O				●	O	O	O	O
Electronic engineering		O	O	◐	O	O	●	O	O		O	O	O	O	O				O	O			O
Bolton																							
Electronic and computer engineering	Y 2					●	●	●	●	●	●		◑	◑	●	●							●
Bournemouth																							
Electronics				O		O			O	O			O		●	●	O			O	●		O	O
Bradford																							
Electrical and electronic engineering ①			◑	◑	◑							◑		◑					◑	◑		◑	◑
Electronic, telecommunications and internet engineering ②			◑	◑	◑							◑		◑		◑			◑			◑	
Brighton																							
Audio electronics ①	Y 1			●		●													●				
Digital electronics, computing and communication ②	Y 1													●	●	●			●	●			
Electrical and electronic engineering	Y 1	●		●		●						●							●				
Bristol																							
Communications and multimedia engineering ①			O			O							O	O	●	●			●		O	O	●
Computer systems engineering						O	O		O				●	●	O	O	●	O	O	O	O	O	O

Final-year course content

Key:
- ○ optional; ● compulsory; ◐ compulsory + options; ◗ available for MEng only
- ① ②: see notes after table
- Time of specialisation: S = semester; Y = year

Institution / Course title	Time of specialisation	Power distribution and machines	RF circuits/antenna	Microwave circuits/radar	Power electronics	Optoelectronics/photonics	Advanced analogue circuits	Device fabrication	VLSI design/CAD	Circuit simulation	Production/manufacture/CIM	High voltage engineering	Software engineering	Computer architectures	Telecommunications	Networks/datacommunications	Control engineering	Automation/robotics	DSP	Image processing/vision	Speech processing	AI/IKBS/Neural nets	Management/economics/law
Bristol (continued)																							
Electrical and electronic engineering ②		○	○	○	●	○	●		○				○		○	○	●	○	○				●
Electronic and communications engineering ③				○	○	○	●						○	○	●	○			○				●
Electronic engineering ④				○		○	●	●	●				○		●	○			○		○		●
Bristol UWE																							
Computer systems engineering ①													●			○			○	○	○	○	○
Digital systems engineering ②						○			○	○			◐	◐		○			○		○	○	◗
Electrical and electronic engineering		○	○	○	○		○		○	○	○			○		○			○		○	○	◗
Electronic engineering ③		○				○	○		○	○			○		○	○	○	○	○		○	○	◗
Brunel																							
Computer systems engineering ①									○				◐	◐		○			○			○	○
Electronic and electrical engineering ②		○			○		●		○				○	○		○	○		○			○	○
Electronic and microelectronic engineering ③							●	●					○	○		○	○		○			○	○
Electronic/electrical engineering (communication systems) ④					○		●		○				○	○	●	○			●			○	○
Electronic/electrical engineering (control systems)					○		●		○				○	○		○	●					○	○
Electronic/electrical engineering (power electronics systems) ②		◐			◐		●		○				○	○		○	●		○			○	○
Internet engineering ⑤							○						○	◐		◐			○			○	○
Cambridge																							
Electrical and electronic engineering	Y 3	○	○	○	○	○	○	○	○	○	○	○	○	○	○	○	○	○	○	○	○		
Electrical and information sciences	Y 3																						
Information and computer engineering	Y 3																						
Cardiff																							
Computer systems engineering					○	○	○		○	○			●	●	○	●	○	○	○			○	●
Electrical and electronic engineering		○	○	○	○	○	○					○	○	○	○	○	○	○	◗		◗	◗	
Electronic engineering			○	○	○	○	○		○				○	○	○	○	○	○	◗		◗	◗	
Central Lancashire																							
All courses	Y 2			○					○	●			●	○	○	○		●	○			○	
City																							
Computer systems engineering ①													●	○	○	○	○	○	○			○	●
Electrical and electronic engineering ②			○			○	○						○	○	○	○	○	○	○			○	●
Coventry																							
Computers, networking and communications technology							○	●					●	●	●	●			○				○
Electronics engineering							●	●					○		○	○	○	●	○	○			
De Montfort																							
Electronic engineering			○		○	○			○	○			○		○	○	○		○				

(continued) Table 3b

Final-year course content

○ optional; ● compulsory; ◑ compulsory + options; ◗ available for MEng only

① ②③ : see notes after table

Time of specialisation: S = semester; Y = year

Institution / Course title	Time	Power distribution and machines	RF circuits/antennae	Microwave circuits/radar	Power electronics	Optoelectronics/photonics	Advanced analogue circuits	Device fabrication	VLSI design/CAD	Circuit simulation	Production/manufacture/CIM	High voltage engineering	Software engineering	Computer architectures	Telecommunications	Networks/datacommunications	Control engineering	Automation/robotics	DSP	Image processing/vision	Speech processing	AI/KBS/Neural nets	Management/economics/law
Derby																							
Computer networks													○	●		●							○
Electrical and electronic engineering	Y 2	●			●	●			●				●			●							
Electronics	Y 2					○			○	○			○		○				○				
Dundee																							
Electronic and electrical engineering	Y 2	○	○		○	○			○	◗			○	○	○	○	○	○	○				○
Electronic engineering				○					○	◗			●	○	○	○	○	○	○				○
Electronics and computing				○	○				○	○			○	○			○	○					
Microelectronics and photonics	Y 2					●		●	●														
Durham																							
Communications engineering	Y 3																						
Computer engineering	Y 3																						
Electronic engineering ①	Y 3				○				●	●			○		○								○
East London																							
Electrical and electronic engineering ①	Y 2						●						●	●		●		●					
Edinburgh																							
Electrical engineering	Y 2	●				●	●	○	○	○		◗	●	●	●				●				
Electronics	Y 2	○	○	○	●	●	●	●	●	●	○	◗	●	●	●	●			●		○		
Electronics and electrical engineering	Y 2	○	●	○	○	●	●	●	●	●			●	●	●	●	○		●				
Electronics and electrical engineering (communications)	Y 2		●	●		●							●	●	●	●			●				
Essex																							
Audio engineering	Y 2		○				●		●						●	○			●	○			
Computer engineering ①	Y 2					○			○				●	●	○	○			○	○		●	
Computers and networks ②	Y 2												●	○	○	●							
Computers and telecommunications	Y 2													●	●	●							
Electronic engineering	Y 2		○	○		○	○						○	○	○	○			○	○		○	
Electronics and computers	Y 2				○								○	○	○	○	○						
Network and internet technology ③	Y 2									○				○		●							
Optoelectronics and communications systems	Y 2		●			●									●								
Telecommunications engineering ④	Y 2		●	●		○									●	●			●	●			
Exeter																							
Electronic engineering ①	Y 2					○			○	○					○	○			○	◗	◗		●
Glamorgan																							
Computer systems engineering									◑	◑	○	◑		○	○	◑	○		○	○	○	○	◗
Electrical and electronic engineering		○	○	○	○	○			○	○	○	○											
Electronic and communication engineering			●			●			●					●	●								
Electronic engineering			○	○	○	○			○	○			○	○	○	○	○	○	○	○			
Electronics							●		○	○			◑	◑	◑	◑	◑	◑	◑	◑	◑	◑	●

Final-year course content

Key: ○ optional; ● compulsory; ◐ compulsory + options; ▶ available for MEng only
Time of specialisation: S = semester; Y = year

Institution / Course title	Time	Power distribution and machines	RF circuits/antennae	Microwave circuits/radar	Power electronics	Optoelectronics/photonics	Advanced analogue circuits	Device fabrication	VLSI design/CAD	Circuit simulation	Production/manufacture/CIM	High voltage engineering	Software engineering	Computer architectures	Telecommunications	Networks/datacommunications	Control engineering	Automation/robotics	DSP	Image processing/vision	Speech processing	AI/IKBS/Neural nets	Management/economics/law
Glasgow																							
Electronic and software engineering				○		○	○	○	○	○			○	○	○		○		○	○		○	○
Electronics and electrical engineering			○	○	○	○	○	○	○	○			○	○	○	○	○	○	○	○		○	○
Electronics with music ①						○	○	○	○	○				○	○		○		○	○		○	○
Microcomputer systems engineering						◐	○	○	○	○			●	○		●	○		○	○		○	○
Glasgow Caledonian																							
Computer engineering ①													●	●		●				●			
Electrical power engineering	Y 3	●			●					●		●	●	●		◐			●				●
Electronic engineering (BEng)	Y 3		○	○	○		●	●	●	●			●	●	●	●	●			●			●
Electronic engineering (BSc)			○	●	◐	○	●	●	●	●	◐	○	●	●	●	●	●			●			●
Telecommunications engineering	Y 2		○		○		●	●	●	●			●	●	●	●	●			●			●
Greenwich																							
Communications systems and software engineering													○		○	○							
Computer networking													○		●	●							●
Computer systems and software engineering													●	●									●
Control and instrumentation engineering	Y 2	○			○		○	○	○				○	○	○		●	●	○				●
Electrical and electronic engineering	Y 2				○		○	○	○				○	○	○		○	○	○				●
Electrical engineering	Y 2	●			●		○						○		○		○	○					●
Electronic engineering					○		●	●	●				○	○		○	○		○			●	●
Electronics and computer systems							●	○		○			○	○		○	○					○	●
Internet technologies ①													○	○	●	○						●	●
Heriot-Watt																							
Computing and electronics			○	○		○	○	○	○	○			○	○	○	○	○		○	○	○	○	○
Electrical and electronic engineering		○	○	○		○	○	○	○	○		○	○	○	○	○	○		○	○		○	○
Electronic and photonic engineering						○	○		○				○	○	○	○	○		○	○		○	○
Robotics and cybertronics		○			○	○	○				○		○	○		○	○	○	○	○	○	○	○
Hertfordshire																							
Digital communications and electronics	S 2																						
Digital systems and computer engineering	S 2																						
Electrical and electronic engineering	S 2		●												●	●							
Huddersfield																							
Computer control systems		○	○	○	○	○	○		○	○		○	○	○	○	○	●						●▶
Electronic and communication engineering ①		◐	●	●	◐	●	◐		●	●			○	○	●	●			●	●			●▶
Electronic and electrical engineering ③		●	○		●		○	○	●	●		●	○	○	●	●			○	○			
Electronic design ③					○		○		○	●			●	●	○				○	○	○		

Electrical and Electronic Engineering

(continued) Table 3b — Final-year course content

Key: O optional; ● compulsory; ◐ compulsory + options; ▼ available for MEng only
Time of specialisation: S = semester; Y = year

Institution / Course title	Time	Power distribution and machines	RF circuits/antennae	Microwave circuits/radar	Power electronics	Optoelectronics/photonics	Advanced analogue circuits	Device fabrication	VLSI design/CAD	Circuit simulation	Production/manufacture/CIM	High voltage engineering	Software engineering	Computer architectures	Telecommunications	Networks/datacommunications	Control engineering	Automation/robotics	DSP	Image processing/vision	Speech processing	AI/IKBS/Neural nets	Management/economics/law
Huddersfield (continued)																							
Electronic engineering ④			O	O	O	O	●		●	●	O		O	●	O	O			O	O			▼
Electronic engineering and computer systems ⑤			O	O	O	O	O		●	O		O	●	●	O	●	O		●	●			
Hull																							
Computer systems engineering			O	O		O	O	O	O	O			O	O	O	O	O	O	O	O			
Electronic engineering			O	O	●								O	O	O	O	O	O	O	O		O	
Information and computer control technology	S 2					O	O	O	O	O	O		O	O	O	O	O	O	O	O			O
Mobile telecommunications technology				O		O	O						O	O	O	O							O
Imperial College London																							
Electrical and electronic engineering			O	O	O	O	O	O	O	O			O	O	O	O	O	O	O	O	O	O	O
Information systems engineering						O	O		O	O			O	O	▼	O	O	O	O	▼	O	O	O
Kent																							
Computer systems engineering ①			O						▼						▼								▼
Electronic and communications engineering ②							●		▼						▼								▼
King's College London																							
Computer systems and electronics	Y 2		O			O							O		O	●			●				●
Electronic engineering	Y 2					O	●	O					O		O	●	O	O	●				●
Telecommunication engineering	Y 2					O	O		O				O		●	●	O		O				●
Lancaster																							
Computer systems engineering	Y 2																						
Leeds																							
Electronic and communications engineering ①		●	O	O	O	◐	O	O	◐			◐	◐	◐	◐	O	O	◐	◐	◐	◐	◐	
Electronic and electrical engineering ②			O	O	●	O	O	O	●	O		◐	●	◐	O	●	O	●	O	O	O	●	
Electronic engineering ③			O	O	●	O	●	O	●	O		◐	O	O	O	●	O	●	O		O	●	
Leicester																							
Communications and electronic engineering ①	S 3			●		O	●										◐	◐	●				
Electrical and electronic engineering	S 4	O	O	●		O	●					O	●	●	◐		O						●
Liverpool																							
Computer science and electronic engineering ①						O	O	O	O	O	O	O	O	●	O	O	O	O	●	O			
Electrical and electronic engineering ②	Y 2	O	O	O	O	O	O	O	O	O	O	O	O	O	O	O	O	O	O	O	O	O	O
Electrical engineering		O	O	O	●	O	O	O	O	O	O	O	O	O	O	●	O	O	O	O	O	O	O
Electrical engineering and electronics ③		O	O	O	O	O	O	O	O	O	O	O	O	O	●	O	O	O	O	O	O	O	O
Electronic and communication engineering		O	●	O	O	O	O	O	O	O	O	O	O	O	●	O	O	O	●	O	O	O	O
Electronics ④		O	O	O	O	O	O	O	●	O	O	O	O	O	O	O	O	O	O	O	O	O	O
Wireless communications and 3G technology ③		O	O	O	O	O	O	O	●	O	O	O	O	O	O	O	O	O	●	O	O	O	O

(continued) Table 3b

Final-year course content

Key: O optional; ● compulsory; ◑ compulsory + options; ▾ available for MEng only

Time of specialisation: S = semester; Y = year

Institution / Course title	Time	Power distribution and machines	RF circuits/antennae	Microwave circuits/radar	Power electronics	Optoelectronics/photonics	Advanced analogue circuits	Device fabrication	VLSI design/CAD	Circuit simulation	Production/manufacture/CIM	High voltage engineering	Software engineering	Computer architectures	Telecommunications	Networks/datacommunications	Control engineering	Automation/robotics	DSP	Image processing/vision	Speech processing	AI/IKBS/Neural nets	Management/economics/law
Liverpool John Moores																							
Broadcast engineering	Y 2		●			O	O			●					●				O	O	O	O	O
Broadcast technology	Y 2												O	O	●	O O				●	●		O O
Computer engineering ①	Y 2					O	O						●	●	●	●				O		O	O
Design electronics ②	Y 2						●	●	●										O	O		O	O
Electrical and electronic engineering	Y 2	●			●				O							●			O	O		O	O
Networks and telecommunications engineering ③													●	●	●				O	O		O	O
London Metropolitan																							
Audio electronics ①							●	O	O	O			O						●				O
Communications systems			O	O		O	O	O	O		O		O	O	O	O	O		O				O
Computer networking			O	O		O	O	O	O		O		O	O	●	O			O				O
Electronic and communications engineering			O	O		O	O				●		O		O	O	O		O				O
Electronics						O	O	O	O	O			O	O	O	O	O		O				O
London South Bank																							
Electrical and electronic engineering	Y 2	O	O	O	◑	O	O	O	O	◑	O		◑	◑	O	O	◑	◑					◑
Loughborough																							
Computer network and internet engineering			O	O	O	O		O	O	O		O	O	O	O	O	O	O	O				◑
Computer systems engineering			O	O	O	O		O					O	O	O	O	O	O	O				O
Electrical and electrical engineering			O	O	O	O		O	O				O	O	O	O	O	O	O				◑
Electronics and software engineering			O	O	O	O		O	O				●	●	O		O	O	O				O
Manchester																							
Computer systems engineering ①	Y 2						O						●		●					O		O	●
Computing and communications systems engineering ②	Y 2		●	●				O							O				●	▾	O		●
Electrical and electronic engineering ③	Y 2	O		O	●								O	O	●	O	O		O				●
Electronic systems engineering ④	Y 2				O	O		O					●	▾	O				●	▾			●
Manchester Metropolitan																							
Communication and electronic engineering ①					O		O					O	O	●	●	●	O		O				●
Computer and electronic engineering ②					O			●				O	●	O	O	O	O						●
Electrical and electronic engineering ③					O		O					O	O	O	O	O	O						●
Middlesex																							
Computer networks													O	O	●								
Napier																							
Electronic and computer engineering					O		O		O	●			●	●		●			O				
Electronic and electrical engineering		O	O	O	O	O	O		O	O			O	O	O	O			O				

(continued) Table 3b

Final-year course content

○ optional; ● compulsory; ◑ compulsory + options; ◐ available for MEng only

① ②: see notes after table

Time of specialisation: S = semester, Y = year

Institution / Course title	Time of specialisation	Power distribution and machines	RF circuits/antennae	Microwave circuits/radar	Power electronics	Optoelectronics/photonics	Advanced analogue circuits	Device fabrication	VLSI design/CAD	Circuit simulation	Production/manufacture/CIM	High voltage engineering	Software engineering	Computer architectures	Telecommunications	Networks/datacommunications	Control engineering	Automation/robotics	DSP	Image processing/vision	Speech processing	AI/KBS/Neural nets	Management/economics/law
Newcastle																							
Computer systems engineering ①						○	◑	◑	◑				○	○	○	○			○	○			◑
Electrical and electronic engineering		○	○	○	○	○	○	◑	○				○	○	○	○	○	○	○	○			○
Electronic communications				○	○	○	◑		○				○		○	○			○	○			○
Electronic engineering		○		○	○	○	◑		○				○		○	○	○		○	○			
North East Wales I																							
Electrical and electronic engineering	S 2	○	○	○	○	○	○	○	○	○	○	○	○	○	○	○	○	○	○	○	○	○	○
Northumbria																							
Communication and electronic engineering ①	Y 2		●	●			●							●		●	●	●					
Electrical and electronic engineering	Y 2	○	○		○		○		○					●	●	○	○		○				
Nottingham																							
Electrical and electronic engineering ①				○	○		●		○	○			○				●		○				●
Electrical engineering ②		○			●				○	○			○				●		○				●
Electronic and communications engineering ③			●	○	○				○				○		●	○							●
Electronic and computer engineering ④				○	○	○	●		●			●	●	○									●
Electronic engineering ⑤				○	○	○	●		○				○		○		○						●
Oxford																							
Electrical engineering	Y 3	○	○	○	○	○	○		○				○	○	○	○	○		○			○	
Information engineering	Y 3		○	○		○			○	○			○	○	○		○		○			○	
Oxford Brookes																							
Telecommunications ①										○			○	●	○				○				○
Plymouth																							
Communication engineering			●	●												●			●				
Computer engineering						●							●	●		●			●				
Computer systems engineering						○	◑						●	●		●			●				
Electrical and electronic engineering	Y 3			●		●			●						●		○		●				●
Electrical and electronic systems		◑	◑	◑	◑	◑		◑	◑					◑	◑	◑	◑	◑	◑				
Electronic communication systems		●	●				○								●	●							●
Electronic engineering	Y 3					●	●		●	●						●							
Portsmouth																							
Communication systems engineering ①			●	●					○						●	●			●	○		○	●
Communications engineering ②			○	●					○					○	●	●			●			○	●
Computer engineering ③									○				●	●	●	○						○	●
Computer technology ④					●				○				●	●							○		●
Electronic and electrical engineering ③		○	○	○		●			◑	●			○	●	○	○	○	○		●	●		●
Electronic engineering ⑥		○	○	○	○		●		●	●	○		○	●	●							○	●
Internet technology ⑦									○					●	●	●			○				●
Queen Mary																							
Communication engineering			○	○		○			○				○		○	○			○	○		○	○

Key: ○ optional; ● compulsory; ◐ compulsory + options; ➤ available for MEng only — ① ②: see notes after table. Time of specialisation: S = semester; Y = year.

Institution / Course title	Time of specialisation	Power distribution and machines	RF circuits/antenna	Microwave circuits/radar	Power electronics	Optoelectronics/photonics	Advanced analogue circuits	Device fabrication	VLSI design/CAD	Circuit simulation	Production/manufacture/CIM	High voltage engineering	Software engineering	Computer architectures	Telecommunications	Networks/datacommunications	Control engineering	Automation/robotics	DSP	Image processing/vision	Speech processing	AI/IKBS/Neural nets	Management/economics/law
Queen Mary (continued)																							
Computer engineering			○	○	○	○			●	○			○	○	○	○			○	○		○	●
Electrical and electronic engineering			○	○	○	○			○				○	○	○	○	○		○			○	●
Electronic engineering	Y 2		○	○		○				○				○	○	○			○			○	●
Internet engineering													○	○	○								
Physics and electronics ①	Y 1		○	○		●							●	◐	●				●	◐		◐	
Telecommunications			○	○		○			○					○	●	○			○				○
Reading																							
Cybernetics	Y 2															●	●	●				●	
Electronic engineering					●		●		●	●			●	○	●		●			○			
Robert Gordon																							
Communications and computer engineering ①													●		●	●			●				
Communications and computer network engineering ②	Y 2														●	●	●		●				
Electrical and energy engineering ③	Y 2	●			●												●						
Electronic and communications engineering ④	Y 2			○			●		●				○		●	●	●						○
Electronic and computer engineering ⑤	Y 2						●		●				●	●								●	
Electronic and electrical engineering ⑥	Y 2	●			●		●		●	●							○	●					●
St Andrews																							
Microelectronics and photonics	Y 2					●		●	●														
Sheffield																							
Data communications engineering																◐							
Electrical engineering		○	○	○	○	○	○	○	○	○	○	○	○	○	○	○	○	○	○	○	○	○	○
Electronic and communications engineering		○	○	●	○	○	○	○	○	○	○	○	○	○	●	○	○	○	○	○	○	○	○
Electronic engineering		○	○	○	●	○	○	○	○	○	○	○	○	○	○	○	○	○	●	○	○	○	○
Sheffield Hallam																							
Computer and network engineering	S 2															●							
Southampton																							
Computer engineering							●						●	●	●	●			○	○	➤	➤	○
Electrical engineering ①	Y 2				●	●						●	●	●		●	○	○					
Electronic engineering			○	○		○	●	●	●	○			●	●	●	○	○			➤	➤	➤	
Southampton Solent																							
Electronic engineering	S 2		●	●		○	○						○		○	●			○			●	●
Staffordshire																							
Computer systems													●	●	○	○	●		○			○	○
Electronic engineering	Y 2				○	●	●							○	○	○			○			○	○
Electronic systems design	Y 2	○	○	○	○	○		○	○					○	○	○	○		○			○	○
Strathclyde																							
Computer and electronic systems ①			○	○	○	○			○	○			○	○	○	○	○	○	○	○	○	○	○
Electrical energy systems ②		○			○							○											○

(continued) Table 3b

Final-year course content

Legend: O optional; ● compulsory; ◑ : see notes after table; O optional/● compulsory; ◐ compulsory + options; ◑ available for MEng only

Time of specialisation: S = semester; Y = year

Institution / Course title	Time	Power distribution and machines	RF circuits/antennae	Microwave circuits/radar	Power electronics	Optoelectronics/photonics	Advanced analogue circuits	Device fabrication	VLSI design/CAD	Circuit simulation	Production/manufacture/CIM	High voltage engineering	Software engineering	Computer architectures	Telecommunications	Networks/datacommunications	Control engineering	Automation/robotics	DSP	Image processing/vision	Speech processing	AI/IKBS/Neural nets	Management/economics/law
Strathclyde (continued)																							
Electronic and digital systems			O	O		O	O		O	O			O	O	O	O	O	O	O				O
Electronic and electrical engineering		O	O	O	O	O	O		O	O		O	O	O	O	O	O	O	O				O
Surrey																							
Digital media engineering						O	O	O	O	O			O	O	O	O			O		O	O	
Electronic engineering			O	O	O	O	O	O	O	O	O		O	O	O	O			O	O	O	O	
Electronics and computer engineering				O	O	O	O	O	O				O	O	O	O			O	O		O	
Electronics with satellite engineering			O	O	O	O	O	O	O			O	O	O	O	O		O	O	O	O	O	
Telecommunication systems		O	O	O	O	O		O	O	O			O	O	O	O		O	O	O		O	
Sussex																							
Computer systems engineering	Y 2	O		O		O	O		O				O	O	O	●				O		O	O
Electrical and electronic engineering	Y 2	O	O		O	O	O							O			O	O	O			O	O
Electronic and communication engineering	Y 2		O	O	O	O	O		O				O	O			O	O		O		O	O
Swansea																							
Communication systems	Y 1		●	◐		O	O			●			●		●	●	●			●	●		◑
Electronic and electrical engineering	Y 1		O	O	●	O	●			●			●		O	●				●	●		◑
Electronics with computing science/management	Y 1									●			●	●	O	●				●	●	O	●
Internet technology	Y 1								O				●	●		●				O	O		●
Mobile communications and Internet technology	Y 1									●			O	O	●	●							●
Swansea IHE																							
Computer networks ①															●	●							O
Computer systems and electronics										●			●			●							
Internet technology and networks ②													●			●							
Teesside																							
Electrical and electronic engineering	Y 2	O			O		●		O	●	O		O				●		●	●	●		●
Instrumentation and control engineering ①	Y 2	O			O												●	●	●	●		O	●
UCL																							
Electronic and electrical engineering ①			◑	◑	O	O	O	O	O	O			O	O	O	O	O		O				O
Electronic engineering with communications engineering ②			O	O	O	O	O	O	O	O			O	O	O	O			O				O
Electronic engineering with computer science/nanotechnology ③			O	O	O	O	O	O	O			◐	O	O					O				O
Wales (Newport)																							
Electrical engineering ①					O	O											●	O					
Electronic engineering ②					O	O									O	●							
Wales (UWIC)																							
Electronic communication systems						●	●	●		●					●	●				●			

(continued) Table 3b — Final-year course content

Institution / Course title	Time of specialisation (S = semester; Y = year)	Power distribution and machines	RF circuits/antennae	Microwave circuits/radar	Power electronics	Optoelectronics/photonics	Advanced analogue circuits	Device fabrication	VLSI design/CAD	Circuit simulation	Production/manufacture/CIM	High voltage engineering	Software engineering	Computer architectures	Telecommunications	Networks/datacommunications	Control engineering	Automation/robotics	DSP	Image processing/vision	Speech processing	AI/IKBS/Neural nets	Management/economics/law
Wales (UWIC) (continued)																							
Electronic control systems		●			●		●		●								●	●	●				●
Electronic microcomputer systems							●		●				●	●					●				●
Music and audioelectronic systems						●			●				●						●				●
Warwick																							
Electronic engineering	Y 1/2	○	●	◖	○	◖	●	●	●	●	◖	○		●		◖	●	○	●	◖	◖	◖	◖
Westminster																							
Computer systems engineering													●										●
Digital communications engineering			●				●									●			●				●
Digital signal processing																●			●				●
Electronic engineering			○	○			●		●						○	●	◐		●				●
Electronics																●			●				●
Mobile communications			●													●			●				●
Networks and communications engineering											○	●				●				○	○		●
York																							
Electronic and communication engineering		◖	◖		◖	○	○		○				○	○	◖	○			◖		○		◐
Electronic and computer engineering		◖	◖			○	○		○				○	○	◖	○			◖		○		◐
Electronic engineering		◖	◖		◖	○	○		○				○	○	◖	○			◖		○		◐

Legend: ○ optional; ● compulsory; ◐ compulsory + options; ◖ available for MEng only; ①②: see notes after table

Aston ①Instrumentation ○; VHDL design ●; internet systems ●; network management ● ②Instrumentation ○; compiler writing ○; real-time systems ○; VHDL design ●; object-oriented design ○; internet systems ○; network management ● ③Instrumentation ○; VHDL design ●; internet systems ●; network management ● ④Internet and telecommunications ◐; signals and systems ◐; e-commerce ◐; Java ◐; project management ◐

Bath ①Finite element analysis ◖; electromagnetic compatibility ◖ ②Numerical methods ◖; finite element analysis ◖

Belfast ①Data acquisition and instrumentation; energy studies

Bradford ①②Real-time systems; signal processors and filters; signals and systems; transmission systems; measurement and control

Brighton ①Audio electronics ● ②Digital television ●

Bristol ①Mobile communications; mathematics; networks and protocols; multimedia systems ②Embedded and real-time systems; mathematics; European language ③Coding theory; mathematics; digital communications ④Embedded and real-time systems; mathematics

Bristol UWE ①Systems administration ●; concurrent and parallel systems ○; distributed and parallel databases ○; computing and music ○; compiler design ○; operating systems ○, multimedia technology ○ ②Multimedia technology ③Avionics; multimedia technology

Brunel ①②Digital communications ○ ③Digital communications ○; analogue chip design ● ④Digital communications ● ⑤Digital communications ◐ ⑥Digital communications ○

City ①Signal processing ●; embedded and real-time systems ● ②Signal processing ●; systems engineering and design ●

Durham ①Microprocessor systems ○; digital communications ○; microelectronic materials and devices ●

East London ①Financial and industrial systems; embedded systems

Essex ①Information systems ○ ②Information systems ● ③Middleware ●; internet computing ● ④Numerical methods ○

Exeter ①Microprocessor systems ●; electromagnetic compatibility ◖; data storage and display technologies ◖

Glasgow ①Music/music technology ●; acoustics ●

Glasgow Caledonian ①Multimedia ●; internet design ●

Greenwich ①Communications systems and processes ●; internet engineering ●
Huddersfield ①Distributed computing ○ ②Distributed computing ③Design for manufacture and manufacturing ④Distributed computing ⑤Distributed computing ○; digital system design ○
Kent ①Parallel programming ●; risk management ● ②Instrumentation ●; signals and systems ▼
Leeds ①②③Career development ●; industrial management ●
Leicester ①Risk, reliability and quality ○; technology and society ●; modelling and classification of data ○; digital communications ○; radio communications ○
Liverpool ①Program design and verification; instrumentation ②③Instrumentation ④Instrumentation ●; electronics for communications ● ⑤Human, computer and multimedia interaction ●
Liverpool John Moores ①Programmable electronics ○ ②Programmable electronics ● ③Programmable devices ○
London Metropolitan ①Acoustics ●
Manchester ①Technology transfer ●; real-time systems ●; computer graphics ○; multimedia and virtual reality ○; parallel systems ○; machine learning ○; internet security ○ ②Electronic systems ●; digital communications ●; real-time systems ○; communication engineering; high-speed devices ▼; technology transfer ●; codes and cryptography ○; multimedia communications ▼ ③Electronic systems ○; instrumentation ●; economics of power systems ▼; technology transfer ●; electrical drive systems ▼; tomography ▼; image engineering ▼; multimedia communications ▼ ④Electronic systems ●; instrumentation ○; technology transfer ●; real-time systems ●; digital communications ○; tomography ▼; system on-chip design ▼; multimedia communications ▼
Manchester Metropolitan ①②③Electronic instrumentation ○

Newcastle ①Digital electronics ▼; algorithm design and analysis ▼; databases ▼
Northumbria ①Communication systems; system design; sub-systems
Nottingham ①Engineering mathematics ○ ②Power quality and EMC ○; induction motor drives ○; engineering mathematics ○ ③④⑤Instrumentation ○; engineering mathematics ○
Oxford Brookes ①Space science; satellite systems; optical communication systems
Portsmouth ①Digital electronics and microprocessors ○ ②Digital audio and video ●; RF engineering ● ③Programming ●; digital electronics ●; microprocessors ● ④Formal specification ●; digital audio and video ○; advanced programming languages ● ⑤⑥Digital electronics and microprocessors ● ⑦Web programming ●; network security ●
Queen Mary ①Statistical physics; atom and photon physics; molecular physics; condensed matter
Robert Gordon ①②Safety, risk and reliability engineering ● ③Safety, risk and reliability management ●; energy conservation ●; renewable energy systems ● ④⑤Safety, risk and reliability engineering ● ⑥Safety, risk and reliability management ●
Southampton ①Industrial electrostatics ○; electrical materials ●; microprocessor applications ○; power system engineering ○; electrical machines dynamics ○; intelligent control ○; digital control ○
Strathclyde ①Human computer interface; real-time systems; distributed computing; concurrency and parallelism ②Industry-based project in year 3
Swansea IHE ①Internet security ●; web-based information systems ○; WANs ○ ②Routing; WANs; Java
Teesside ①Measurement systems ●
UCL ①Microwave semiconductor devices ▼; mathematics ○ ②Optical communications ◑ ③Range of computer science topics
Wales (Newport) ①②Project management

Glossary

This section gives some brief notes on the topic headings in TABLE 3b. Note that these may not be precisely the names used by individual institutions, and most institutions will teach other topics in addition to those listed. However, they are reasonably representative of the major areas covered in most of the courses.

Advanced analogue circuits Analogue circuits are used in signal processing, control and audio frequency circuits. Digital techniques are increasingly taking over in what have traditionally been analogue applications, but there are still many areas where analogue circuitry is simpler and produces better performance.

AI/IKBS/neural nets Intelligent knowledge-based systems (also known as expert systems) and neural networks are two very different artificial intelligence techniques for constructing computerised decision-making systems. The former uses a set of rules derived from experts in the particular field for which the system is being built. A neural network is a system that learns from experience. Initially there is a training period during which it is given input data and suggests a result, which you have to say is right or wrong. As the training period progresses, the network structure adapts until eventually it can be used for making real decisions on new data.

Audio/hearing   Audio frequency circuits have to be designed to optimise the signal as it is perceived by the listener, so it is important to have an understanding of the physiology and psychology of hearing.

Automation/robotics   This topic deals with the use of robots and other control systems in factories.

Biomedical electronics   Electronic sensors, monitors and data presentation devices such as computed tomography scanners (body scanners) are used increasingly in medical applications.

Circuit simulation   The cost of producing small numbers of integrated circuits is very large, so, as far as possible, computer simulations are used to test the circuits before they are fabricated on silicon for final testing.

Computer architectures   A computer's architecture is its large-scale system structure. It shows the relationship between major components such as the central processor, arithmetic logic unit, memory and input/output devices. Demand for increased performance and specialist applications has led to a variety of new architectures. In particular, parallel architectures have been designed to avoid the bottleneck of a single processor. However, the associated hardware and software problems are formidable.

Control engineering   In control engineering, a variety of motion, pressure, light and other sensors are linked through a feedback system to drive a variety of actuators, such as electric motors, servos, solenoids or pneumatic valves, to control machinery or other devices. The principles of feedback control are fundamental to all electrical and electronic engineering courses. Modern control systems can be extremely sophisticated and use advanced digital signal processing circuitry. Examples are found in fly-by-wire aircraft, which are often intrinsically unstable and can only be kept in the air by continually adjusting the control surfaces automatically through feedback circuits.

Cryptography/security   As the information held on computers and transmitted across networks increases in volume and value, there is an increasing need to ensure that the data is not read or changed by anyone who does not have appropriate authorisation. Cryptography is one approach to the general problem of security. In this approach, data is stored or transmitted in a coded form (using software or hardware), which (hopefully) can only be read by someone with the appropriate decryption key.

Device fabrication   These are the processes involved in manufacturing electronic components or devices such as integrated circuits.

DSP   Digital signal processing is used in a wide variety of applications where complex information signals have to be encoded/decoded, compressed/decompressed or treated to remove errors. Examples range from DVD players and mobile phones to radar systems. In some cases the required algorithms (mathematical procedures) can be implemented on general purpose computers, but a particular concern of DSP is the design of high-speed specialised computing architectures and integrated circuits.

High-voltage engineering   This topic deals with the special engineering techniques required for handling high voltages, which are typically required for efficient power transmission.

Image processing/vision   Video signals are used in a wide variety of applications ranging from entertainment to control systems. The signals carry a large amount of information and are subject to noise. Processing can improve the quality of the signal, and extract and enhance the required information. Attention is now focused on making computers understand what they see in an image.

Industrial/consumer electronics   Electronics courses are mainly concerned with low-level technology, but this topic is concerned with the use of this technology to make actual products, and is closely related to manufacturing engineering. Naturally, the cost of production is an important consideration, and the demanding challenge is to design for lowest cost.

Management/economics/law   Electrical and electronic engineers work within indus-trial and commercial organisations, and many move on to management as their careers progress, so they must have a good understanding of the business context.

Microwave circuits/radar   This topic covers the generation, transmission and detec-tion of microwave frequency signals. With the explosion in the use of mobile tele-phones, now operating at frequencies above 1GHz, the distinction between microwave and other radio frequency circuits is becoming blurred, and the design of radio circuits operating at these frequencies is of even greater commercial importance.

Networks/datacommunications   Communication between computers locally (LANs – local area networks) and globally (WANs – wide area networks) is becoming increas-ingly important as the use of computers spreads and the pattern of computer use becomes more decentralised. Different types of information need to be handled in dif-ferent ways. For example, video signals have to be transmitted and received in the same order and in real time: if there is a fault, the data should be thrown away as it is more important to keep pace with the timing of the image than to ensure the image is perfect. On the other hand, if there is a fault when a bank transaction is being transmitted, it must be sent again or millions of pounds could get lost in the computer system.

Optoelectronics/photonics   This topic deals with the use of devices to convert signals between optical and electrical media. Photonics is the use of devices to control and switch signals in optical media such as fibre-optic cables. This is becoming increasingly important as the use of fibre optics spreads rapidly. Photonics also holds out the possibility of very rapid and powerful alternatives to electronic devices.

Power distribution and machines   This topic deals with the power lines and transformers used to distribute electrical power from the power station to the consumer, and with the theory and practice of electrical generators and motors.

Power electronics   This topic deals with the control and high-frequency transformation and conditioning of electrical power using high-frequency switching techniques. Applications include the power supplies and electronic motor drives found in most domestic appliances.

Production/manufacture/CIM   This topic deals with the processes involved in manufacturing electrical and electronic products. Computers are increasingly used to plan, monitor and control the manufacturing process through computer-integrated manufacture.

RF circuits and antennae   This covers the generation, transmission, detection and processing of radio frequency signals, and forms the key to most modern mobile telephones and radio transmitters/receivers.

Software engineering   All software systems are run on electronic systems, and other electronic devices are increasingly controlled by external or embedded software systems. In either case, the electronic engineer frequently has to write programs or work closely with specialist software engineers.

Speech processing   Considerable progress has been made in both artificial speech production and recognition. The potential applications are immense and the technology is approaching the point at which it can be used on a widespread basis.

Telecommunications   This covers the transmission of signals over long distances through a variety of media including copper wire, microwave, radio and optical fibre. However, the method of transmission is only part of the problem; the other and often more challenging part comes from the fact that many signals have to be transmitted together but kept separate to avoid interference, and then these signals have to be switched and routed through networks to their correct destination.

VLSI design/CAD   The design of very large-scale integrated circuits uses sophisticated computer-aided design systems to develop and test the logic, and then plan the layout of the circuits on the surface of the chip.

General studies and additional subjects   Specialised engineering courses run the risk of focusing the student's attention entirely on technical subjects. To prevent this, some institutions include topics on liberal, complementary or general studies in their specialised courses. TABLE 3a contains information about language teaching; you will need to look at prospectuses or ask institutions direct for information about other subjects.

Where a course is part of a modular degree scheme (see TABLE 2a), there will probably be more opportunities to take a wide range of other subjects, though the requirements for professional qualification may mean that the choice is more restricted than it would be if you were following a course in a non-engineering subject.

MEng courses   TABLE 2a lists many courses that can lead to the award of either a BEng or an MEng degree. For these courses, most of the tables in this Guide give information specifically for the BEng stream (the tables show information for MEng courses that are MEng only). Where both degrees are available, much of the information will also apply to the MEng stream, but TABLE 3c shows you where there are differences for the MEng stream, and at what point the MEng course separates from the BEng.

Table 3c	MEng course differences						
Institution	Course title	MEng separates from BEng	More engineering	More management	More languages	Other differences in MEng course
Aberdeen	Electrical and electronic engineering	Year 3	●	●		More extensive group project; project management
	Electronics and computer/software engineering	Year 3	●	●		More extensive group project; project management
Bath	All courses	Year 3				
Birmingham	All courses	Year 3				Major group project
Bradford	Electrical and electronic engineering	Year 3	●	●		Enhanced design skills and deeper knowledge, including marketing and finance
	Electronic, telecommunications and internet engineering	Year 3	●	●		Telecommunications subjects studied in greater depth
Brighton	Electrical and electronic engineering	Year 3	●	●		Major team and industry-related projects
Bristol	Both courses	Year 3	●			Greater breadth and depth; group project

(continued) Table 3c	MEng course differences					
Institution	Course title	MEng separates from BEng	More engineering	More management	More languages	Other differences in MEng course
Bristol UWE	Digital systems engineering	Year 3	●	●		3-module project dissertation; industrial case studies; wider range of optional modules
	Electrical and electronic engineering	Year 3	●	●		Substantial project dissertation; industrial case studies; wider range of optional modules
	Electronic engineering	Year 3	●	●		3-module project dissertation; industrial case studies; wider range of optional modules
Cambridge	*All courses*	Year 3	●			
Cardiff	Electrical and electronic engineering	Year 3	●			Deeper coverage of core subjects
	Electronic engineering		●			Deeper coverage of core subjects
City	Electrical and electronic engineering	Year 2				
Dundee	Electronic and electrical engineering	Year 3		●	●	Professional development courses
	Electronic engineering	Year 3		●	●	Professional development courses
	Microelectronics and photonics	Year 4	●			
Exeter	Electronic engineering	Year 2	●			
Glamorgan	Electrical and electronic engineering	Year 3	●			Management element; wider technical coverage
Glasgow	Electronic and software engineering	Year 4		●	●	6 months in Europe
	Electronics and electrical engineering	Year 4	●	●	●	6 months abroad
	Electronics with music	Year 4		●	●	6 months in Europe
	Microcomputer systems engineering	Year 4	●	●	●	6 months abroad
Heriot-Watt	*All courses*	Year 4	●	●		5-month industrial placement; industrial project in final year
Huddersfield	*Both courses*	Year 4		●		Greater depth in technical subjects; teamworking and leadership; major group project
Hull	Electronic engineering	Year 3	●	●		Additional year
	Mobile telecommunications technology	Year 3	●	●		
Imperial College London	Electrical and electronic engineering	Year 3	●	●		
	Information systems engineering	Year 3	●	●		Optional further language study
Kent	*Both courses*	Year 4				
King's College London	*All courses*	Year 3	●			
Leeds	*All courses*	Year 3	●			Greater emphasis on group project work, CAD and IT
Leicester	Electrical and electronic engineering	Year 3	●	●		Further options; major group design project
Liverpool	*All courses*	Year 3	●			
Manchester	Computer systems engineering	Year 3	●	●		Group project; enterprise studies
	Computing and communications systems engineering	Year 3	●	●		Group project; enterprise studies
	Electrical and electronic engineering	Year 3	●	●		Group project; enterprise studies
	Electronic systems engineering	Year 3	●	●		Group project; enterprise studies
Newcastle	*All courses*	Year 3	●	●		Industrial/research project; technical review; group design project
Nottingham	*All courses*	Year 3	●	●		Group project; industrial awareness module
Portsmouth	*All courses*	Year 3	●			Additional topics; design, build, test and research group project
Queen Mary	Computer engineering	Year 3				More group project work
	Electronic engineering	Year 3				More group project work
Reading	Electronic engineering	End Part 2				Students must obtain more than 60% in Part 2; industry-based project
Robert Gordon	Electronic and electrical engineering	Year 3	●	●		'Fast track' MEng has compulsory assessed industrial placement

MEng course differences

Institution	Course title	MEng separates from BEng	More engineering	More management	More languages	Other differences in MEng course
St Andrews	Microelectronics and photonics	Year 4		●		
Sheffield	Computer systems engineering	Year 2				
	Data communications engineering	Year 3	●			
	Electrical engineering	Year 3	●			
	Electronic engineering	Year 3	●			
	Electronic, control and systems engineering	Year 2				
Sheffield (continued)	Systems and control engineering	Year 2				
Southampton	Computer engineering	Year 3	●			
	Electrical engineering	Year 3	●	●	●	Individual project year 3; group project year 4
	Electronic engineering	Year 3	●			More professional options
Staffordshire	Computer systems	Year 4		●		
	Electronic engineering	Year 4				
	Electronic systems design	Year 4				
Strathclyde	Computer and electronic systems	Year 3	●	●		Major group project in year 5
	Electrical energy systems	Year 3	●	●		Major group project in year 5
	Electronic and digital systems	Year 3	●	●		Major group project in year 5
	Electronic and electrical engineering	Year 3	●	●	●	Major group project in year 5
Surrey	Digital media engineering	Year 2	●			
	Electronic engineering		●			More advanced modules during course; 2.2 standard required throughout
	Electronics and computer engineering	Year 2	●			
	Telecommunication systems	Year 2	●			
Sussex	*Both courses*	Year 3				
Swansea	*All courses*	Level 3	●			
UCL	Electronic and electrical engineering	Year 3	●			
Warwick	Electronic engineering	Year 3	●			Optional study abroad; substantial multidisciplinary group project in final year; optional intercalated year in industry or research; elective in final year
York	*All courses*	Year 3				

See Chapter 4 in the *Introduction* (page 16) for general information about teaching and assessment methods used in all types of engineering course, including those covered in this part of the Guide. It also explains how to interpret TABLE 4, which gives information about projects and assessment methods used on individual courses.

Table 4 — Assessment methods

Institution	Course title	Frequency of assessment (Key: ◐ term; ◐ semester; ○ year)	Years of exams contributing to final degree (years of exams not contributing to final degree)	Coursework: minimum/maximum %	Project/dissertation: minimum/maximum %	Time spent on projects in: first/intermediate/final years %			Group projects: ● compulsory; ○ optional	Orals: ◐ if borderline; ● everyone; ○ for projects
Aberdeen	Electrical and electronic engineering	◐	(1),(2),**3,4,5**	15/**20**	39/**42**	20	20	50	●	○
	Electronics and computer/software engineering	◐	(1),(2),**3,4**	15/**20**	39/**39**	20	20	50	●	○
Anglia Ruskin	Electronics	◐	(1),**2,3**	40/**60**	10/**20**		8	17	○	○
Aston	Communications engineering	○	**1,2,3**	20/**20**	20/**20**		10	20		◐○
	Electrical and electronic engineering	○	**1,2,3,4**	20/**20**	20/**20**		10	20		◐○
	Electronic engineering		(1),**2,3,4**	20/**20**	20/**20**		10	20	●	◐○
	Electronic systems engineering	○	**2,3,4**	20/**20**	20/**20**		20	20	●	◐○
	Internet engineering	○	(1),**2,3,4**	20/**20**	20/**20**		10	20		◐○
Bangor	Electronic engineering	◐	(1),**2,3,4**		**25**		10	25	●	○
Bath	Computers, electronics and communications	◐	(1),**2,3,4**	10	**25**			25		◐○
	Electrical and electronic engineering	◐	(1),**2,3,4**	10	**25**	8	8	25	●	◐○
	Electrical engineering and applied electronics	◐	(1),**2,3,4**	10	**25**	8	8	25		◐○
	Electronic and communication engineering	◐	(1),**2,3,4**	10	**25**	8	8	25		◐○
	Electronics with space science and technology	◐	(1),**2,3,4**	10	**25**	8	8	25		◐○
Belfast	Electrical and electronic engineering	◐	(1),**2,3,4**	10/**30**	25/**25**			25		
Birmingham	Communication systems engineering	◐	(1),**2,3,4**	10/**25**	20/**40**	5	10	40	●	◐○
	Computer systems engineering	◐	(1),**2,3,4**	10/**25**	20/**40**	5	10	40	●	◐○
	Electronic and electrical engineering	◐	(1),**2,3,4**	10/**25**	20/**40**	5	10	40	●	◐○
	Electronic engineering	◐	(1),**2,3,4**	10/**25**	20/**40**	5	10	40	●	◐○
Bolton	Electronic and computer engineering	○	(1),**2,3**	20/**50**	20/**50**	20	20	30	●	○
Bournemouth	Electronics	○	(1),**2,3,4**	20/**40**	25/**30**	15	30	40	●	○
Bradford	Electrical and electronic engineering	◐	(1),**2,3,4,5**		**26**			30		
	Electronic, telecommunications and internet engineering	◐	(1),**2,3,4,5**		**26**			30	●	
Brighton	Audio electronics	◑	(1),**2,3**	19/**27**	25/**25**	25	25	35	●	○
	Digital electronics, computing and communication	◑	(1),**2,3**	20/**30**	25/**25**	25	25	35	●	○
	Electrical and electronic engineering	◑	(1),**2,3**	20/**30**	25/**25**	25	25	35	●	○
Bristol	Communications and multimedia engineering	○	(1),**2,3,4**		20/**20**			25	●	◐○
	Computer systems engineering	○	(1),**2,3,4**		30/**30**		15	33	●	◐○
	Electrical and electronic engineering	○	(1),**2,3,4**		20/**20**			25	●	◐○
	Electronic and communications engineering	○	(1),**2,3**		20/**20**			25	●	◐○

Electrical and Electronic Engineering

Electrical and Electronic Engineering

Key for frequency of assessment column: ◑ term; ◐ semester; ○ year

Group projects: ● compulsory, ○ optional
Orals: ◐ if borderline; ● everyone; ○ for projects

Institution	Course title	Frequency of assessment	Years of exams contributing to final degree (years of exams not contributing to final degree)	Coursework: minimum/maximum %	Project/dissertation: minimum/maximum %	Time spent on projects in: first/intermediate/final years %			Group projects	Orals
Bristol (continued)	Electronic engineering	○	(1),2,3		20/20			25		◐○
Bristol UWE	Digital systems engineering	◐	(1),2,3	20/30	20/25	15	20	25	●	○
	Electrical and electronic engineering	◐	(1),2,3	20/30	20/25	15	20	25	●	◐○
	Electronic engineering	◐	(1),2,3	20/30	20/25	15	20	25	●	○
Brunel	Computer systems engineering	◐●	(1),2,3,4,(5)		26/26		16	33	●	○●
	Electronic and electrical engineering	◐●	(1),2,3,4,(5)		26/26		16	33	●	○●
	Electronic and microelectronic engineering	◐●	(1),2,3,4,(5)		26/26		16	33	●	○●
	Electronic/electrical engineering (communication systems)	◐●	(1),2,3,4,(5)		26/26		16	33	●	○●
	Electronic/electrical engineering (control systems)	◐●	(1),2,3,4,(5)		26/26		16	33	●	○●
	Electronic/electrical engineering (power electronics systems)	◐●	(1),2,3,4		26/26		16	33	●	○●
	Internet engineering	○	(1),2,3,4		26/26		16	33		○●
Cambridge	Electrical and electronic engineering	○	(1),(2),3,4	20	50/50	5	20	50		○
	Electrical and information sciences	○	(1),(2),3,4	0/20	50/50	5	25	50		○
	Information and computer engineering	○	(1),(2),3,4	20	50/50	5	20	50		○
Cardiff	Computer systems engineering	◐	(1),2,3	10/15	25/25			25	●	◐○
	Electrical and electronic engineering	◐	(1),2,3,4,5	10/15	25/25			25	●	◐○
	Electronic engineering	◐	(1),2,3,4,5	10/15	25/25			25	●	◐○
Central Lancashire	Computer engineering	○	(1),2	20/20	20/20	10	25	34		○
	Digital communications	○	(1),2,4	20/20	20/20	10	25	34		○
	Digital signal and image processing	○	(1),2,4	20/20	20/20	10	25	34	●	○
	Electronic engineering	○	(1),2,4	20/20	20/20	10	25	34	●	○
City	Computer systems engineering		(1),2,3	20/25	10/15	20	20	20		◐○
	Electrical and electronic engineering	○	(1),2,3,4	20/25	10/15					◐○
Coventry	Computers, networking and communications technology	◐●	(1),2,3	30/40	30/45	25	30	30		◐○
	Electronics engineering	○	(1),2,3	20/50	25/50	25	25	40	●	◐○
De Montfort	Broadcast technology		2,4							
	Electronic engineering	◐	(1),2,4	20/100	/20			25		
Derby	Computer networks	◐	(1),2,3	30/30	20/20	5	15	25	●	
	Electrical and electronic engineering	◐	(1),(2),3	30	20			25	○	◐○
	Electronics	◐	(1),2,3	30	20/35	10	10	25	●	○●
Dundee	Electronic and electrical engineering	◐	(1),(2),3,4,5	20	25	15	20	40	●	◐○
	Electronic engineering	◐	(1),(2),3,4,5	20	25	15	20	40	●	◐○
	Electronics and computing	◐	(1),(2),3,4	20/40	25/25	0	0	50	●	○
	Microelectronics and photonics	◐	(1),(2),3,4,5	15/20	25/25	10	15	30	●	◐○
Durham	*All courses*	○	(1),2,3,4	5/10	25/25	5	10	50	●	◐○
East London	Electrical and electronic engineering	◐	(1),2,3	20	25			30		○
Edinburgh	Electrical engineering	◐	(1),(2),(3),4,5	25/25	35/40	25	30	50	●	◐○
	Electronics	◐	(1),(2),(3),4,5	25/25	50/50	25	30	50		◐○
	Electronics and electrical engineering	◐	(1),(2),3,4	25/25	50/50	25	30	50		◐○
	Electronics and electrical engineering (communications)	○	(1),(2),(3),4	25/25	50/50	25	30	50		◐○
Essex	Audio engineering	○	(1),2,3	10/20	22/22	10	10	30	●	◐○
	Computer engineering	○	(1),2,3	10/20	22/22	10	10	30	●	◐○
	Computers and networks	○	(1),2,3	10/20	22/22	10	10	30		◐○

Institution	Course title	Frequency of assessment	Years of exams contributing to final degree (years of exams not contributing to final degree)	Coursework: minimum/maximum %	Project/dissertation: minimum/maximum %	Time spent on projects in: first/intermediate/final years %			Group projects: ● compulsory; ○ optional	Orals: ◐ if borderline; ○ for projects; ● everyone
Essex (continued)	Computers and telecommunications	◐	(1),**2,3**	10/**20**	22/**22**	10	10	30	●	◐ ○
	Electronic engineering	◐	(1),**2,3**	10/**20**	22/**22**	10	10	30	●	◐ ○
	Electronics and computers	◐	(1),**2,3**	10/**20**	22/**22**	10	10	30	●	○
	Network and internet technology	◐	(1),**2,3**	10/**20**	22/**22**	10	10	30	●	○
	Optoelectronics and communications systems	◐	(1),**2,3**	10/**20**	22/**22**	10	10	30	●	○
	Telecommunications engineering	◐	(1),**2,3**	10/**20**	22/**22**	10	10	30	●	◐ ○
Exeter	Electronic engineering	◑	(1),**2,3,4**	20/**30**	20/**20**		25	25	●	○ ●
Glamorgan	Computer systems engineering	◑	(1),(2),**3,4**		20/**20**		20	40		◐
	Electrical and electronic engineering	◑	(1),(2),**4,5**	20/**20**	30/**30**	10	20	30	●	◐
	Electronic and communication engineering	◑	(1),(2),**4**	20/**20**	30/**30**	10	20	30	●	◐
	Electronic engineering	◑	(1),(2),**4**	20/**20**	20/**20**	10	20	30	●	◐
	Electronics	◑	(1),**4**	20/**20**	20/**20**	10	20	30		◐
Glasgow	Electronic and software engineering	◐	(1),(2),**3,4,5**	8/**8**	20/**20**	5	10	20	●	◐ ○
	Electronics and electrical engineering	◐	(1),(2),**3,4,5**	8/**8**	20/**20**	5	10	20	●	◐ ○
	Electronics with music	◐	(1),(2),**3,4,5**	8/**8**	20/**20**	5	10	20	●	◐ ○
	Microcomputer systems engineering	◐	(1),(2),**3,4,5**	8/**8**	20/**20**	5	10	20	●	◐ ○
Glasgow Caledonian	Computer engineering	◑	(1),(2),(3),**4**	30/**50**	30/**30**	15	15	30		◐ ○
	Electrical power engineering	◑	(1),(2),**3,4,5**	20/**30**	25	20	10	33	●	◐ ○
	Electronic engineering (BEng)	◑	(1),(2),**3,4,5**	20/**30**	10/**15**	20	10	15	●	◐ ○
	Electronic engineering (BSc)	◑	(1),(2),**3,4**	30/**43**	15/**35**	10	20	30		
	Telecommunications engineering	◑	(1),(2),(3),**4**	20/**30**	10/**15**			15		◐ ○
Greenwich	Communications systems and software engineering		(1),**2,3**	25/**40**	25/**40**		10	30		
	Computer networking	◐	(1),**2,3**	25/**40**	25/**40**		10	30		
	Computer systems and software engineering	◐	(1),**2,3**	25/**40**	25/**40**		10	30		
	Control and instrumentation engineering	◐	(1),**2,3**	20/**20**	25/**25**	15	15	25	●	○
	Electrical and electronic engineering	◐	(1),**2,3**	20/**20**	25/**25**	15	15	25	●	○
	Electrical engineering	◐	(1),**2,3**	20/**20**	25/**25**	15	15	25	●	○
	Electronic engineering	◐	(1),**2,3**	20/**20**	25/**25**	15	15	25	●	○
	Electronics and computer systems	◐	(1),**2,3**	20/**20**	25/**25**	15	15	25	●	○
	Internet technologies	◐	(1),**2,3**	25/**40**	25/**40**		10	30	●	○
Heriot-Watt	Computing and electronics	◐	(1),(2),(3),**4**	0/**20**	33/**33**	30	30	33	●	◐ ○
	Electrical and electronic engineering	◐	(1),(2),(3),**4,5**	40/**70**	30/**40**	30	30	30	●	◐ ○
	Electronic and photonic engineering	◐	(1),(2),**3,4,5**	0/**10**	20/**30**	30	30	30	●	◐ ○
	Robotics and cybertronics	◐	(1),(2),(3),**4,5**	20/**20**	30/**30**	30	30	30	●	◐ ○
Hertfordshire	*All courses*	◐	(1),**2,4**	10/**30**	20/**25**	25	13	38	●	○
Huddersfield	Computer control systems	◐	(1),**2,4,5**	30/**40**	33/**33**	20	20	30	●	◐
	Electronic and communication engineering	◐	(1),**2,4**	30/**40**	33/**33**	20	20	30	●	◐
	Electronic and electrical engineering	◐	(1),**2,4**	30/**40**	33/**33**	20	20	30	●	◐
	Electronic design	◐	(1),**2,4**	30/**40**	30/**30**	20	20	30	●	◐
	Electronic engineering	◐	(1),**2,4,5**	30/**40**	33/**33**	20	20	30	●	◐
	Electronic engineering and computer systems	◑	(1),**2,4**	30/**40**	33/**33**	20	20	30	●	◐
Hull	Computer systems engineering	◑	(1),**2,3**	5/**10**	30/**35**	15	15	30	●	○
	Electronic engineering	◑	(1),**2,3,4**	15/**15**	30/**30**	15	15	30	●	○

Electrical and Electronic Engineering

(continued) Table 4

Assessment methods

Institution	Course title	Frequency of assessment	Years of exams contributing to final degree (years of exams not contributing to final degree)	Coursework: minimum/**maximum** %	Project/dissertation: minimum/**maximum** %	Time spent on projects in: first/intermediate/final years %			Group projects: ● compulsory; ○ optional	Orals: ◐ if borderline; ● everyone; ○ for projects
Hull (continued)	Information and computer control technology		(1),**2,3**	30/**50**	20/**35**	10	30	50	●	◐○●
	Mobile telecommunications technology	◑	(1),**2,3,4**	30/**50**	20/**35**	10	30	50	●	◐○●
Imperial College London	Electrical and electronic engineering	○	**1,2,3,4**	40/**40**	12/**12**	10	25	50	●	◐○●
	Information systems engineering	○	**1,2,3,4**	18/**18**	23/**23**	10	25	50	●	◐○●
Kent	Computer systems engineering	○	(1),**2,3**	20/**25**	12/**12**	12	12	25	○	◐○
	Electronic and communications engineering	○	(1),**2,3**	12/**12**	18/**18**	10	10	25	●	◐○
King's College London	Computer systems and electronics	◑	**1,2,3,4**	32/**38**	21/**21**	6	10	30	●	
	Electronic engineering	◑	**1,2,3,4**	32/**38**	21/**21**	6	10	30		
	Telecommunication engineering	◑	**1,2,3,4**	32/**38**	21/**21**	6	10	30		
Lancaster	Computer systems engineering		(1)							
	Electronic systems engineering		(1)							
Leeds	*All courses*	◑	(1),**2,3,4**	12/**12**	25/**40**	17	25	40	●	◐○
Leicester	Communications and electronic engineering	◑	**1,2,3,4**	10/**20**	10/**20**	10	20	35	●	○
	Electrical and electronic engineering	◑	**1,2,3,4**	10/**20**	10/**20**	10	20	35	●	○
Liverpool	Computer science and electronic engineering	◑	(1),**2,3,4**	15/**15**	25/**25**	20	20	25		◐○
	Electrical and electronic engineering	◑	(1),(2),**3,4**	15/**15**	25/**25**	20	20	25		◐○
	Electrical engineering	◑	(1),**2,3**	15/**15**	25/**25**	20	20	25		◐○
	Electrical engineering and electronics	◑	(1),**2,3,4**	15/**15**	25/**25**	20	20	25		◐○
	Electronic and communication engineering	◑	(1),**2,3**	15/**15**	25/**25**	20	20	25		◐○
	Electronics	◑	(1),**2,3**	15/**15**	25/**25**	20	20	25		◐○
	Wireless communications and 3G technology	◑	(1),**2,3,4**	15/**15**	25/**25**	20	20	25		◐○
Liverpool John Moores	Broadcast engineering	◑	(1),**2,3**	20/**40**	30/**30**	10	20	30	●	○
	Broadcast technology	◑	(1),**2,3**	20/**30**	25/**35**	20	20	30	●	○
	Computer engineering	◑	(1),**2,3**	20/**30**	25/**35**	20	20	30	●	○
	Design electronics	◑	(1),**2,3**	20/**30**	25/**35**	20	20	30	●	○
	Electrical and electronic engineering	◑	(1),**2,3,4**	20/**30**	25/**35**	20	20	30	●	○
	Networks and telecommunications engineering	◑	(1),**2,3**	15/**20**	25/**30**	10	20	30		○
London Metropolitan	Audio electronics	◑	(1),**2,3,4**	40/**60**	12/**25**			25		
	Communications systems	◑	(1),**2,3,4**	40/**60**	12/**25**			25	○	◐○
	Computer networking	◑	(1),**2,3,4**	20/**40**	8/**10**			25		◐○
	Electronic and communications engineering	◑	(1),**2,3,4**	40/**50**	12/**25**			25		○
	Electronics	◑	(1),**2,3,4**	40/**60**	12/**25**			25		
London South Bank	Electrical and electronic engineering	◑	(1),**2,3**	20/**40**	10/**25**	20	20	20	●	◐
Loughborough	Computer network and internet engineering	◑	(1),**2,3,4**	5/**10**	20/**25**			25		
	Computer systems engineering	◑	(1),**2,3,4**	15/**20**	23/**23**	17	17	33	●	○●
	Electronic and electrical engineering	◑	(1),**2,3,4**	5/**10**	20/**25**			25	●	◐○
	Electronics and software engineering	◑	(1),**2,3,4**	5/**10**	20/**25**			25	●	
Manchester	Computer engineering	◑	(1),**2,3,4**	5/**5**	20/**20**					
	Computer systems engineering	◑	(1),**2,3,4**	20/**20**	15/**50**	10	20	50	●	

Institution	Course title	Frequency of assessment	Years of exams contributing to final degree (years of exams not contributing to final degree)	Coursework: minimum/maximum %	Project/dissertation: minimum/maximum %	Time spent on projects in: first/intermediate/final years %			Group projects: ● compulsory; ○ optional	Orals: ◐ if borderline; ● for projects; ○ everyone
Manchester (continued)	Computing and communications systems engineering	◐	(1),2,3,4	20/**20**	15/**50**	10	20	50		
	Electrical and electronic engineering	◐	(1),2,3,4	20/**20**	15/**50**	10	20	50	●	◐
	Electronic systems engineering	◐	(1),2,3,4	20/**20**	15/**50**	10	20	50		
Manchester Metropolitan	Communication and electronic engineering	○	(1),2,3	26/**36**	23/**23**		10	25		◐○
	Computer and electronic engineering	○	(1),2,3	26/**36**	23/**23**		10	25		◐○
	Electrical and electronic engineering	○	(1),2,3	26/**36**	23/**23**		10	25	●	◐○
Middlesex	Computer networks	◐	**2,3**	25/**25**	25/**35**	5	5	30	●	
Napier	Electronic and computer engineering	◐	(1),(2),**3,4**	20/**25**	25/**25**			25		◐○
	Electronic and electrical engineering	◐	(1),(2),**3,4**	15/**20**	25/**25**			25	●	◐○
Newcastle	Computer systems engineering	◐	(1),2,3,4	10/**20**	30/**50**	10	10	30		○
	Electrical and electronic engineering	◐	(1),2,3,4	10/**20**	30/**50**	10	10	30		○
	Electronic communications	◐	(1),2,3,4	10/**20**	27/**50**	10	10	30	●	
	Electronic engineering	◐	(1),2,3,4	10/**20**	27/**50**	10	10	30	●	○
North East Wales I	Electrical and electronic engineering	◐	**1,2,3**	40/**60**	40/**60**	15	30	60		
Northumbria	Communication and electronic engineering	◐	(1),2,3	**10**	**20**			20		
	Electrical and electronic engineering	◐	(1),2,3	10/**10**	10/**20**			20		
Nottingham	Electrical and electronic engineering	◐	(1),2,3,4	10/**15**	20/**25**	15	10	30	●	○●
	Electrical engineering	◐	(1),2,3,4	10/**15**	20/**25**	15	10	30	●	○●
	Electronic and communications engineering	◐	(1),2,3,4	10/**15**	20/**25**	15	10	30	●	○●
	Electronic and computer engineering	◐	(1),2,3,4	10/**15**	20/**25**	15	10	30	●	○●
	Electronic engineering	◐	(1),2,3,4	10/**25**	20/**25**	15	10	30	●	○●
Oxford	*Both courses*		(1),**3,4**	13/**13**	25/**25**	10	15	50	●	◐
Oxford Brookes	Telecommunications	◐	(1),2,3	**30**				25		
Plymouth	Communication engineering	◐	(1),2,3	**30**	**20**	10	10	20	●	
	Computer engineering	◐	(1),2,3	**30**	**20**	10	10	20		
	Computer systems engineering	◐	(1),2,3	**30**	**20**	10	10	20	●	
	Electrical and electronic engineering	◐	(1),2,3,4,5	**30**	**20**	10	10	20		◐
	Electrical and electronic systems	◐	(1),2,3	**30**	**20**	10	10	20		
	Electronic communication systems	◐	(1),2,3	**70**	**20**	10	10	20	●	
	Electronic engineering	◐	(1),2,3	**30**	**20**	10	10	20		◐
Portsmouth	Communication systems engineering	◐	(1),(2),**3,4,5**	16/**25**	33/**33**	15	15	33	●	◐
	Communications engineering	◐	(1),(2),**3**	15/**25**	33/**33**	15	20	33	●	◐○
	Computer engineering	◐	(1),(2),**3,4,5**	15/**25**	33/**33**	15	15	33	●	◐○
	Computer technology	◐	(1),(2),**3**	15/**20**	33/**33**	15	20	33	●	◐○
	Electronic and electrical engineering	◐	(1),(2),**3,4**	16/**20**	33/**33**	15	15	33	○	◐○
	Electronic engineering	◐	(1),(2),**3**	15/**25**	33/**33**	15	20	33	●	
	Internet technology	◐	(1),(2),**3,4**	15/**25**	33/**33**	15	25	33		◐○
Queen Mary	Communication engineering	○	**1,2,3,4**	12/**15**	20/**25**	5	20	25	●	○
	Computer engineering	○	**1,2,3,4**	15/**20**	20/**20**	5	20	25	●	○
	Electrical and electronic engineering	○	**1,2,3**	15/**20**	20/**20**	5	20	25	●	○
	Electronic engineering	○	**1,2,3,4**	15/**20**	20/**20**	5	20	25	●	○●
	Internet engineering	○	**1,2,3**	15/**20**	20/**20**	5	20	25	●	○
	Physics and electronics	◐	(1),**2,3,4**	10/**20**	15/**25**			27	○	○
	Telecommunications	○	**1,2,3**	15/**20**	20/**20**	5	20	25	●	○

Electrical and Electronic Engineering

Assessment methods

Key for frequency of assessment column: ● term; ◐ semester; ○ year
Group projects: ● compulsory; ○ optional
Orals: ◐ if borderline; ○ for projects; ● everyone

Institution	Course title	Frequency of assessment	Years of exams contributing to final degree (years of exams not contributing to final degree)	Coursework: minimum/maximum %	Project/dissertation: minimum/maximum %	Time spent on projects in: first/intermediate/final years %			Group projects: ● compulsory; ○ optional	Orals: ◐ if borderline; ○ for projects; ● everyone
Reading	Computer engineering		**3**							
	Cybernetics	○	(1),**2,3,4**	25	30		25	30	○	◐●
	Electronic engineering	○	(1),**2,3**	20/**25**	20/**25**	10	20	25	●	◐●
Robert Gordon	Communications and computer engineering	◐	(1),(2),**3,4**	25/**25**	25/**25**	5	12	25	●	◐
	Communications and computer network engineering	◐	(1),(2),**3,4**	25/**25**	25/**25**	5	12	25	●	◐
	Electrical and energy engineering	◐	(1),(2),**3,4**	20/**20**	25/**25**	5	12	25	●	◐
	Electronic and communications engineering	◐	(1),(2),**3,4**	25/**25**	25/**25**	5	12	25	●	◐
	Electronic and computer engineering	◐	(1),(2),**3,4**	30/**30**	25/**25**	5	12	25	●	◐
	Electronic and electrical engineering	◐	(1),(2),**3,4**	25/**25**	25/**25**	5	12	25	●	◐
St Andrews	Microelectronics and photonics	◐	(1),**2,3,4,5**	15/**20**	25/**25**	10	15	30	●	○
Sheffield	Computer systems engineering	◐	(1),**2,3,4**						●	
	Data communications engineering	◐	(1),**2,3,4**	2/**4**	18/**22**	15	15	20	●	
	Electrical engineering	◐	(1),**2,3,4**	2/**4**	18/**22**	15	15	20	●	◐
	Electronic and communications engineering	◐	(1),**2,3,4**	2/**4**	18/**22**	15	15	20	●	○
	Electronic engineering	◐	(1),**2,3,4**	2/**4**	18/**22**	15	15	20	●	◐
	Electronic, control and systems engineering	◐	(1),**2,3,4**							
	Engineering (computing, electronics, systems and control)	◐	(1),**2,3,4**							
	Systems and control engineering	◐	(1),**2,3,4**							
Sheffield Hallam	Computer and network engineering		(1),**2,3,4**	20/**40**	20/**40**	10	30	40		○
Southampton	Computer engineering	◐	(1),**2,3,4**	10/**10**	20/**20**	10	15	20	●	○
	Electrical engineering	◐	(1),**2,3,4**	10/**10**	20/**20**	5	15	25	●	◐○
	Electronic engineering	◐	(1),**2,3,4**	10/**10**	20/**20**	10	15	20		○
Southampton Solent	Electronic engineering	◐	(1),**2,3**	40/**50**	33	5	10	40	●	○
Staffordshire	Computer systems	◐	(1),**2,3,4**	30/**60**	38/**38**	0	0	38		○
	Electronic engineering	◐	(1),**2,3**	20/**30**	40/**40**	12	25	40	●	◐○
	Electronic systems design	◐	(1),**2,3**	20/**30**	40/**40**	12	25	40		○
Strathclyde	Computer and electronic systems	◐	(1),(2),**3,4,5**	10	25	10	20	20		◐○
	Electrical energy systems	◐	(1),(2),**3,4,5**	10/**15**	25/**25**		15	25	●	◐○
	Electronic and digital systems	◐	(1),(2),**3,4,5**	0/**15**	25/**25**		15	25	●	◐○
	Electronic and electrical engineering	◐	(1),(2),**3,4,5**	2/**13**	25/**25**		15	25	●	◐○
Surrey	Digital media engineering	◐	(1),**2,3,4**	20/**30**	20/**20**	5	10	25		○
	Electronic engineering	◐	(1),**2,3,4,5**	20/**30**	20/**20**	10	10	20		◐○
	Electronics and computer engineering	◐	(1),**2,3,4,5**	20/**30**	20/**20**	5	10	25		
	Electronics with satellite engineering	◐	(1),**2,3,4,5**	20/**30**	20/**20**	5	10	25		○
	Telecommunication systems	◐	(1),**2,3,4,5**	20/**30**	20/**20**	10	10	20	○	◐○
Sussex	Computer systems engineering	○	(1),**2,3**	13/**21**	20/**20**		15	20	○	◐○
	Electrical and electronic engineering	○	(1),**2,3,4**	5/**6**	20/**20**			20	○	◐○
	Electronic and communication engineering	○	(1),**2,3,4**	9/**36**	18/**20**			20	○	◐○
Swansea	Communication systems	◐	(1),**2,3,4,5**	9/**10**	/**20**	15	20	25	●	◐○
	Electronic and electrical engineering	◐	(1),**2,3,4,5**	9/**10**	/**20**	15	20	25	●	◐○

Assessment methods

Institution	Course title	Frequency of assessment (Key: ◑ term; ◐ semester; ○ year)	Years of exams contributing to final degree (years of exams not contributing to final degree)	Coursework: minimum/maximum %	Project/dissertation: minimum/maximum %	Time spent on projects in: first/intermediate/final years %			Group projects: ● compulsory; ○ optional	Orals: ◐ if borderline; ● everyone ○ for projects
Swansea (continued)	Electronics with computing science/management	◑	(1),**2,3,4,5**	10/**12**	/**20**	5	20	25	●	◐ ○
	Internet technology	◑	(1),**2,3**	10/**12**	/**20**	5	20	25	●	◐ ○
	Mobile communications and internet technology	◑	(1),**2,3**	10/**12**	/**20**	5	20	25	●	◐ ○
Swansea IHE	Computer systems and electronics	○	(1),**2,3**	40/**40**	30/**30**	40	40	50	●	◐ ○ ●
	Internet technology and networks	○	(1),**2,3**	40/**40**	30/**30**	40	40	50	●	◐ ○ ●
Teesside	*Both courses*	◑	(1),**2,3**	30/**30**	25/**25**	10	10	30		◐ ○
UCE Birmingham	Computing and electronics	◑	(1),**2,3**							
	Electronic engineering	◑	(1),**2,3**							○
	Telecommunications and networks	◑	(1),**2,3**							○
UCL	Electronic and electrical engineering	○	**1,2,3,4**	6/**10**	15/**18**		45	25	●	○
	Electronic engineering with communications engineering	○	**1,2,3,4**	6/**10**	15/**18**		12	25	○	○
	Electronic engineering with computer science/nanotechnology	○	**1,2,3,4**	6/**10**	15/**18**		12	37	○	○
Ulster	Electronics and computer systems	◑	(1),(2),**4**	13/**26**	/**33**	2	10	33	●	◐ ○ ●
Wales (Newport)	Electrical engineering		(1),**2,3**		**18**			33		
	Electronic engineering	○	(1),**2,3**		**18**			33		
Wales (UWIC)	Electronic communication systems	○	(1),**2,3**	30	30	0	20	40		○
	Electronic control systems	○	(1),**2,3**		30	0	20	40		○
	Electronic microcomputer systems	○	(1),**2,3**	30	30	0	20	40		○
	Music and audioelectronic systems	○	(1),**2,3**	30	30	0	20	40		○
Warwick	Electronic engineering	○	**1,2,3,4**	8/**28**	18/**18**	12	12	25	●	○ ●
Westminster	Computer systems engineering	◑	(1),**2,3**	10/**15**	30/**30**	12	50	38	●	○ ●
	Digital communications engineering	◑	(1),**2,3**	10/**15**	30/**30**	12	50	38		○ ●
	Digital signal processing	◑	(1),**2,3**	10/**10**	25/**25**	15	50	25		○ ●
	Electronic engineering	◑	(1),**2,3**	10/**15**	30/**30**	12	50	38	●	○ ●
	Electronics	◑	(1),**2,3**	10/**10**	25/**25**	15	50	25		○ ●
	Mobile communications	◑	(1),**3**	10/**10**	25/**25**	15	50	25	●	○ ●
	Networks and communications engineering	◑	(1),**2,3**	10/**10**	25/**25**	15	50	25		○
York	Electronic and communication engineering	◕	**1,2,3,4**	75/**80**	20/**25**	5	10	25	●	◐ ○ ●
	Electronic and computer engineering	◕	**1,2,3,4**	75/**80**	20/**25**	5	10	25	●	◐ ○ ●
	Electronic engineering	◕	**1,2,3,4**	75/**80**	20/**25**	5	10	25	●	○ ●

Chapter 5: Entrance requirements

See Chapter 5 in the *Introduction* (page 18) for general information about entrance requirements that applies to all types of engineering course, including those covered in this part of the Guide. It also explains how to interpret TABLE 5, which gives information about entrance requirements for individual courses.

Table 5 — Entrance requirements

Institution	Course title	Number of students (includes other courses)	Typical offers UCAS points	A-level grades	SCQF Highers grades	● compulsory; ○ preferred A-level Mathematics	A-level Physics
Aberdeen	Electrical and electronic engineering	(170)		CCD	ABBBC	●	●
	Electronic and computer engineering			CCD	ABBBC		
	Electronic engineering with communications			CCD	ABBBC		
	Electronics and computer/software engineering	(170)		CCD	ABBBC	●	●
Anglia Ruskin	Electronics	20	160		BBCC	○	○
Aston	Communications engineering	15	240–280	BCC	BBBBC	●	●
	Electrical and electronic engineering	30	240–280	BCC	BBBBC	●	○
	Electronic engineering	15	240–280	BCC	BBBBC		
	Electronic systems engineering	12				●	●
	Internet engineering	10	240–280	BCC	BBBBC	●	●
Bangor	Communications and computer systems		240–260				
	Electronic engineering	30	240–260			●	●
Bath	Computers, electronics and communications	(25)		BBC		●	○
	Electrical and electronic engineering	80		BBC		●	○
	Electrical engineering and applied electronics	10		BBC		●	○
	Electrical power engineering			BBC			
	Electronic and communication engineering	25		BBC		●	○
	Electronics with space science and technology	15		BBC		●	●
Bedfordshire	Computer networking		120				
Belfast	Electrical and electronic engineering	80		BCC	BBBC	●	●
	Electronic and software engineering			BCC	BBBC		
Birmingham	Communication systems engineering	15		ABB	ABBBB	●	○
	Computer systems engineering	20		ABB	ABBBB	●	
	Electronic and communications engineering			ABB	ABBBB		
	Electronic and electrical engineering	70		ABB	ABBBB	●	○
	Electronic engineering	15		ABB	ABBBB	○	○
Bolton	Electronic and computer engineering	35	200			●	●
	Internet communications and networks		200				
Bournemouth	Electronics		160–240			○	○
Bradford	*Both courses*	(60)	200–240			○	○
Brighton	Audio electronics	(50)	260			●	●
	Digital electronics, computing and communication	(50)	260			●	○
	Electrical and electronic engineering	(50)	260			●	○
Bristol	Communications and multimedia engineering	60				●	●
	Computer systems engineering	20				●	○
	Electrical and electronic engineering	(60)		AAB	AAABB	●	●

Entrance requirements

Institution	Course title	Number of students (includes other courses)	UCAS points	A-level grades	SCQF Highers grades	A-level Mathematics (● compulsory, ○ preferred)	A-level Physics
Bristol (continued)	Electronic and communications engineering	(60)		AAB	AAABB	●	●
	Electronic engineering	(60)		AAB	AAABB	●	●
Bristol UWE	Computer systems engineering		240–260			○	○
	Computing and telecommunications		200–240				
	Digital systems engineering	12	180–220			●	○
	Electrical and electronic engineering	12	180–220			●	○
	Electronic engineering	12	180–220			●	○
	Music systems engineering		220–240				
Brunel	Computer systems engineering	50	260			○	○
	Electronic and electrical engineering	50	260	BCC		●	○
	Electronic and microelectronic engineering	50	260	BCC		●	○
	Electronic/electrical engineering (communication systems)	50	260	BCC		●	○
	Electronic/electrical engineering (control systems)	50	260	BCC		●	○
	Electronic/electrical engineering (power electronics systems)	50	260	BCC		●	○
	Internet engineering	50	260			○	
Buckinghamshire Chilterns UC	Internet technology		120–240		CCCC		
Cambridge	Electrical and electronic engineering	(300)		AAA		●	●
	Electrical and information sciences	(300)		AAA		●	●
	Information and computer engineering	(300)		AAA		●	●
Cardiff	Computer systems engineering	15	280	BBC		●	
	Electrical and electronic engineering	30	280	BBC		●	○
	Electronic engineering	25	280	BBC		●	○
Central Lancashire	Computer engineering	25	200			●	○
	Digital communications	25	200			●	○
	Digital signal and image processing	25	200			●	○
	Electronic engineering	25	200			●	○
City	Communications	40	220–300				
	Computer systems engineering	30	240–330			●	○
	Electrical and electronic engineering	40	220–300		BBBBB	●	●
	Systems and control engineering	25	240–330			●	○
Coventry	Communications engineering		240–280			○	○
	Computer hardware and software engineering		200			○	○
	Computers, networking and communications technology		200				○
	Electrical systems engineering		200			○	○
	Electronics engineering	30	160–200			○	○
De Montfort	Broadcast technology		160–240			○	○
	Electronic engineering	40	160–240			●	●
Derby	Computer networks	25	180–200				
	Electrical and electronic engineering	15	160–180			●	○
	Electronics	5	160–180			●	○
Dundee	Electronic and electrical engineering	40	260			●	○
	Electronic engineering	5	240			●	○
	Electronic engineering and microcomputer systems		240				

(continued) Table 5

Entrance requirements

Institution	Course title	Number of students (includes other courses)	Typical offers / UCAS points	A-level grades	SCQF Highers grades	● compulsory; ○ preferred A-level Mathematics	A-level Physics
Dundee (continued)	Electronics and computing	10	240–300			●	●
	Microelectronics and photonics	5		ABB	AABB	●	●
Durham	Communications engineering					●	○
	Computer engineering	(130)				●	○
	Electronic engineering	(140)				●	○
East Anglia	Computer systems engineering		280				
East London	Electrical and electronic engineering	50	220			●	○
Edinburgh	Electrical engineering	5				●	○
	Electronics	10		BBB	BBBB	●	○
	Electronics and electrical engineering	40		BBB	BBBB	●	○
	Electronics and electrical engineering (communications)	10		BBB	BBBB	●	○
Essex	Audio engineering	10	280		ABBB	●	○
	Computer engineering	10	280–300		ABBB	○	○
	Computers and networks	15	280–300		ABBB	○	○
	Computers and telecommunications	5	280		ABBB	●	○
	Electronic engineering	20	280		ABBB	●	○
	Electronics and computers	25	200–220		CCCC	○	○
	Network and internet technology	10	280–300		ABBB	○	○
	Optoelectronics and communications systems	5	280		ABBB	●	○
	Telecommunications engineering	10	280		ABBB	●	○
Exeter	Electronic engineering	30	240			●	○
	Internet engineering	20	260–300			●	
Glamorgan	Computer systems engineering	12	240				
	Electrical and electronic engineering	40	260			●	○
	Electronic and communication engineering	15	260			●	○
	Electronic engineering	15	260			●	○
	Electronics	15	220				○
	Mobile telecommunications		260				
Glasgow	Audio and video engineering			BCC	BBBBC	●	○
	Electronic and software engineering	(80)		BCC	BBBBC	●	●
	Electronics and electrical engineering	(80)		BCC	BBBBC	●	●
	Electronics with music	25		BBC	ABBBB	●	●
	Microcomputer systems engineering	(80)		BCC	BBBBC	●	●
Glasgow Caledonian	Computer engineering	20		CD	BCCC	●	●
	Electrical power engineering	10		BC	BBBC	○	○
	Electronic engineering (BEng)	10		AC	BBBC	●	●
	Electronic engineering (BSc)			CD	BBC	○	○
	Telecommunications engineering	30		CD	BBC	○	○
Greenwich	Communications systems and software engineering		240	CCC	BBCC	○	○
	Computer networking	50	240	CCC	BBCC	●	○
	Computer systems and software engineering	50	240	CCC	CCCC	●	○
	Control and instrumentation engineering	10	60			●	○
	Electrical and electronic engineering	30	240	CCC	CCCC	●	○
	Electrical engineering		240	CCC	BBCC	●	○
	Electronic engineering	30	240	CCC	CCCC	●	○
	Electronics and computer systems		240	CCC	CCCC	○	○

Electrical and Electronic Engineering

	(continued) **Table 5** Entrance requirements						
Institution	**Course title**	**Number of students** (includes other courses)	**Typical offers** UCAS points	A-level grades	SCQF Highers grades	● compulsory; ○ preferred **A-level Mathematics**	**A-level Physics**
Greenwich (continued)	Internet technologies	15	240	CCC	CCCC	○	○
Heriot-Watt	Computing and electronics	20		CCD	BBBC	●	○
	Electrical and electronic engineering	(100)		CDD	BBBC	●	○
	Electronic and photonic engineering	15		CDD	BBBC	●	●
	Robotics and cybertronics	30		CDD	BBBB	●	●
Hertfordshire	Digital communications and electronics	(60)	220			●	○
	Digital systems and computer engineering	(60)	240–280			●	○
	Electrical and electronic engineering	(60)	240–280			●	○
Huddersfield	Computer control systems	30	220		BBBB	●	○
	Electronic and communication engineering	(50)	220		BBBB	●	○
	Electronic and electrical engineering	(50)	220		BBBB	●	○
	Electronic design	20	220		BBBB		○
	Electronic engineering	(50)	220		BBBB	●	○
	Electronic engineering and computer systems	(50)	220		BBBB	●	○
Hull	Computer systems engineering	30	220–280			●	○
	Electronic engineering	30	240			●	○
	Information and computer control technology	10	240			○	○
	Mobile telecommunications technology	10	240			●	●
Imperial College London	Electrical and electronic engineering	130		AAA		●	●
	Information systems engineering	50		AAA		●	●
Kent	*Both courses*		260			●	●
King's College London	Computer systems and electronics	25		BBB	AABBB	●	○
	Electronic engineering	45		BBB	AABBB	●	○
	Telecommunication engineering	25		BBB	AABBB	●	○
Kingston	Communication systems		160–220				
Lancaster	Computer systems engineering	(90)		BCC	AAAB	●	○
	Electronic communication systems			BBC	AAAB	●	○
	Electronic systems engineering	(90)		BBC	AAAB	●	○
Leeds	*All courses*	(95)		BBB		●	○
Leicester	Communications and electronic engineering	30	260			●	○
	Electrical and electronic engineering	(90)	260			●	○
	Embedded systems engineering		260				
Liverpool	Computer science and electronic engineering	10	280	BBC	AAAB	●	○
	Electrical and electronic engineering	20	200		CCCC	○	○
	Electrical engineering	10	280	BBC	AAAB	●	●
	Electrical engineering and electronics	20	280	BBC	AAAB	●	●
	Electronic and communication engineering	10	280	BBC	AAAB	●	●
	Electronics	10	280	BBC	AAAB	●	●
	Wireless communications and 3G technology	5	280	BBC	AAAB	●	○
Liverpool John Moores	Broadcast engineering		240			○	○
	Broadcast technology	25	200			○	○
	Computer engineering	25	140–200			○	○
	Design oloctronics	10	200			○	○
	Electrical and electronic engineering	(50)	240			○	○
	Networks and telecommunications engineering	30	240			○	○

(continued) Table 5

Entrance requirements

Institution	Course title	Number of students (includes other courses)	Typical offers UCAS points	A-level grades	SCQF Highers grades	● compulsory; ○ preferred A-level Mathematics	A-level Physics
London Metropolitan	Audio electronics	20	160			●	○
	Communications systems	10	120–180				○
	Computer networking	75	160			○	○
	Electronic and communications engineering	55	160			●	○
	Electronics	20	120–160				○
London South Bank	Electrical and electronic engineering	50		CC	BBB	●	○
	Internet and multimedia engineering			CC	BBB		
	Telecommunications and computer networks engineering			CC	BBB		
Loughborough	Computer network and internet engineering	(150)				●	●
	Computer systems engineering	(150)	260			●	●
	Electrical and electrical engineering	(150)	260			●	●
	Electronics and software engineering	(150)				●	●
	Wireless communication engineering	(150)				●	●
Manchester	Computer engineering	(130)		AAB	AAABB	●	
	Computer systems engineering	30	320	ABB	AABBB	●	●
	Computing and communications systems engineering	30	320	ABB	AABBB	●	●
	Electrical and electronic engineering	50	320	ABB	AABBB	●	●
	Electronic systems engineering	40	320	ABB	AABBB	●	●
Manchester Metropolitan	Communication and electronic engineering	(70)	240		BBBB	●	○
	Computer and electronic engineering	(70)	240		BBBB	●	○
	Computer and network technology		220			○	
	Electrical and electronic engineering	(70)	240		BBBB	●	○
Middlesex	Computer networks	50	160–280			○	○
Napier	Computer networks and distributed systems		200				
	Electronic and communication engineering		220				
	Electronic and computer engineering	15	220			○	○
	Electronic and electrical engineering	45	220			○	○
Newcastle	Computer systems engineering	15		BBB		●	○
	Electrical and electronic engineering	45		BBB		●	○
	Electronic communications	10		BBB		●	○
	Electronic engineering	15		BBB		●	○
North East Wales I	Computer networks		160				
	Electrical and electronic engineering	10	140			○	○
Northumbria	Communication and electronic engineering		240		CCCCC	●	○
	Computer and network technology		240				
	Electrical and electronic engineering	115	240		CCCCC	●	○
Nottingham	Electrical and electronic engineering	(150)		AAB		●	●
	Electrical engineering	(150)		AAB		●	●
	Electronic and communications engineering	(150)		AAB		●	●
	Electronic and computer engineering	(150)		AAB		●	●
	Electronic engineering	(150)		AAB		●	●
Oxford	Electrical engineering	(170)				●	●
Oxford Brookes	Electronic systems design			CDD			
	Telecommunications	20		CDD		●	○
Paisley	Internet technologies			DE	BCC	○	○

Entrance requirements

Institution	Course title	Number of students (includes other courses)	UCAS points	A-level grades	SCQF Highers grades	● compulsory; ○ preferred	A-level Mathematics	A-level Physics
Plymouth	Communication engineering	30	240				●	○
	Computer engineering	30	240				●	○
	Computer systems and networks		240					
	Computer systems engineering	20	180				●	○
	Electrical and electronic engineering	30	240				●	○
	Electrical and electronic systems	40	180				●	○
	Electronic communication systems	15	180				●	○
	Electronic engineering	30	240				●	○
	Internet technologies and applications		180					
Portsmouth	Communication systems engineering	25	240				●	○
	Communications engineering	20	200				●	○
	Computer engineering	40	240				○	○
	Computer technology	40	200				○	○
	Electronic and electrical engineering	30	240				●	○
	Electronic engineering	30	200				○	○
	Internet technology	50	240				○	○
Queen Mary	Audio systems engineering		260				●	●
	Communication engineering	5					●	●
	Computer engineering	10	260–300		BBBBC		●	○
	Electrical and electronic engineering	15	260–300		BBBBC		●	●
	Electronic engineering	20	260–300		BBBBC		●	●
	Internet engineering	5	260–300				●	●
	Physics and electronics	5	280		BBBCC		●	●
	Telecommunications	10	260–300		BBBBC		●	●
Reading	Computer engineering		300				●	●
	Cybernetics	40					●	●
	Electronic engineering	35	280				●	●
Robert Gordon	Communications and computer engineering	30	220–240		BBCC		●	○
	Communications and computer network engineering	30	220–240		BBCC		●	○
	Electrical and energy engineering	30	220–240		BBCC		●	○
	Electronic and communications engineering	30	220–240		BBCC		●	○
	Electronic and computer engineering	30	220–240		BBCC		●	○
	Electronic and electrical engineering	30	220–240		BBCC		●	○
St Andrews	Microelectronics and photonics	5		AAB	AAAB		●	●
Sheffield	Computer systems engineering	30		AAB	AAAA		●	○
	Data communications engineering	(90)	280	BBB	BBBB		●	●
	Electrical engineering	80		AAB	AAAA		●	●
	Electronic and communications engineering	80					●	●
	Electronic engineering	80		AAB	AAAA		●	●
	Electronic, control and systems engineering	10		AAB	AAAA		●	○
	Engineering (computing, electronics, systems and control)			ABB	AABB		●	○
	Systems and control engineering	15		AAB	AAAA		●	○
Sheffield Hallam	Computer and network engineering	20	200				○	○
	Computer networks		200					
	Electrical and electronic engineering		240				●	
	Electronic engineering		180					

Electrical and Electronic Engineering

(continued) Table 5

Entrance requirements

Institution	Course title	Number of students (includes other courses)	Typical offers UCAS points	A-level grades	SCQF Highers grades	● compulsory, O preferred A-level Mathematics	A-level Physics
Southampton	Computer engineering	25	350	AAB		●	O
	Electrical engineering	20	350	ABB		●	●
	Electronic engineering	80	370	AAB		●	O
Southampton Solent	Computer network communications		120				
	Computer network management		120				
	Computer systems and networks		120				
	Electronic engineering	(55)	120			O	O
Staffordshire	Broadcasting technology		200–240				
	Computer systems	(300)	200–240				
	Electronic engineering	30	200–240			●	●
	Electronic systems design	(50)	200–240			●	●
	Internet technology	(300)	200–240				
	Mobile and wireless business systems	20	200–240				
	Mobile device technology	20	200–240				
	Wireless networking technology	20	200–240				
Strathclyde	Computer and electronic systems	40		CCC	BBBBC	●	●
	Digital communication and multimedia systems			CCC	BBBBC		
	Electrical energy systems	10		CCC	BBBBC	●	O
	Electronic and digital systems	10		CCC	BBBBC	●	O
	Electronic and electrical engineering	70		CCC	BBBBC	●	O
Surrey	Audio media engineering		300			●	O
	Digital media engineering	10	300			●	O
	Electronic engineering	110	300			●	O
	Electronics and computer engineering		300			●	O
	Electronics with satellite engineering					●	O
	Telecommunication systems	5	300			●	O
Sussex	Computer systems engineering	25		BBB	BBBBB	●	O
	Electrical and electronic engineering	30		BBB	BBBBB	●	O
	Electronic and communication engineering	35		BBB	BBBBB	●	O
	Electronic engineering			BBB	BBBBB		
Swansea	Communication systems	5	260			●	O
	Electronic and electrical engineering	35	260			●	O
	Electronics with computing science/management	5	260			●	O
	Internet technology	20	260			O	
	Mobile communications and internet technology	10	260			O	
Swansea IHE	Computer networks		80–340				
	Computer systems and electronics	15	80–340				
	Internet technology and networks	15	80–340				
Teesside	Electrical and electronic engineering	20	180–240			O	O
	Instrumentation and control engineering	25	180–240			O	O
UCE Birmingham	Computing and electronics		220				
	Electronic engineering		220			●	O
	Telecommunications and networks		220–240			●	O
UCL	Electronic and electrical engineering	(80)		AAA		●	●
	Electronic engineering with communications engineering	(80)				●	●

Entrance requirements

Institution	Course title	Number of students (includes other courses)	Typical offers UCAS points	A-level grades	SCQF Highers grades	● compulsory; ○ preferred A-level Mathematics	A-level Physics
UCL (continued)	Electronic engineering with computer science/nanotechnology	(80)				●	●
Ulster	Electronics and computer systems	30	260			○	○
	Electronics, communications and software		220			○	○
	Engineering (electrical/electronic)		260			●	○
Wales (Newport)	*Both courses*		160				
Wales (UWIC)	Electrical systems engineering		140				
	Electronic communication systems	10	140			●	
	Electronic control systems	12	140			●	
	Electronic microcomputer systems	12	140			●	
	Music and audioelectronic systems	15	140			●	
Warwick	Electronic engineering	70		BBB		○	○
Westminster	Computer systems engineering	(80)	260	BCC	BBCC	●	
	Computer systems technology	30		CC	CCCC		
	Digital communications engineering	(80)	260	BCC	BBCC	●	
	Digital signal processing	20	200	CDD	CCCC		
	Electronic engineering	(80)	260	BCC	BBCC	●	
	Electronics			CDD	CCCC		
	Mobile communications			CDD	CCCC	○	○
	Networks and communications engineering	20	200	CDD	CCCCC		
Wolverhampton	Electronics and communications engineering		160–220				
York	Electronic and communication engineering	(110)		BBB		●	●
	Electronic and computer engineering	(110)		BBB		●	●
	Electronic engineering	(110)		BBB		●	●

See Chapter 6 in the *Introduction* (page 20) for general information about professional qualification that applies to all types of engineering course, including those covered in this part of the Guide.

For most of the courses covered by this Guide, the Institution of Engineering and Technology (formed by the merger of the Institution of Electrical Engineers and the Institution of Incorporated Engineers) is the relevant institution, but the Institute of Measurement and Control (InstMC) also gives accreditation to several courses. TABLE 6 shows whether courses have been accredited by these institutions at Chartered or Incorporated Engineer level. Where TABLE 6 shows that a course leading to a BEng degree only, or a BEng or an MEng degree, is accredited at Chartered level, this accreditation may in future be withdrawn, or apply only to the MEng degree, so if you are aiming to become a Chartered Engineer you should check with the professional body or the institution offering the course before applying.

A number of other institutions have given accreditation to courses: the British Computer Society (BCS); the Institution of Mechanical Engineers (IMechE); the Institute of Physics (IoP); the Royal Aeronautical Society; and the Chartered Institution of Building Services Engineers (CIBSE).

Remember that the process of accreditation goes on continuously, so you should check with the professional institutions for the current position on specific courses. Information has not been supplied for courses that are not included in TABLE 6 or the list above. Where courses are accredited by several institutions, the accreditation by each one may depend on your having taken certain options, which may not coincide with those required by another institution.

Chapter 7 in the *Introduction* (page 24–26) lists the professional institutions' addresses and websites.

Institution	Course title	IET CEng	IET IEng	InstMC CEng	InstMC IEng	Others
Aberdeen	Electrical and electronic engineering	●				
	Electronics and computer/software engineering	●				
Aston	Communications engineering	●				
	Electrical and electronic engineering	●				
	Electronic engineering	●				
	Electronic systems engineering	●				BCS
Bangor	Electronic engineering	●				
Bath	Computers, electronics and communications	●	●			
	Electrical and electronic engineering	●	●			

Table 6 — Accreditation by professional institutions

Accreditation by professional institutions						
Institution	Course title	IET CEng	IET IEng	InstMC CEng	InstMC IEng	Others
Bath (continued)	Electrical engineering and applied electronics	●	●			
	Electronic and communication engineering	●	●			
Belfast	Electrical and electronic engineering	●				
Birmingham	Communication systems engineering	●				
	Computer systems engineering	●				
	Electronic and electrical engineering	●				
	Electronic engineering	●				
Bolton	Electronic and computer engineering	●				
Bradford	*Both courses*	●				
Brighton	*All courses*	●				
Bristol	Computer systems engineering	●				BCS
	Electrical and electronic engineering	●				
	Electrical and communications engineering	●				
	Electronic engineering	●				
Bristol UWE	Digital systems engineering	●				
	Electrical and electronic engineering	●				
	Electronic engineering	●				
Brunel	Computer systems engineering	●				
	Electronic and electrical engineering	●				
	Electronic and microelectronic engineering	●				
	Electronic/electrical engineering (communication systems)	●				
	Electronic/electrical engineering (control systems)	●				
	Electronic/electrical engineering (power electronics systems)	●				
Cambridge	*All courses*	●				
Cardiff	Computer systems engineering	●				BCS
	Electrical and electronic engineering	●				
	Electronic engineering	●				
City	Computer systems engineering	●		●		
	Electrical and electronic engineering	●		●		
Coventry	Computers, networking and communications technology		●			
	Electronics engineering		●			
De Montfort	Electronic engineering	●				
Derby	Computer networks					MBCS IEng
	Electrical and electronic engineering				●	
	Electronics				●	
Dundee	Electronic and electrical engineering	●				
	Electronic engineering	●				
	Microelectronics and photonics	●				
Durham	Electronic engineering	●				
Edinburgh	*All courses*	●				
Essex	Audio engineering	●				
	Computer engineering	●				
	Computers and networks	●				
	Computers and telecommunications	●				
	Electronic engineering	●				

Accreditation by professional institutions

Institution	Course title	IET CEng	IET IEng	InstMC CEng	InstMC IEng	Others
Essex (continued)	Telecommunications engineering	●				
Exeter	Electronic engineering	●	●			
Glamorgan	Computer systems engineering	●				
	Electrical and electronic engineering	●				
	Electronic and communication engineering	●				
	Electronic engineering	●				
Glasgow	Electronic and software engineering	●				BCS
	Electronics and electrical engineering	●				
	Electronics with music	●				
	Microcomputer systems engineering	●				
Glasgow Caledonian	Computer engineering		●			
	Electrical power engineering	●				
	Electronic engineering (BEng)	●	●		●	
	Electronic engineering (BSc)		●	●		
	Telecommunications engineering		●	●	●	
Greenwich	Electrical and electronic engineering		●			
	Electronic engineering	●		●		
Heriot-Watt	Computing and electronics	●				BCS
	Electrical and electronic engineering	●				
	Electronic and photonic engineering	●				IoP
	Robotics and cybertronics	●				IMechE
Hertfordshire	All courses	●				
Huddersfield	Computer control systems			●		
	Electronic and communication engineering	●				
	Electronic and electrical engineering	●				
	Electronic engineering	●				
	Electronic engineering and computer systems	●				
Hull	Computer systems engineering	●				
	Electronic engineering	●				
	Mobile telecommunications technology	●				
Imperial College London	Electrical and electronic engineering	●		●		
	Information systems engineering	●				BCS
Kent	Both courses	●				
King's College London	All courses	●				
Lancaster	Computer systems engineering	●		●		
	Electronic systems engineering	●				
Leeds	All courses	●	●			
Leicester	Electrical and electronic engineering	●		●		
Liverpool	All courses	●				
Liverpool John Moores	Electrical and electronic engineering	●				
London South Bank	Electrical and electronic engineering	●				CIBSE
Loughborough	Computer network and internet engineering	●				
	Computer systems engineering			●		RAeS
	Electronic and electrical engineering	●				
	Electronics and software engineering	●				

Institution	Course title	IET CEng	IET IEng	InstMC CEng	InstMC IEng	Others
Manchester	Computer systems engineering	●				
	Computing and communications systems engineering	●				
	Electrical and electronic engineering	●				
	Electronic systems engineering	●				
Manchester Metropolitan	Communication and electronic engineering	●				
	Computer and electronic engineering	●				
	Electrical and electronic engineering	●				
Middlesex	Computer networks					BCS
Napier	Electronic and computer engineering	●				
	Electronic and electrical engineering	●				
Newcastle	*All courses*	●				
Northumbria	Communication and electronic engineering	●				
	Electrical and electronic engineering	●				
Nottingham	Electrical and electronic engineering	●				
	Electrical engineering	●				
	Electronic and computer engineering	●				
	Electronic engineering	●				
Oxford	Electrical engineering	●				
	Information engineering	●		●		
Plymouth	Communication engineering	●				
	Computer engineering	●				
	Computer systems engineering		●			
	Electrical and electronic engineering	●				
	Electrical and electronic systems		●			
	Electronic communication systems		●			
	Electronic engineering	●				
Portsmouth	Communication systems engineering	●				
	Computer engineering	●				
	Electronic and electrical engineering	●				
	Electronic engineering	●				
Queen Mary	Communication engineering	●				
	Computer engineering	●				
	Electrical and electronic engineering	●				
	Electronic engineering	●				
	Internet engineering	●				
	Physics and electronics					IoP
	Telecommunications	●				
Reading	Computer engineering					BCS
	Cybernetics	●	●	●	●	
	Electronic engineering	●				
Robert Gordon	Communications and computer engineering	●	●			
	Communications and computer network engineering	●	●			
	Electrical and energy engineering	●	●			
	Electronic and communications engineering	●		●		
	Electronic and computer engineering	●		●		
	Electronic and electrical engineering	●	●	●		

(continued) Table 6

Accreditation by professional institutions

Institution	Course title	IET CEng	IET IEng	InstMC CEng	InstMC IEng	Others
St Andrews	Microelectronics and photonics	●				
Sheffield	Computer systems engineering	●		●		
	Electrical engineering	●	●			
	Electronic and communications engineering	●	●			
	Electronic engineering	●	●			
	Electronic, control and systems engineering	●		●		
	Systems and control engineering	●		●		
Southampton	*All courses*	●				
Southampton Solent	Electronic engineering	●				
Staffordshire	Computer systems					BCS
Strathclyde	Computer and electronic systems	●				BCS
	Electrical energy systems	●				
	Electronic and digital systems	●				
	Electronic and electrical engineering	●				
Surrey	Electronic engineering	●				
	Electronics and computer engineering	●				
	Electronics with satellite engineering	●				
	Telecommunication systems	●				
Sussex	Computer systems engineering	●				BCS
	Electrical and electronic engineering	●				
	Electronic and communication engineering	●				BCS
Swansea	Communication systems	●				
	Electronic and electrical engineering	●				
	Electronics with computing science/management	●				
Teesside	Electrical and electronic engineering		●			
UCE Birmingham	Electronic engineering	●		●		
UCL	Electronic and electrical engineering	●	●			
	Electronic engineering with communications engineering	●				
	Electronic engineering with computer science/nanotechnology	●				
Wales (UWIC)	*All courses*		●			
Warwick	Electronic engineering	●		●		
Westminster	Computer systems engineering	●				
	Digital communications engineering	●				
	Digital signal processing			●		
	Electronic engineering	●				
	Electronics			●		
	Mobile communications			●		
	Networks and communications engineering			●		
York	Electronic and communication engineering	●				
	Electronic and computer engineering	●				
	Electronic engineering	●				

See Chapter 7 in the *Introduction* (page 24) for a list of sources of information that apply to all types of engineering course, including those covered in this part of the Guide. It also lists a number of books that can give some general background to engineering and the work of engineers. Visit the websites of the professional institutions for information on careers, qualifications and course accreditation, as well as information about publications, many of which you can download. Chapter 7 in each of the other individual subject parts of the Guide contains further suggestions for background reading in particular engineering disciplines.

The following books illustrate the type of material you would study in the first and second years of an electrical/electronic engineering degree course (alongside subjects such as mathematics). If you want to look at books that reflect the huge range of exciting subjects in the later years of an electronics course, have a look at university websites where, along with the course syllabuses, there are extensive booklists.

Circuits, Devices and Systems R J Smith & R C Dorf. Wiley, 1992 (5th edition), £81.50

Electronics for Today and Tomorrow T Duncan. Hodder Arnold, 1997 (2nd edition), £15.99

Hughes Electrical and Electronic Technology J Hiley, K Brown & I McKenzie Smith. Pearson Higher Education, 2004 (9th edition), £44.99

Telecommunications Engineering J Dunlop & D G Smith. Taylor & Francis, 1998 (3rd edition), £31.99

Introduction to Computer Science and Programming L Vilms. Harper Collins, 1993

The courses This Guide gives you information to help you narrow down your choice of courses. Your next step is to find out more about the courses that particularly interest you. Prospectuses cover many of the aspects you are most likely to want to know about, but some departments produce their own publications giving more specific details of their courses. University and college websites are shown in TABLE 2a.

You can also write to the contacts listed below.

Aberdeen Student Recruitment and Admissions Service (sras@abdn.ac.uk), University of Aberdeen, Regent Walk, Aberdeen AB24 3FX

Anglia Ruskin Contact Centre (answers@anglia.ac.uk), Anglia Ruskin University, Bishop Hall Lane, Chelmsford CM1 1SQ

Aston Dr Geof Carpenter (g.f.carpenter@aston.ac.uk), School of Engineering and Applied Science, Aston University, Aston Triangle, Birmingham B4 7ET

Bangor Admissions Tutor, School of Electronic Engineering and Computer Systems, University of Wales, Bangor, Dean Street, Bangor, Gwynedd LL57 1UT

Bath Professor N J Mitchell (elec-eng@bath.ac.uk), Admissions Tutor, Department of Electronic and Electrical Engineering, University of Bath, Claverton Down, Bath BA2 7AY

Bedfordshire Admissions Department (admission@beds.ac.uk), University of Bedfordshire, Park Square, Luton LU1 3JU

Belfast Admissions Officer, The Queen's University of Belfast, Belfast BT7 1NN

Birmingham Dr Sandra Woolley (uga-eece@bham.ac.uk), Department of Electronic and Electrical Engineering, University of Birmingham, PO Box 363, Birmingham B15 2TT

Bolton Admissions Tutor (pkl1@bolton.ac.uk), Department of Computing and Electronic Technology, Bolton University, Deane Road, Bolton BL3 5AB

Bournemouth Mrs H Impett, School of Design, Engineering and Computing, Bournemouth University, Talbot Campus, Fern Barrow, Poole BH12 5BB

Bradford Mr Jack Bradley (ug-eng-enquiries@bradford.ac.uk), Admissions Tutor, School of Engineering, Design and Technology, University of Bradford, Bradford BD7 1DP

Brighton Dr D S Gill (d.s.gill@brighton.ac.uk), School of Engineering, University of Brighton, Cockcroft Building, Lewes Road, Brighton BN2 4GJ

Bristol Computer systems engineering Undergraduate Admissions Co-ordinator, Department of Computer Science; All other courses Admissions Tutor, Department of Electrical and Electronic Engineering; both at University of Bristol, Merchant Venturers Building, Woodland Road, Bristol BS8 1UB

Bristol UWE Pat Cottrell (admissions.cems@uwe.ac.uk), Faculty of Computing, Engineering and Mathematics, University of the West of England Bristol, Coldharbour Lane, Frenchay, Bristol BS16 1QY

Brunel T Kissack (thomas.kissack@brunel.ac.uk), School of Engineering and Design, Brunel University, Uxbridge UB8 3PH

Buckinghamshire Chilterns UC Admissions Office (admissions@bcuc.ac.uk), Buckinghamshire Chilterns University College, Queen Alexandra Road, High Wycombe HP11 2JZ

Cambridge Cambridge Admissions Office (admissions@cam.ac.uk), University of Cambridge, Fitzwilliam House, 32 Trumpington Street, Cambridge CB2 1QY

Cardiff Admissions Office (griffithsh@cardiff.ac.uk), School of Engineering, Cardiff University, PO Box 917, Cardiff CF24 3XF

Central Lancashire Martin Varley (mrvarley@uclan.ac.uk), Department of Technology, University of Central Lancashire, Preston PR1 2HE

City Undergraduate Admissions Office (ugadmissions@city.ac.uk), City University, Northampton Square, London EC1V 0HB

Coventry Electrical systems engineering Electronics engineering N R Poole (n.poole@coventry.ac.uk); All other courses Mr R Jinks (r.jinks@coventry.ac.uk); both at School of Engineering, Coventry University, Priory Street, Coventry CV1 5FB

De Montfort Admissions Unit (cse@dmu.ac.uk), Faculty of Computing Sciences and Engineering, De Montfort University, The Gateway, Leicester LE1 9BH

Derby Student Recruitment Unit (enquiries-admissions@derby.ac.uk), University of Derby, Kedleston Road, Derby DE22 1GB

Dundee Electronics and computing Dr Iain R Murray (irmurray@computing.dundee.ac.uk), Department of Applied Computing; All other courses Dr D I Jones (d.i.jones@dundee.ac.uk), Electronic Engineering and Physics Division; both at University of Dundee, Dundee DD1 4HN

Durham Dr Tim Short (engineering.admissions@durham.ac.uk), School of Engineering, University of Durham, South Road, Durham DH1 3LE

East Anglia Admissions Office (admissions@uea.ac.uk), University of East Anglia, Norwich NR4 7TJ

East London Student Admissions Office (admiss@uel.ac.uk), University of East London, Docklands Campus, 4–6 University Way, London E16 2RD

182

Edinburgh Undergraduate Admissions Office (sciengug@ed.ac.uk), College of Science and Engineering, University of Edinburgh, The King's Buildings, West Mains Road, Edinburgh EH9 3JY

Essex Undergraduate Admissions Office (admit@essex.ac.uk), University of Essex, Wivenhoe Park, Colchester CO4 3SQ

Exeter Electronic engineering Admissions Secretary (eng-admissions@exeter.ac.uk), Department of Engineering; Internet engineering Admissions Secretary (dcs-admissions@exeter.ac.uk), Department of Computer Science; both at School of Engineering, Computer Science and Mathematics, University of Exeter, North Park Road, Exeter EX4 4QF

Glamorgan Electronic and communication engineering Electronic engineering Alex Beaujean; All other courses Akram Hammoudeh (amhammou@glam.ac.uk); both at School of Electronics, University of Glamorgan, Pontypridd, Mid Glamorgan CF37 1DL

Glasgow Dr G Harrison (g.harrison@elec.gla.ac.uk), Department of Electronics and Electrical Engineering, Glasgow University, Glasgow G12 8LT

Glasgow Caledonian Computer engineering Telecommunications engineering Dr B J Beggs (bjbe@gcal.ac.uk); Electrical power engineering Electronic engineering (BEng) Angela Geddes (age@gcal.ac.uk); Electronic engineering (BSc) Dr H Fernandez-Conque (hfe@gcal.ac.uk); all at School of Engineering, Science and Design, Glasgow Caledonian University, Cowcaddens Road, Glasgow G4 0BA

Greenwich Computer systems and software engineering Control and instrumentation engineering Admissions Co-ordinator (eng-courseinfo@gre.ac.uk), Medway School of Engineering, University of Greenwich, Central Avenue, Chatham Maritime ME4 4TD; Electrical engineering Electronic engineering Electronics and computer systems Internet technologies Information Officer (eng-courseinfo@gre.ac.uk), University of Greenwich, Recruitment and Marketing, Wellington Street, Woolwich SE18 6PF; All other courses Enquiry Unit (courseinfo@gre.ac.uk), University of Greenwich, Maritime Greenwich Campus, Old Royal Naval College, London SE10 9LS

Heriot-Watt Dr John Hiley (j.hiley@hw.ac.uk), Department of Computing and Electrical Engineering, Heriot-Watt University, Riccarton, Edinburgh EH14 4AS

Hertfordshire Dr F A Muhammad (admissions@herts.ac.uk), Department of Electrical and Electronic Engineering, University of Hertfordshire, College Lane, Hatfield AL10 9AB

Huddersfield Department of Engineering and Technology (engtech@hud.ac.uk), University of Huddersfield, Queensgate, Huddersfield HD1 3DH

Hull Admissions Tutor (engineering-admissions@hull.ac.uk), Department of Engineering, University of Hull, Hull HU6 7RX

Imperial College London Dr S Lucyszyn (s.lucyszyn@imperial.ac.uk), Department of Electrical and Electronic Engineering, Imperial College London, South Kensington Campus, London SW7 2AZ

Kent Registry (recruitment@kent.ac.uk), University of Kent, Canterbury, Kent CT2 7NZ

King's College London Admissions Tutor (ugadmissions.engineering@kcl.ac.uk), Division of Engineering, King's College London, Strand, London WC2R 2LS

Kingston Student Information and Advice Centre, Cooper House, Kingston University, 40–46 Surbiton Road, Kingston upon Thames KT1 2HX

Lancaster Dr R V Chaplin, Engineering Department, Lancaster University, Lancaster LA1 4YR

Leeds Admissions Team (electronics@leeds.ac.uk), Department of Electronic and Electrical Engineering, University of Leeds, Leeds LS2 9JT

Leicester Mr I M Jarvis (ms263@le.ac.uk), Department of Engineering, University of Leicester, University Road, Leicester LE1 7RH

Liverpool Admissions Tutor (admiss.ug.eee@liv.ac.uk), Department of Electrical Engineering and Electronics, University of Liverpool, Brownlow Hill, Liverpool L69 3GJ

Liverpool John Moores Student Recruitment Team (recruitment@ljmu.ac.uk), Liverpool John Moores University, Roscoe Court, 4 Rodney Street, Liverpool L1 2TZ

London Metropolitan Admissions Office (admissions@londonmet.ac.uk), London Metropolitan University, 166–220 Holloway Road, London N7 8DB

London South Bank Admissions Office, London South Bank University, 103 Borough Road, London SE1 0AA

Loughborough Mr J G Hooper (j.g.hooper@lboro.ac.uk), Department of Electronic and Electrical Engineering, Loughborough University, Loughborough LE11 3TU

Manchester Computer engineering Undergraduate Admissions Office (ug-admissions@cs.man.ac.uk), School of Computer Science, University of Manchester, Manchester M13 9PL; All other courses Admissions Tutor (ug.elec.eng@manchester.ac.uk), School of Electrical and Electronic Engineering, University of Manchester, PO Box 88, Manchester M60 1QD

Manchester Metropolitan Ms Joan Jackson, Assistant Administrator, Department of Engineering and Technology, Manchester Metropolitan University, John Dalton Building, Chester Street, Manchester M1 5GD

Middlesex Admissions Enquiries (admissions@mdx.ac.uk), Middlesex University, North London Business Park, Oakleigh Road South, London N11 1QS

Napier Roy Goodwin, School of Engineering, Napier University, 219 Colinton Road, Edinburgh EH14 1DJ

Newcastle Admissions Officer (eeeng.ugadmin@ncl.ac.uk), School of Electrical, Electronic and Computer Engineering, University of Newcastle upon Tyne, Newcastle upon Tyne NE1 7RU

North East Wales I C Fordwhalley, School of Engineering, North East Wales Institute, Mold Road, Wrexham LL11 2AW

Northumbria Admissions (er.educationliaison@northumbria.ac.uk), University of Northumbria, Trinity Building, Northumberland Road, Newcastle upon Tyne NE1 8ST

Nottingham Ms Helen Ireson (ug.admin@eee.nottingham.ac.uk), School of Electronic and Electrical Engineering, University of Nottingham, University Park, Nottingham NG7 2RD

Oxford Deputy Administrator (Academic), Department of Engineering Science, Oxford University, Parks Road, Oxford OX1 3PJ

Oxford Brookes P T Moran, School of Engineering, Oxford Brookes University, Headington, Oxford OX3 0BP

Paisley Admissions Officer, Department of Electronic Engineering and Physics, University of Paisley, High Street, Paisley PA1 2BE

Plymouth Dr Peter White (pwhite@plymouth.ac.uk), Department of Communication and Electronic Engineering, University of Plymouth, Drake Circus, Plymouth PL4 8AA

Portsmouth Dr Nick Savage (nick.savage@port.ac.uk), Department of Electronic and Computer Engineering, University of Portsmouth, Anglesea Building, Anglesea Road, Portsmouth PO1 3DJ

Queen Mary Physics and electronics Department of Physics (physics@qmul.ac.uk); All other courses Kate Dunster (enquiries@elec.qmul.ac.uk), Department of Electronic Engineering; both at Queen Mary University of London, Mile End Road, London E1 4NS

Reading Computer engineering Dr Michael Evans (csug.admissions@reading.ac.uk), Department of Computer Science; Cybernetics Mrs C Leppard (cybernetics@reading.ac.uk), Department of Cybernetics; Electronic engineering Dr S A Shirsavar (ee-admissions@lists.reading.ac.uk), Department of Engineering; all at University of Reading, PO Box 225, Whiteknights, Reading RG6 6AY

Robert Gordon Ken Gow, School of Electronic and Electrical Engineering, The Robert Gordon University, Schoolhill, Aberdeen AB10 1FR

St Andrews Admissions Applications Centre (admissions@st-andrews.ac.uk), University of St Andrews, St Andrews KY16 9AX

Sheffield Computer systems engineering Electronic, control and systems engineering Engineering (computing, electronics, systems and control) Systems and control engineering Admissions (ugacse@sheffield.ac.uk), Department of Automatic Control and Systems Engineering; All other courses Dr D A Stone (infoeee@sheffield.ac.uk), Department of Electronic and Electrical Engineering; both at University of Sheffield, Mappin Street, Sheffield S1 3JD

Sheffield Hallam Computer and network engineering Computer networks Frank Rippon; Electrical and electronic engineering Student Administration; Electronic engineering Sharon Wilson, Marketing Officer, School of EIT; all at Sheffield Hallam University, Pond Street, Sheffield S1 1WB

Southampton Admissions Secretary (ucas@ecs.soton.ac.uk), Department of Electronics and Computer Science, University of Southampton, Southampton SO17 1BJ

Southampton Solent Faculty of Technology Admissions, Southampton Solent University, East Park Terrace, Southampton SO14 0RD

Staffordshire Computer systems Mauren Hindhaugh, Faculty of Computing, Engineering and Technology; All other courses Ann Grainger, School of Engineering and Advanced Technology; both at Staffordshire University, Beaconside, Stafford ST18 0AD

Strathclyde Mr Dougie Grant (d.grant@eee.strath.ac.uk), Department of Electronic and Electrical Engineering, University of Strathclyde, George Street, Glasgow G1 1XW

Surrey Dr E Chilton, Department of Electronic and Electrical Engineering, University of Surrey, Guildford GU2 7XH

Sussex Admissions Tutor (ug.admissions@engineering.sussex.ac.uk), Department of Engineering and Design, University of Sussex, Falmer, Brighton BN1 9QT

Swansea Mrs Lynnette Jones (eng.recruitment@swansea.ac.uk), School of Engineering, University of Wales Swansea, Singleton Park, Swansea SA2 8PP

Swansea IHE Computer networks Sue Maw (sue.maw@sihe.ac.uk); Computer systems and electronics Internet technology and networks I Wells (i.wells@sihe.ac.uk); both at Faculty of Applied Design and Engineering, Swansea Institute of Higher Education, Mount Pleasant, Swansea SA1 6ED

Teesside Sandra Joyce, School of Science and Technology, University of Teesside, Middlesbrough TS1 3BA

UCE Birmingham Information Officer (enquiries@tic.ac.uk), Technology Innovation Centre, University of Central England in Birmingham, Millennium Point, Curzon Street, Birmingham B4 7XG

UCL Dr Richard B Jackman (r.jackman@ee.ucl.ac.uk), Department of Electronic and Electrical Engineering, University College London, Torrington Place, London WC1E 7JE

Ulster Electronics and computer systems Dr L McDaid, School of Electrical and Mechanical Engineering, University of Ulster, Magee College, Londonderry BT48 7JL; Electronics, communications and software Engineering (electrical/electronic) N D Hunter, Faculty of Informatics, University of Ulster, Jordanstown, Newtownabbey BT37 0QB

Wales (Newport) Admissions (admissions@newport.ac.uk), University of Wales College, Newport, PO Box 101, Newport NP18 3YH

Wales (UWIC) Marketing and Student Recruitment (admissions@uwic.ac.uk), University of Wales Institute, Cardiff, PO Box 377, Western Avenue, Llandaff CF5 2SG

Warwick Director of Undergraduate Admissions, Department of Engineering, University of Warwick, Coventry CV4 7AL

Westminster Admissions and Marketing Office (cav-admissions@wmin.ac.uk), Cavendish Campus, University of Westminster, 115 New Cavendish Street, London W1W 6UW

Wolverhampton Admissions Unit (enquiries@wlv.ac.uk), University of Wolverhampton, Compton Road West, Wolverhampton WV3 9DX

York Computer systems and software engineering W Freeman, Department of Computer Science; All other courses K L Todd, Department of Electronics; both at University of York, Heslington, York YO10 5DD

See the *Introduction* to this Guide (page 3) for an overview of how to use each chapter in the individual subject parts, including this one, together with general information about engineering and the structure of engineering courses.

The scope of this Guide Materials science and materials engineering involve the study of a wide range of traditional and new materials used by modern technology. These include metals, glasses, ceramics, concrete, wood, textiles, polymers, composite materials and functional materials such as semiconductors, superconductors and magnets. Materials science and engineering form the basis of our understanding of the properties of materials and are used to develop methods for their production and fabrication.

This part of the Guide gives information about a wide variety of courses dealing with materials. These include general materials courses, courses that deal with a particular type of material, and applied courses specialising in an area such as aerospace, automotive, biomedical or sports materials.

General materials courses These courses cover a wide range of materials. They allow you to compare and contrast the properties of different materials and the processes associated with them. You will learn about properties and techniques that are common to a range of different types of material, and ones that are specific to particular materials. Although the early parts of these courses give a broad coverage, in later years you can usually concentrate your study, at least to some extent, on a particular type of material.

Specific materials courses These courses specialise in a particular type or group of materials, such as metals, ceramics, glasses, polymers, composites, textiles, paper and wood. Each of these materials has its own special properties and associated techniques for production and processing. The individual characteristics of the materials mean that they find application in a wide variety of contexts. For example, ceramics are used in the chemical, steel and more general engineering industries, and are used for electrical insulators and magnets, as well as for the more familiar domestic applications.

Applied materials courses These include courses in aerospace, automotive, biomedical and sports materials. You will study some general materials science, but with particular reference to the materials important for the application. For example, you will probably concentrate on metals, ceramics and polymers on a biomedical materials course. You will also need to study aspects of the application area in order to understand and analyse the properties required from the materials. For example, you will study topics like biomechanics and the design of sports equipment on a sports materials course, and anatomy and physiology on a biomedical materials course.

187

Materials scientists and engineers Materials scientists are employed in making improvements to existing materials and in the development of new materials up to the point at which they can go into production. They may then be involved in trouble-shooting and solving problems that may arise during production and in the quality control of finished products. After the production stage, they may play a role in supporting customers, helping those who have any problems with the products. Materials scientists also contribute to the choice of suitable materials for safe, effective and economic engineering designs.

Materials engineers require an overview of the whole production process, which means they must have an integrated knowledge of materials, engineering and design. Materials engineers choose the most appropriate material to use for a given application, taking into account the cost of the overall design and the environment in which the materials will be used, and if there is no suitable material, they may need to develop new materials. They may also be involved in finding applications for newly developed materials. Each material has its own characteristics, so when engineers replace one material with another they have to re-examine the whole design to take account of any weaknesses and to exploit the strengths of the new material. Factors they need to consider include the stress and strain that will be imposed, the temperature and pressure, the chemical environment and the amount of vibration to which they will be subjected. Quality, safety and reliability are also major considerations.

Materials and engineering Materials lie at the heart of all technology, and materials specialists contribute to solving problems across the whole range of engineering. Mechanical engineers need to know that the materials they use can withstand the wear and forces they are subjected to. Aeronautical engineers continually demand stronger and lighter materials. The chemical engineer has to design reaction vessels that can withstand high temperatures and pressures, and corrosive environments. The nuclear engineer needs to know that a containment vessel is going to maintain its integrity for decades despite high levels of radiation and in the knowledge that it may be impossible to make repairs at a later stage. Civil and structural engineers rely on the strength and reliability of the materials that they use to ensure the safety of buildings, bridges and tunnels. The huge progress made in electronic engineering could not have been made without the parallel development of semiconductor materials.

New materials and improvements in old ones Given the central role played by materials, it should come as no surprise that many of the outstanding advances in engineering in recent years have been inspired either by the development of new materials or by improvements to the properties of existing materials. On the other hand, the demands of specific engineering applications have often been the driving force behind the development of improved and new materials. Jet engines and space research have required a whole range of special alloys, polymers and ceramics, combining strength, oxidation resistance at high temperatures and lightness. The need to cram even more components onto integrated circuits to meet the demand for ever-increasing computer power at ever-decreasing cost is pushing the development of new processes and the search for new materials. Recent years have seen the development of ceramic materials with zero electrical resistance at temperatures approaching room temperature. These 'warm' superconductors have enormous potential in a vast range of applications, but this will only be realised once materials scientists and engineers have found ways of refining their properties so that they become useful engineering materials. Materials developments are having an increasing impact on society. For example: medical materials allow more sophisticated and long-term implants to be used; degradable polymers reduce environmental pollution; and new sports materials increase performance and enjoyment from leisure activities.

The science and engineering of materials   The first materials to be studied were metals – a reflection of their dominant role as engineering materials – but the knowledge and techniques developed for metals have been transferred and extended to the study of other materials such as glasses, ceramics, textiles and plastics, which are now all major engineering materials in their own right. For example, new ceramics have played a significant part in the latest steel-making processes, and, following extensive research that has led to improvements in their toughness, ceramics are increasingly used to replace metals in wear-resistant applications. The UK has been particularly prominent in the development of carbon fibre materials to produce low-mass, high-stiffness composites for demanding applications in the aerospace, sports and other industries. The rapidly increasing use of large flat-screen displays for televisions and computers has only been possible because of the immense progress made in the development of both the semiconductor materials required and the processes needed to manufacture them economically. New plastics with improved and specialised properties are taking over from traditional materials in an ever-increasing number of applications; in fact, the tonnage production of plastics now exceeds that of non-ferrous metals.

Techniques and processes   Metals are shaped and joined using a wide variety of techniques including moulding, rolling, pressing, stamping, drawing through a die and welding. Some of these metal fabrication techniques can also be used for other materials. For example, plastic sections can be produced by forcing the materials through a die in a process known as extrusion; some synthetic materials can be welded together by heat; and concrete can be cast into a mould. However, the nature of new materials often means that new processes have to be developed for their production and fabrication. These processes depend on a knowledge of the relationship between the structure and properties of materials, and require the skills and knowledge of the materials scientist and engineer to develop, monitor and control.

Beyond a degree   The skills of metallurgists, materials scientists and materials engineers are rather specialised, and only a few firms, such as those in the steel industry and large chemical companies, employ them in large numbers. However, a wide range of other organisations are involved in the development of new materials and the improvement of old ones, in materials production, in the analysis of the failure of materials, and in designing the equipment to make use of materials. Many of these companies recruit a few graduates each year, which means that there is a reasonable demand for the relatively small numbers of students graduating each year in the materials sciences and engineering.

Materials producers include steel and other metals companies, and glass, textiles, plastics, rubber and ceramics manufacturers and processors.

The electronics industry is an important employer. Most of the electronic components in common use are built on silicon, but there is much research into alternatives, such as gallium arsenide, which have other properties that allow the construction of more powerful and cost-effective devices.

Materials are extremely important in the energy industries. For example, the oil industry needs metallurgists and materials experts to investigate how to minimise corrosion on North Sea platforms, and to improve the insulation of ships designed to carry liquefied gases. The nuclear power and nuclear reprocessing industries also require specialists in materials because of the need to maintain the safety of vessels and pipes, often at high temperatures and pressures. The electricity and gas industries use a wide variety of materials, often in extreme conditions, for turbines, pipelines, conductors, insulators and a host of other applications, so they also recruit graduates from this area to avoid materials problems and to deal with any that do occur. Engineering companies in the automotive and aerospace industries, turbine manufacturers and defence contractors are also recruiters.

Polymer scientists work principally in chemical companies, researching, developing and manufacturing polymers, which can later be transformed into a huge variety of moulded products, fibres or sheet materials. Many textile technologists find work in the research, development and manufacturing arms of clothing and other textile retail organisations.

Some civil service laboratories employ materials specialists, and a small number are employed by organisations undertaking contract research for their clients.

Most graduates in materials science and engineering use their qualification directly in their jobs. However, you should remember that around 40% of all vacancies for graduates are open to people from any subject background, and for these graduates are recruited for their intellectual abilities, experience and personal skills, rather than their subject knowledge. Skills such as communication, teamwork, organisation, numeracy and problem-solving are valued by many employers. You will have opportunities to develop such skills on your course, and also through work experience and involvement in sports or student societies while at university.

For more information on career opportunities in materials, see the website of the Institute of Materials, Minerals and Mining at www.iom3.org.uk.

Further study   Quite a large proportion of graduates from materials science and engineering courses go on to further study. Some of them take Master's courses in order to specialise in a particular field of materials science or engineering, while a few choose vocational courses, such as IT conversion courses, to enhance their chances of entry into some other career area. The remainder study for a research degree, such as a PhD. A PhD is an advantage, and in many cases essential, if you are considering a career in research and development, whether in industry or academia. Further study can increase your chances of gaining employment and open up new opportunities, but a good class of degree is necessary, and funding can be difficult to obtain.

Chapter 2: The courses

TABLE 2a lists the specialised and combined courses at universities and colleges in the UK that lead to the award of an honours degree in materials engineering, materials science or metallurgy. When the table was compiled it was as up to date as possible, but sometimes new courses are announced and existing courses withdrawn, so before you finally fill in your application you should check the UCAS website, www.ucas.com, to make sure the courses you plan to apply for are still on offer. See Chapter 2 in the *Introduction* (page 6) for advice on how to use TABLE 2a and for an explanation of what the various columns mean.

Table 2a — First-degree courses in **Materials Engineering, Materials Science and Metallurgy**

Institution / Course title	①②③ combined	Degree	Duration	Foundation year	At this institution	At franchised inst.	Second-year entry	Modes of study (full/part time)	(time abroad)	(sandwich)	Modular scheme	Course type	No of combined courses
Birmingham www.bham.ac.uk													
Biomedical materials science ①		BMedSc	3		●			●				●	1
Materials science and engineering with business management		BEng/MEng	3, 4	●	●			●	○			●	0
Materials science and technology/materials engineering		BEng/MEng	3, 4	●	●			●	○	○		●	0
Mechanical and materials engineering ②		BEng/MEng	3, 4	●	●			●	○	○		◑	1
Metallurgy/materials engineering		BEng/MEng	3, 4	●	●			●	○			●	0
Sports science and materials technology ③		BSc	3		●			●				◑	1
Bolton www.bolton.ac.uk													
Textile technology		BSc	3	●	●			● ▬	○	◑		●	0
Cambridge www.cam.ac.uk													
Materials science and metallurgy		BA/MSci	3, 4		●			●			✿	●	0
Imperial College London www.imperial.ac.uk													
Aerospace materials		MEng	4							◑		●	0
Biomaterials and tissue engineering		MEng	4							◑		●	0
Materials science and engineering		BEng/MEng	3, 4		●			●	○	◑		●	0
Materials with management ①		BEng	3, 4		●			●	○			● ◑	1
Leeds www.leeds.ac.uk													
Biomaterials		BSc	3		●			●				●	0
Materials science and engineering		BEng/MEng	3, 4	●	●			●	○			●	0
Sports materials technology		BSc	3, 4		●			●	○	◑		●	0
Liverpool www.liv.ac.uk													
Aerospace materials		BSc	3	●	●			●				●	0
Biomaterials science and engineering		BEng	3		●			●				●	0
Materials engineering		MEng	4	●	●			●				●	0
Materials science ①		BEng	3		●	○		●				● ◑	1
Materials science and engineering		BEng	3		●	○		●				●	0
Materials, design and manufacture		BEng	3		●	○		●				●	0
Mechanical and materials science		BSc	3		●			●				●	0
London Metropolitan www.londonmet.ac.uk													
Polymer engineering		BEng	3, 4	●	●			● ▬		◑		●	0
Loughborough www.lboro.ac.uk													
Automotive materials		BEng/MEng	3, 4, 5	●	●			●	○	◑		●	0

First-degree courses in **Materials Engineering, Materials Science and Metallurgy**

Institution / Course title	① ② ③ see combined subject list – Table 2b	Degree	Duration (Number of years)	Foundation year: ● at this institution	○ at franchised institution	◐ second-year entry	Modes of study: ● full time; ➝ part time	○ time abroad	◐ sandwich	❖ Modular scheme	Course type ● specialised; ◐ combined	No of combined courses
Loughborough (continued)												
Materials engineering		BEng/MEng	3, 4, 5	●			●	○	◐		●	0
Materials with management studies ①		BEng	3, 4	●			●	○	◐		● ◐	1
Manchester www.man.ac.uk												
Biomedical materials science		BSc/MEng	3, 4	●			●	○	◐		●	0
Materials science and engineering		BSc/MEng	3, 4	●			●	○	◐	❖	●	0
Textile science and technology		BSc	3, 4	●			●	○	◐		●	0
Textile technology (business management)		BSc	3, 4	●			●	○	◐		●	0
Manchester Metropolitan www.mmu.ac.uk												
Plastics and rubber materials		BSc	3, 4	●	○		● ➝		◐		●	0
Textile technology for fashion		BSc	3, 4	●	○		●	○	◐		●	0
Napier www.napier.ac.uk												
Polymer engineering		BEng	4			◐			◐	❖	●	0
Newcastle www.ncl.ac.uk												
Materials and process engineering		BEng/MEng	3, 4	●			●	○			●	0
Mechanical and materials engineering ①		MEng	4	●			●	○			◐	1
Northampton www.northampton.ac.uk												
Materials technology (leather)		BSc	3	●			●				●	0
Nottingham www.nottingham.ac.uk												
Biomedical materials science		BSc	3	●			●				●	0
Oxford www.ox.ac.uk												
Materials science		MEng	4				●				●	0
Materials, economics and management		MEng	4				●				●	0
Plymouth www.plymouth.ac.uk												
Composite materials engineering		BEng	3, 4	●	○		● ➝	○	◐		●	0
Marine and composite technology		BSc	3, 4				●		◐		●	0
Queen Mary www.qmul.ac.uk												
Aerospace materials technology		BEng/MEng	3, 4	●			●	○			●	0
Biomaterials		BSc	3	●			●				●	0
Biomedical materials science and engineering		BEng	3	●			●	○	◐		●	0
Environmental materials technology		BEng/MEng	3, 4	●			●	○	◐		●	0
Materials and mechanical engineering ①		MEng	4	●			●	○	◐	❖	◐	1
Materials engineering in medicine		MEng	4	●			●	○	◐		●	0
Materials science ②		MEng	3, 4	●			●	○	◐		● ◐	3
Materials science and engineering		BSc/BEng	3, 4	●			●	○	◐		●	0
Polymer technology		BEng	3	●			●	○	◐		●	0
Sports materials		BEng/MEng	3, 4	●			●	○	◐		●	0
St Andrews www.st-and.ac.uk												
Materials science		MSci	4, 5				●				●	0
Sheffield www.sheffield.ac.uk												
Biomaterial science and tissue engineering		BEng/MEng	3, 4	● ●			●				●	0
Materials chemistry		BSc/MChem	3, 4	● ●			●	○			●	0
Materials science and engineering ①		BEng/MEng	3, 4	● ●			●				● ◐	2
Metallurgy ②		MEng	4	●			●				● ◐	1
Southampton www.soton.ac.uk												
Aerospace engineering/advanced materials ①		MEng	4				●				◐	1
Mechanical engineering/advanced materials ②		MEng	4				●				◐	1
Swansea www.swan.ac.uk												
Materials engineering with management ①		BSc	3				●				● ◐	1

(continued) Table 2a					Foundation year	Modes of study	Modular scheme	Course type	No of combined courses
Institution Course title	① ② ③ see combined subject list – Table 2b	Degree		Duration Number of years	● at this institution ○ of franchised institution ◑ second-year entry	● full time; ▼ part time ○ time abroad ◐ sandwich	◑ Modular scheme	● specialised; ◕ combined	No of combined courses
Swansea (continued)									
Materials science and engineering		BEng/MEng		3, 4	●	●　　○		●	0
Resistant materials design and technology		BEng		3		●		●	0

Subjects available in combination with materials engineering, materials science or metallurgy

TABLE 2b shows those subjects that can make up at least one-third of your degree programme when taken with materials engineering, materials science or metallurgy in the combined degrees listed in TABLE 2a. See Chapter 2 in the *Introduction* (page 6) for general information about combined courses and for an explanation of how to use TABLE 2b.

Table 2b	Subjects to combine with **Materials Engineering, Materials Science or Metallurgy**

Aerospace engineering　Southampton①
Business studies　Loughborough①, Queen Mary②
European languages　Sheffield②
Forensic science　Queen Mary②
Management studies　Imperial College London①, Liverpool①, Sheffield①, Swansea①

Mechanical engineering　Birmingham②, Newcastle①, Queen Mary①, Southampton②
Modern languages　Sheffield①
Physics　Queen Mary②
Sports science　Birmingham① ③

Other courses that may interest you

Queen Mary offers a course in dental materials, and at Aberdeen and Strathclyde, materials engineering can be studied as a minor along with mechanical engineering.

You may also find courses in the following parts of this Guide of interest:

- *Integrated Engineering and Mechatronics:* for courses covering materials engineering with other engineering subjects
- *Mechanical and Manufacturing Engineering:* for many courses that include a substantial component of work on materials.

Finally, you may also be interested in applied physics courses, which are listed in the *Physics and Chemistry* Guide.

Industrial experience and time abroad Many courses provide a range of opportunities for spending a period of industrial training in the UK or abroad, or of study abroad. TABLE 3a gives information about the possibilities for the courses in this Guide. See Chapter 3 in the *Introduction* (page 11) for further information.

Table 3a — Time abroad and sandwich courses

Location key: ● Europe; ○ North America; ➤ industry; ◑ academic institution
Language study: ○ optional; ● compulsory; * contributes to assessment
Sandwich courses: ● thick; ○ thin
Arranged by: ● institution; ○ student
① ② : see notes after table

Institution / Course title	Named 'international' variant of the course	Location	Maximum time abroad (months)	Time abroad assessed	Language study	Socrates–Erasmus	Sandwich courses	Arranged by
Birmingham								
Biomedical materials science ②								
Materials science and engineering with business management		● ○ ➤ ◑	12	●	○*	●		
Materials science and technology/materials engineering		● ○ ➤ ◑	12	●	○*	●		
Mechanical and materials engineering		● ○ ➤ ◑	12	●	○*	●		
Metallurgy/materials engineering		● ○ ➤ ◑	12	●	○*			
Bolton								
Textile technology ②		● ○ ➤	12	●	○*		●	● ○
Cambridge								
Materials science and metallurgy ③					○			
Imperial College London								
Aerospace materials ②		● ○ ➤ ◑	3	●	○	●	○	●
Biomaterials and tissue engineering ②		● ○ ➤ ◑	3	●	○*		○	●
Materials science and engineering ③	●	● ○ ➤ ◑	12	●	○*	●	● ○	●
Materials with management ②	●	● ○ ➤ ◑	12	●	○*			●
Leeds								
Materials science and engineering ③	●	◑	9	●	○*	●		
Liverpool								
Materials science	●							
London Metropolitan								
Polymer engineering ②	●	➤	12			●	● ○	● ○
Loughborough								
Automotive materials ① ③		● ○ ➤ ◑	12	●	○*	●	●	● ○
Materials engineering ① ③		● ○ ➤ ◑	12	●	○*	●	●	● ○
Materials with management studies ① ③		● ○ ➤ ◑	12	●	○*	●	●	● ○
Manchester								
Biomedical materials science ②		➤	12		○	●	●	● ○
Materials science and engineering ②	●	➤			○ ○	●	●	● ○
Textile science and technology ②	●	● ○ ➤ ◑	12		○*	●	●	● ○
Textile technology (business management) ②	●	● ○ ➤ ◑	12		○*	●	●	● ○
Manchester Metropolitan								
Plastics and rubber materials	●	➤	12		○	●	●	●
Textile technology for fashion		● ○ ➤ ◑	12	●			●	●
Napier								
Polymer engineering		● ○ ➤ ◑	12	●	○	●	●	● ○
Newcastle								
Materials and process engineering ②		● ○ ➤ ◑	12		○	●		

(continued) Table 3a

Time abroad and sandwich courses

Institution / Course title ① ② : see notes after table	Named 'international' variant of the course	Location: ● Europe; ○ North America; ◖ industry; ◗ academic institution	Maximum time abroad (months)	Time abroad assessed	Language study: ○ optional; ● compulsory; * contributes to assessment	Socrates–Erasmus	Sandwich courses: ● thick; ○ thin	Arranged by: ● institution; ○ student
Newcastle (continued)								
Mechanical and materials engineering ① ②		● ◖	9	●	○*			
Nottingham								
Biomedical materials science					○*	●		
Oxford								
Materials science		● ○ ◖ ◗	6	●			●	
Materials, economics and management		● ○ ◖ ◗	6	●			●	
Plymouth								
Composite materials engineering ②		● ○ ◖ ◗	12		○		● ●	● ○
Queen Mary								
Biomedical materials science and engineering ②		● ○ ◗	6		○*		● ●	● ○
Environmental materials technology ②		● ○ ◗	6		○*		● ●	● ○
Materials and mechanical engineering ② ③		● ○ ◖ ◗	12		○*		● ●	● ○
Materials engineering in medicine ② ③		● ○ ◖ ◗	12		○*		● ●	● ○
Materials science ② ③		● ○ ◖ ◗	12		○*		● ●	● ○
Materials science and engineering ②		● ○ ◗	6		○*		● ●	● ○
Polymer technology ②		● ○ ◖ ◗	12		○*		● ●	○
Sports materials ②		● ○ ◖ ◗	12		○*		● ●	○
St Andrews								
Materials science ②		● ○ ◖ ◗	12	●	○		●	●
Sheffield								
Materials chemistry ②		● ○ ◗		●		●		
Materials science and engineering ②		● ○ ◖ ◗	12	●	○*	●		
Metallurgy ②		● ○ ◖ ◗	12	●	○*	●		
Swansea								
Materials science and engineering ②	●	● ○ ◗	12	●	○*	●		

① A year of industrial experience before the course starts is recommended for students on non-sandwich courses

② Students on non-sandwich courses can take a year out for industrial experience

③ Minimum period of vacation industrial experience for non-sandwich students (weeks) : Cambridge <u>Materials science and metallurgy</u> 10 ; Loughborough <u>Automotive materials</u> 6 <u>Materials engineering</u> 6 <u>Materials with management studies</u> 6 ; Queen Mary <u>Materials and mechanical engineering</u> 12 <u>Materials engineering in medicine</u> 12 <u>Materials science</u> 12

Basic content All the courses in this Guide build on a solid foundation in the physical sciences, with a variable engineering component depending on the orientation of the course. TABLE 3b shows where courses have an orientation towards sciences or engineering, or, if there is an entry in both columns, where there is a roughly even balance between the science and engineering content. Many of the other engineering disciplines have similar foundations, so at several institutions engineering students from a wide range of courses share a common introductory period: see TABLE 3b for where this is the case and at what point specialisation occurs.

In the first year of most courses you will be given a grounding in mathematics, physics and chemistry. However, some of the applied courses, such as biomedical materials and sports materials, place rather less emphasis on basic science and mathematics, and are less demanding in this respect, but more demanding in other areas, such as the need to understand the application area.

The basic science is usually taught with specific reference to materials science, though in some courses you may not begin the study of materials proper until a later stage. For instance, you will probably concentrate more on physical chemistry and thermodynamics than on organic chemistry (unless it is a polymer course); your studies in physics are more likely to be on the properties of matter and the physics of the solid state rather than on special relativity; and the mathematics will be more concerned with topics like differential equations than with abstract algebra. The first-year work will usually include some work that is more specifically concerned with materials, such as investigations of their internal or microstructure, and how their strength can be measured and assessed.

Practical work   Practical work forms an essential part of all the courses. You will be instructed in the microscopical examination of materials, and carry out experiments on the physical, chemical and mechanical properties of solids, using modern laboratory and industrial equipment designed for the testing and manufacture of materials. For example, a second-year practical in forensic metallurgy might involve the investigation and analysis of a failed component, such as a mountain bike frame that has cracked around its circumference. You might begin by examining the fracture surface using a scanning electron microscope, and then go on to section the material and look at its microstructure using an optical microscope. You might then test the hardness of the material and determine its composition. The object of the experiment is to identify the failure mechanism (for example, fatigue) and determine the cause (for example, a crack starting at a weld that has grown by repeated loading, and has led to premature failure due to an overload). In the process of carrying out the practical, you will have learned about a number of materials science techniques that can be applied in many different contexts, and placed the course's theoretical work in a practical context.

At nearly all institutions you will carry out group projects, case studies and a major individual project in your final year, and in some of the MEng courses you will also be involved in a multidisciplinary project with other engineering students.

Advanced and specialised content   The later years of the courses build on the earlier physical science foundations with specific work on: the structure and properties of materials; methods of production and fabrication; the detection, analysis and prevention of defects and failures; and the estimation of reliability and life prediction. Most courses, particularly those with an engineering orientation, include related studies such as management, statistics and economics, and most departments have extensive integration of computing techniques into their courses. In particular, courses with an engineering and design orientation will also usually have facilities for computer-aided design (CAD).

Metallurgy   Departments generally offer broadly based courses covering the chemical, physical and industrial aspects of metallurgy. A large majority of the courses also cover other materials such as polymers, ceramics, glasses and composites to a greater or lesser extent. Many institutions allow some degree of specialisation in particular branches of metallurgy or related studies by the selection of options or electives in the final year. In addition to specialisation in chemical, physical and industrial (or process) metallurgy, the options available in certain courses include more specific topics such as: surface metallurgy; extractive metallurgy and process engineering; management or economics; corrosion science and materials conservation. The possible options are wide-ranging and constantly changing, so if you are particularly interested in a specific area you should check prospectuses and contact departments direct: see Chapter 7 for addresses.

Materials science and engineering   Most general materials courses cover the science and technology of a wide variety of materials, which may include metals, ceramics, glasses, synthetic polymers, textiles and wood. Different courses give different emphasis to physical, chemical and mechanical properties, to manufacture and to engineering applications. The structure of these courses is similar to that of metallurgy courses, and in departments that offer both types of course, there is often a common first year. In any case, throughout most courses there is an emphasis on the unifying principles of the behaviour of materials. Characterisation techniques include advanced transmission electron microscopy, scanning electron microscopy, X-rays, defect analysis and infrared spectroscopy.

Courses in specific materials cover topics analogous to those in metallurgy and general materials courses. Naturally, the courses base their examples on the specific materials and are able to go into more detail on their particular properties and uses.

Specialisation   If you are interested in a particular group of materials, you have the choice of taking a specialised course concentrating on that group or taking a more general materials science or engineering course that includes those materials along with a range of others. Both approaches have their advantages and disadvantages. The specialised course allows you to concentrate on your main area of interest from the start, while the broader course gives you a wider perspective. Even if you take a more general course, you may be able to bias it towards a particular group of materials as many of them offer a wide choice of specialist options in later years. TABLE 3b shows where specialised branches of the subject are available as compulsory (●) or optional (○) components, and where there are compulsory components with further optional components (◓). It also shows if a particular branch can form a major specialisation occupying at least 50% of your time in the final year (●* ○* ◓*).

Final-year course content

Legend: ① ②: see notes after table · ○ optional; ● compulsory; · ◑ compulsory + options · * major area of specialisation

Institution	Course title	Engineering orientation	Science orientation	Time of specialisation after a broad introduction	Metals	Polymers	Ceramics	Composites	Powder technology	Functional materials	Materials finishing	Materials processing	Selection/design of materials	Characterisation techniques	Automotive materials	Aerospace materials	Biomedical materials	Sports materials	Management
Birmingham	Biomedical materials science ①		●		○	○	○	○			○	●					○	○	○	
	Materials science and engineering with business management	●	●		●*	●*	●*	●*		○*	○*	◑*	●*	●			○	○	○	
	Materials science and technology/materials engineering ②	●	●		●*	●*	●*	●*	○	○*	○	◑*	●	●			○	○	○	●
	Mechanical and materials engineering ③	●	●				●					○	●				○	○	○	●
	Metallurgy/materials engineering ④	●	●		●*	●*	●*	●*	○	○*	○	◑*	●	●			○	○	○	●
	Sports science and materials technology		●																	
Bolton	Textile technology ①	●				●			○	○	○	●	●	●	●		○	○	○	●
Cambridge	Materials science and metallurgy		●	Year 2	●	●	●	●	●	●	●	●	●	●					
Imperial College London	Aerospace materials	●	●		●	●	○	●	○	○	○	○	●	○	●		●	○		
	Biomaterials and tissue engineering	●	●		○	○	○	○	○	○	○	○	○	○	○			●		●
	Materials science and engineering	●	●		○	○	○	○	○	○	○	○	○	○	○			○		
	Materials with management ①	●	●		○	○	○	○	○	○	○	○	○	○	○		○			
Leeds	Materials science and engineering	●	●		●	●	●	○	●	●	●	●	●	●	●				
Liverpool	Aerospace materials	●		Year 2	◑	◑	◑	◑		◑		◑					●			
	Biomaterials science and engineering		●		◑	◑	◑	◑		◑		◑						●		
	Materials engineering	●	●	Year 2	○	○	●	○		○		○							
	Materials science	●	●		◑	◑	◑	◑		◑		◑							
	Materials science and engineering	●	●	Year 2	◑	◑	◑	◑		◑		◑							
	Materials, design and manufacture	●		Year 2	◑	◑	◑	◑		◑		◑							
London Metropolitan	Polymer engineering ①	●			●	●		●	○	○	●	●	●	●					○
Loughborough	Automotive materials ①	●	●							●		●		●		●			●
	Materials engineering	●	●	Year 2	○	○	○	○	○	○	○	○	○	○					
	Materials with management studies	●	●	Year 2	●*	●*	●	●	○	○	●	●							
Manchester	Biomedical materials science ①	●	●		○	○	○	○		○		○	○	○				●		
	Materials science and engineering	●	●		●	●	●	●	●	●	●	●	●						
	Textile science and technology	●	●			●				●	●	●	●		●		○	○	○	○
	Textile technology (business management)	●	●			●				●	●	●	●		●		○	○	○	●
Manchester Metropolitan	Plastics and rubber materials		●		○	●	○	●				●		●				○		
	Textile technology for fashion		●			●					●	●	●	●				○	○	
Napier	Polymer engineering ①	●		Year 2		●		●			●	●							
Newcastle	Materials and process engineering	●																	
	Mechanical and materials engineering ①	●		Year 3	●	●	●	●		●		●	●		●				●
Northampton	Materials technology (leather) ①	●	●			●					●	●							
Nottingham	Biomedical materials science ①		●		◑	◑	◑	◑	◑			◑		◑				◑		
Oxford	Materials, economics and management		●		○	○	○	○	○	○		○							
Plymouth	Composite materials engineering ①	●		Stage 3		●		●		●	●	●					○		○	●
Queen Mary	Aerospace materials technology						○			○							○			
	Biomaterials						○													

Materials Engineering, Materials Science and Metallurgy

(continued) Table 3b — Final-year course content

① ② : see notes after table
○ optional; ● compulsory; ◐ compulsory + options; ◑ major area of specialisation *

Institution	Course title	Engineering orientation	Science orientation	Time of specialisation after a broad introduction	Metals	Polymers	Ceramics	Composites	Powder technology	Functional materials	Materials finishing	Materials processing	Selection/design of materials	Characterisation techniques	Automotive materials	Aerospace materials	Biomedical materials	Sports materials	Management
Queen Mary (continued)	Biomedical materials science and engineering	●	●		●	○	○	○		○	●	●	●	●			●		●
	Environmental materials technology							○		○									
	Materials and mechanical engineering	●	●		●	○	○	○		○	●	●	●	●		○			●
	Materials engineering in medicine	●	●		●	○	○	○		○	●	●	●	●			●		
	Materials science	●	●			○	○	○		○		●	●	●					
	Materials science and engineering	●	●		●	○	○	○		○	●	●	●	●		○			●
	Polymer technology				○	○	○	○		○		○	○	○		○			○
	Sports materials	●	●		○	○	○	○		○		○	○	○			○	○	
St Andrews	Materials science		●	Year 2	●	●			●	●		●	●	●					
Sheffield	Biomaterial science and tissue engineering	●	●		○	○	○	○		○							○		
	Materials chemistry		●		○	○	○	○		○	○	◑	◑	◑		○	○		
	Materials science and engineering	●	◑		◑*	◑*	◑*	◑		○	○	○	○	◑	◑	◑	○	○	○
	Metallurgy	●	●		◑*	○	○	◑		○	○	○	○	◑	◑	◑	○	○	○
Swansea	Materials science and engineering ①	●	●		◑*	◑	○	◑		○	◑			◑					

Birmingham ① 50% of course is human biology (cell biology, genetics, embryology, biochemistry) and clinical background (anatomy, histology, physiology) ② Biomaterials and materials characterisation towards materials applications for mechanical engineering ④ Biomaterials and materials characterisation
Bolton ① Textile design/management/marketing
Imperial College London ① Managerial economics; project management; marketing
London Metropolitan ① Compulsory structural composites; multiphase polymer systems
Loughborough ① Automotive engineering; vehicle design; engine design; materials recycling

Manchester ① 3 modules of biological and biomedical subjects each year
Napier ① Computer simulation
Newcastle ① Joining technology; new energy and materials technology; advanced materials and processes
Northampton ① Leather
Nottingham ① Biomedical and pharmaceutical science
Plymouth ① Composites design and manufacture; computational methods
Swansea ① Computer modelling and simulation

General studies and additional subjects

Some institutions include topics on liberal, complementary or general studies in their specialised courses. TABLE 3a contains information about language teaching; you will need to look at prospectuses or ask institutions direct for information about other subjects.

Where a course is part of a modular degree scheme, there will probably be more opportunities to take a wide range of other subjects, though the requirements for professional qualification may mean that the choice is more restricted than it would be if you were following a course in a non-engineering subject. TABLE 2a shows where a course is part of a modular degree scheme allowing a relatively free choice of modules from a large number of subjects.

BEng/MEng courses TABLE 2a lists many courses that can lead to the award of either a BEng or an MEng degree. For these courses, most of the tables in this Guide give information specifically for the BEng stream (the tables show information for MEng courses that are MEng only). Much of the information will also apply to the MEng streams, but TABLE 3c shows you where there are differences for the MEng stream. It shows at what point the MEng course separates from the BEng (for those courses where they are not separate from the start), and what proportion of students are expected to leave with an MEng degree.

Table 3c — MEng course differences

Institution	Course title	MEng separates from BEng	Students receiving MEng %	More engineering	More management	More languages	Other differences
Birmingham	Materials science and engineering with business management	Year 3	25	●	●		
	Materials science and technology/materials engineering	Year 3	25	●	●		6-month project in industry
	Mechanical and materials engineering	Year 3	25	●	●		6-month project in industry
	Metallurgy/materials engineering	Year 3	25	●	●		6-month project in industry
Imperial College London	Materials science and engineering	Year 3	50	●	●		Final-year industrial placement/project; group design project
Leeds	Materials science and engineering	Year 3	10	●	●	●	More industrial focus; greater depth of materials teaching
Loughborough	Automotive materials	Year 3	10	●			Group design project
	Materials engineering	Year 3	10	●			Group design project
Manchester	Biomedical materials science	Year 3	50				
	Materials science and engineering	Year 3	50				
Sheffield	Biomaterial science and tissue engineering	Year 3	65		●		Group project work; more substantial final-year project
	Materials chemistry	Year 3	75	●			
	Materials science and engineering	Year 3	25		●		5-month industrial placement; final-year specialisation opportunities
Swansea	Materials science and engineering	Year 3	10	●	●		

Chapter 4: Teaching and assessment methods

See Chapter 4 in the *Introduction* (page 16) for general information about teaching and assessment methods used in all types of engineering course, including those covered in this part of the Guide. It also explains how to interpret TABLE 4, which gives information about projects and assessment methods used on individual courses.

Table 4 — Assessment methods

Key for frequency of assessment column: ● term; ◑ semester; ○ year
Group projects: ● compulsory; ○ optional
Orals: ◑ if borderline; ● for projects; ○ everyone

Institution	Course title	Frequency of assessment	Years of exams contributing to final degree (years of exams not contributing to final degree)	Coursework: minimum/maximum %	Project/dissertation: minimum/maximum %	Time spent on projects in: first/intermediate/final years %			Group projects	Orals
Birmingham	Biomedical materials science	◑	(1),**2,3**	10/**15**	20/**30**	10	15	40	●	◑
	Materials science and engineering with business management	○	(1),**2,3**	15/**20**	20/**20**	5	15	40	●	○
	Materials science and technology/materials engineering	○	(1),**2,3,4**	15/**20**	20/**20**	5	15	40	●	○
	Mechanical and materials engineering	○	(1),**2,3,4**	15/**20**	20/**20**	5	15	40	●	○
	Metallurgy/materials engineering	○	(1),**2,3,4**	15/**20**	20/**20**	5	15	40	●	○
	Sports science and materials technology	○	(1),**2,3**	20/**20**	11/**11**	10	10	35	●	○
Bolton	Textile technology	◑	(1),**2,3**	30/**50**	25/**25**	10	15	33	●	◑○
Cambridge	Materials science and metallurgy	○	(1),(2),**3**	10	10	0	5	15	●	○
Imperial College London	Aerospace materials	◑●	**1,2,3,4**	20/**20**	30/**30**	3	10	60		◑○
	Biomaterials and tissue engineering	◑●	**1,2,3,4**	20/**20**	20/**30**	3	10	50		◑○
	Materials science and engineering	◑●	**1,2,3,4**	20/**20**	20/**30**	3	10	50		◑○
	Materials with management	◑●	**1,2,3,4**	20/**20**	20/**20**	3	10	40		◑○
Leeds	Materials science and engineering	◑	(1),**2,3**	13/**13**	22/**22**	0	8	40	●	◑○
Liverpool	Aerospace materials	◑	(1),**2,3**	25/**40**	10/**20**	0	0	25		○
	Biomaterials science and engineering	◑	(1),**2,3**	25/**40**	10/**20**	0	0	25		○
	Materials engineering	◑	(1),**2,3,4**	25/**35**	10/**20**	0	25	50		○
	Materials science	◑	(1),**2,3**	25/**40**	10/**20**	0	0	25		○
	Materials science and engineering	◑	(1),**2,3**	25/**40**	10/**20**	0	0	25		○
	Materials, design and manufacture	◑	(1),**2,3**	25/**40**	10/**20**	0	0	25		○
London Metropolitan	Polymer engineering	◑	(1),**2,3**	20/**40**	10/**20**			25		
Loughborough	Automotive materials	◑	(1),**2,3,4**	20/**40**	21/**23**	0	0	25	●	◑○
	Materials engineering	◑	(1),**2,3,4**	50/**55**	15/**15**	25	25	50	●	◑○
	Materials with management studies	◑	(1),**2,3,4**	20/**20**	20/**20**	15	15	25	○	◑○
Manchester	Biomedical materials science	◑	(1),**2,3**	8	16	5	10	20	●	◑
	Materials science and engineering	◑	(1),**2,3**	29	12	5	10	20	●	◑
	Textile science and technology	◑	(1),**2,3,4**	11/**11**	17/**17**	30	50	30	●	◑
	Textile technology (business management)	◑	(1),**2,3,4**	11/**11**	17/**17**	30	50	30	●	◑
Manchester Metropolitan	Plastics and rubber materials	○	(1),**2,3,4**	20/**25**	12/**20**			50		◑
	Textile technology for fashion	◑●	(1),**2,4**	60/**60**	20/**20**	20	20	20		
Napier	Polymer engineering	◑	(1),(2),(3),**4,5**	30/**30**	25/**25**	10	15	25	●	○●
Newcastle	Materials and process engineering	◑	(1),**2,3**	25			10	30	●	
	Mechanical and materials engineering	◑	(1),**2,3,4**	7/**7**	24/**24**	5	25	33	●	○
Northampton	Materials technology (leather)	○	(1),**2,3**	25	20	0	0	16		○

Institution	Course title	Frequency of assessment	Years of exams contributing to final degree (years of exams not contributing to final degree)	Coursework: minimum/maximum %	Project/dissertation: minimum/maximum %	Time spent on projects in: first/intermediate/final years %			Group projects: ● compulsory; ○ optional	Orals: ◑ if borderline; ○ for projects; ● everyone
Nottingham	Biomedical materials science	◑	(1),**2,3**	20/**30**	20/**30**	5	0	25	○	◑○
Oxford	Materials science		(1),**3,4**	10	30	0	10	100	●	◑○●
	Materials, economics and management	○	(1),**3,4**	10	15	0	5	50		◑○●
Plymouth	Composite materials engineering	◑	(1),**2,3**	**60**	**40**	40	40	50	●	◑○
Queen Mary	Biomedical materials science and engineering	○	**1,2,3**		25/**25**	25	25	25	●	○
	Environmental materials technology	○	**1,2,3,4**		25/**25**	25	25	25	●	○
	Materials and mechanical engineering	○	**1,2,3,4**		25/**25**	25	25	25	●	○
	Materials engineering in medicine	○	**1,2,3,4**		25/**25**	25	25	25	●	
	Materials science	○	**1,2,3,4**		25/**25**	25	25	25	●	
	Materials science and engineering	○	**1,2,3,4**		25/**25**	25	25	25	●	
	Polymer technology	○	**1,2,3**		25/**25**	25	25	25		○
	Sports materials	○	**1,2,3,4**		25/**25**	25	25	25		○
St Andrews	Materials science	◑	(1),(2),**3,4,5**	**83**	**17**	5	10	33	●	○
Sheffield	Biomaterial science and tissue engineering	◑	(1),**2,3,4**	20/**20**	25/**30**	5	15	40	●	◑○
	Materials chemistry		(1),**2,3,4**	10/**20**	25/**25**	5	15	25	●	◑○
	Materials science and engineering	◑	(1),**2,3,4**	12/**20**	25/**25**	0	20	30	●	◑○
	Metallurgy	◑	(1),**2,3,4**	12/**20**	25/**25**	0	20	30	●	◑○
Swansea	Materials science and engineering	◑	(1),**2,3,4**	30/**35**	20/**20**		10	30		◑○

Chapter 5: Entrance requirements

See Chapter 5 in the *Introduction* (page 18) for general information about entrance requirements that applies to all types of engineering course, including those covered in this part of the Guide. It also explains how to interpret TABLE 5, which gives information about entrance requirements for individual courses.

Table 5 — Entrance requirements

Institution	Course title	Number of students (includes other courses)	Typical offers UCAS points	A-level grades	SCQF Highers grades	● compulsory; ○ preferred A-level Mathematics	A-level Chemistry	A-level Physics
Birmingham	Biomedical materials science	25		BBC	BBBBC		○	
	Materials science and engineering with business management	10		BBB	ABBBB	○	○	
	Materials science and technology/materials engineering	30		BBB	ABBBB	○	○	
	Mechanical and materials engineering	10		BBB	ABBBB	●	○	
	Metallurgy/materials engineering	30		BBB	ABBBB	○	○	
	Sports science and materials technology	30		BBB	ABBBB			
Bolton	Textile technology	30	200			○	○	
Cambridge	Materials science and metallurgy	(600)		AAA		●	○	
Imperial College London	Aerospace materials	15				●		
	Biomaterials and tissue engineering	10				●	○	
	Materials science and engineering	60		BBC		●	○	
	Materials with management	5		BBC		●	○	
Leeds	Biomaterials			BBC				
	Materials science and engineering	(40)		BBC		○	○	
	Sports materials technology			BBC				
Liverpool	Aerospace materials	8	280		AAAB	○	○	
	Biomaterials science and engineering	8	280		AAAB	●	○	
	Materials engineering	8				●	○	○
	Materials science	10	200	CDD	CCCC	○	○	
	Materials science and engineering	10	280	BBC	AAAB	●	○	
	Materials, design and manufacture	6	280	BBC	AAAB	●	○	○
	Mechanical and materials science		280		AAAB			
London Metropolitan	Polymer engineering	20	120			○	○	○
Loughborough	Automotive materials	20	240			○	○	
	Materials engineering	24	240			○	○	
	Materials with management studies	10	240			○	○	○
Manchester	Biomedical materials science	25		ABC	BBBBB		○	
	Materials science and engineering	(50)		ABC	BBBBB	○	○	
	Textile science and technology	12	260	BBB	BBBBB	●	○	○
	Textile technology (business management)	15	260	BBB	BBBBB	●	○	○
Manchester Metropolitan	Plastics and rubber materials	20	160–220				●	
	Textile technology for fashion	15	160			○	○	
Napier	Polymer engineering	(30)	220	CD	BCC	●	○	○

(continued) Table 5 Entrance requirements

Institution	Course title	Number of students (includes other courses)	Typical offers			A-level Mathematics	A-level Chemistry	A-level Physics
			UCAS points	A-level grades	SCQF Highers grades	● compulsory; ○ preferred		
Newcastle	Materials and process engineering	35		BBB		●	●	
	Mechanical and materials engineering	(10)				●	○	
Northampton	Materials technology (leather)	15	180–220				○	
Nottingham	Biomedical materials science	15		BBB	BBBBB	○	●	
Oxford	Materials science	20				○	○	○
	Materials, economics and management	5				○	○	○
Plymouth	Composite materials engineering	40	240			●	○	○
	Marine and composite technology		160			●		
Queen Mary	Aerospace materials technology	10	260		BBBCC	○		
	Biomaterials	25	240		BBBCC	○		
	Biomedical materials science and engineering	10	260			○		
	Environmental materials technology	10	260			○		
	Materials and mechanical engineering	5				○		
	Materials engineering in medicine	10				○		
	Materials science	5				○	○	
	Materials science and engineering	5	260			○		
	Polymer technology	5	260			○	○	
	Sports materials	5	260			○	○	
St Andrews	Materials science	3				●	●	
Sheffield	Biomaterial science and tissue engineering	18		BBC	ABBB	○	○	
	Materials chemistry	5		BBC	AAAA	○	●	
	Materials science and engineering	(50)		BBC	ABBB	○	○	
	Metallurgy	(50)				○	○	
Southampton	Aerospace engineering/advanced materials	(85)						
	Mechanical engineering/advanced materials	(60)						
Swansea	Materials science and engineering	20	240			○	○	○
	Resistant materials design and technology		240					

Chapter 6: Routes to professional qualification

See Chapter 6 in the *Introduction* (page 20) for general information about professional qualification that applies to all types of engineering course, including those covered in this part of the Guide.

TABLE 6 shows courses that the university or college has confirmed have been accredited by the Institute of Materials, Minerals and Mining (IMMM; the main professional organisation for all materials scientists and engineers), the Textile Institute, the Institution of Mechanical Engineers (IMechE), the Institution of Chemical Engineers (IChemE) or the Institution of Engineering and Technology (IET).

Graduates with a non-accredited or ordinary degree in metallurgy or materials studies, or with a degree in physics, chemistry or another engineering discipline, may also qualify for corporate membership provided they demonstrate that they have acquired a knowledge of metallurgy or materials studies equivalent to that of an honours degree by suitable additional academic study, training and experience.

Some courses are accredited by several institutions, but the accreditation by an individual institution may be conditional on your having taken certain options, which may not coincide with those required by another institution.

Remember that the process of accreditation goes on continuously, so you should check with the professional institutions for the current position on specific courses. Full particulars about acquiring professional qualifications may be obtained from the relevant professional bodies: Chapter 7 of the *Introduction* to this Guide (page 24–26) lists their addresses.

Table 6 Accreditation by professional institutions

Institution	Course title	IMMM	Textile Institute	IMechE	Other professional bodies
Birmingham	Materials science and engineering with business management	●			
	Materials science and technology/materials engineering	●			
	Mechanical and materials engineering	●		●	
	Metallurgy/materials engineering	●			
Bolton	Textile technology		●		
Cambridge	Materials science and metallurgy	●			
Imperial College London	*All courses*	●			
Leeds	Materials science and engineering	●			
Liverpool	Biomaterials science and engineering	●			
	Materials engineering	●			
	Materials science and engineering	●			
	Materials, design and manufacture	●			

Accreditation by professional institutions

Institution	Course title	IMMM	Textile Institute	IMechE	Other professional bodies
Loughborough	Automotive materials	●			
	Materials engineering	●		●	
	Materials with management studies	●			
Manchester	Biomedical materials science	●			
	Materials science and engineering	●			
	Textile science and technology		●		
	Textile technology (business management)		●		
Manchester Metropolitan	Textile technology for fashion		●		
Newcastle	Materials and process engineering				IChemE
	Mechanical and materials engineering	●		●	IET
Oxford	Materials, economics and management	●			
Plymouth	Composite materials engineering	●		●	
Queen Mary	Biomedical materials science and engineering	●			
	Materials and mechanical engineering	●			
	Materials engineering in medicine	●			
	Materials science	●			
	Materials science and engineering	●			
	Polymer technology	●			
	Sports materials	●			
Sheffield	Materials science and engineering	●			
	Metallurgy	●			
Strathclyde	Mechanical engineering with materials engineering	●			
Swansea	Materials science and engineering	●			

See Chapter 7 in the *Introduction* (page 24) for a list of sources of information that apply to all types of engineering course, including those covered in this part of the Guide. It also lists a number of books that can give some general background to engineering and the work of engineers. Visit the websites of the professional institutions for information on careers, qualifications and course accreditation, as well as information about publications, many of which you can download. Chapter 7 in each of the other parts of the Guide contains further suggestions for background reading in the individual engineering disciplines.

The Armourers and Brasiers' Company sponsors placements and scholarships for students thinking of studying materials science or engineering. These are available in the final year of A-level study, or equivalent, and then at university. See the website www.armourersandbrasiers.co.uk for further information.

The following books give some background information specifically for materials science/engineering and metallurgy.

Tomorrow's Materials K Easterling. IOM Communications, 1990 (2nd edition), £19.00

Metals in the Service of Man W Alexander & A Street. Penguin, 1998 (11th edition)

Physics of Materials B Cooke & D Sang. Nelson Thornes, 1996, £15.50

The New Science of Strong Materials – Or Why You Don't Fall Through the Floor J E Gordon. Penguin, 1991 (2nd edition), £10.99

The Structure of Matter – From the Blue Sky to Liquid Crystals A Guinier. Butterworth Heinemann, 1984

The courses This Guide gives you information to help you narrow down your choice of courses. Your next step is to find out more about the courses that particularly interest you. Prospectuses cover many of the aspects you are most likely to want to know about, but some departments produce their own publications giving more specific details of their courses. University and college websites are shown in TABLE 2a.

You can also write to the contacts listed below.

Birmingham Biomedical materials science Dr J Barralet (j.e.barralet@bham.ac.uk), Biomedical Materials Science Admissions Tutor, School of Dentistry, St Chad's Queensway, Birmingham B4 6NN; All other courses Dr Mike Jenkins (met-admissions@bham.ac.uk), School of Metallurgy and Materials, University of Birmingham, Edgbaston, Birmingham B15 2TT

Bolton Admissions Tutor (bw2@bolton.ac.uk), Engineering and Design, Bolton University, Deane Road, Bolton BL3 5AB

Cambridge Cambridge Admissions Office (admissions@cam.ac.uk), University of Cambridge, Fitzwilliam House, 32 Trumpington Street, Cambridge CB2 1QY

Imperial College London Dr S Skinner (s.skinner@imperial.ac.uk), Department of Materials, Imperial College London, South Kensington Campus, London SW7 2AZ

Materials Engineering, Materials Science and Metallurgy

Leeds Admissions Secretary (spemeadmissions@leeds.ac.uk), Department of Materials, School of Process, Environmental and Materials Engineering, University of Leeds, Leeds LS2 9JT

Liverpool Admissions Tutor (ugeng@liv.ac.uk), Department of Engineering, University of Liverpool, Brownlow Hill, Liverpool L69 3GH

London Metropolitan Admissions Office (admissions@londonmet.ac.uk), London Metropolitan University, 166–220 Holloway Road, London N7 8DB

Loughborough Dr R L Higginson (r.l.higginson@lboro.ac.uk), IPTME, Loughborough University, Loughborough LE11 3TU

Manchester Biomedical materials science Materials science and engineering Dr S J Eichhorn, Materials Science Centre, University of Manchester, Grosvenor Street, Manchester M1 7HS; Textile science and technology Textile technology (business management) Jacquie Wilson (j.a.wilson@manchester.ac.uk), Undergraduate Admissions Tutor, Textiles and Paper, School of Materials, University of Manchester, Sackville Street, Manchester M14 5RR

Manchester Metropolitan Plastics and rubber materials Dr Nicola Hughes (se.combined@mmu.ac.uk), Admissions Tutor, Combined Studies Scheme, Faculty of Science and Engineering, Manchester Metropolitan University, Chester Street, Manchester M1 5GD; Textile technology for fashion Dr Steve Hayes (se.combined@mmu.ac.uk), Programme Leader, Manchester Metropolitan University, Hollings Campus, Old Hall Lane, Manchester M14 6HR

Napier Colin Hindle, School of Engineering, Napier University, 10 Colinton Road, Edinburgh EH10 5DT

Newcastle Admissions Office (enquiries@ncl.ac.uk), University of Newcastle upon Tyne, 6 Kensington Terrace, Newcastle upon Tyne NE1 7RU

Northampton Admissions Office (admissions@northampton.ac.uk), University of Northampton, Park Campus, Broughton Green Road, Northampton NN2 7AL

Nottingham Dr G S Walker (gavin.walker@nottingham.ac.uk), School of Mechanical, Materials & Manufacturing Engineering, University of Nottingham, University Park, Nottingham NG7 2RD

Oxford Dr C R M Grovenor, Department of Materials, Oxford University, Oxford OX1 3PH

Plymouth Faculty of Technology, University of Plymouth, Drake Circus, Plymouth PL4 8AA

Queen Mary Department of Materials (materials@qmul.ac.uk), Queen Mary University of London, Mile End Road, London E1 4NS

St Andrews Admissions Applications Centre (admissions@st-andrews.ac.uk), University of St Andrews, St Andrews KY16 9AX

Sheffield Materials chemistry Dr N Hyatt (n.hyatt@sheffield.ac.uk); All other courses Dr Emma Wager (e.c.wager@sheffield.ac.uk); both at Department of Engineering Materials, University of Sheffield, Mappin Street, Sheffield S1 3JD

Southampton Admissions Office (admissions@soton.ac.uk), University of Southampton, Highfield, Southampton SO17 1BJ

Swansea Dr S G R Brown, Department of Materials Engineering, University of Wales Swansea, Singleton Park, Swansea SA2 8PP

See the *Introduction* to this Guide (page 3) for an overview of how to use each chapter in the individual subject parts, including this one, together with general information about engineering and the structure of engineering courses.

The scope of this part of the Guide   This part of the Guide covers first-degree courses in the area of mechanical and manufacturing engineering, and product design. This includes courses with a broad coverage and courses in specialised areas such as aeronautical, agricultural, automotive and marine engineering. There are more than 120 different titles listed in TABLE 2a, which reflects the range of different emphases that courses can be given. Given this variety, it is not possible to make very precise statements about the courses in general. However, it is possible to group the courses into a number of categories, and the following sections should give you a reasonable feel for what you can expect from the courses in each of the categories.

Mechanical engineering   Mechanical engineering is the most general of the engineering disciplines. Mechanical engineers are involved in almost anything that is manufactured – things as small as a screw and as large as a diesel locomotive. They are needed in a wide range of industries from motor vehicles to food processing, from pharmaceuticals to the manufacture of hi-fi components. You may learn about different types of handling equipment and computer control devices, and major products such as steam or gas turbines, nuclear reactors and supersonic aircraft: indeed, almost any application involving moving or load-bearing parts. Mechanical engineers learn to understand and apply the science that underpins the design of these products, and how they are mass-produced or manufactured as one-off custom items. The professional mechanical engineer can be involved in many different roles, including research, design, development, manufacture and marketing.

Manufacturing engineering and product design   Manufacturing engineering and product design cover different but related aspects of the total process of producing manufactured goods, so they have a major affinity with each other.

Manufacturing engineering   In recent years a greater emphasis has been placed on automation in manufacturing, and there are now many courses offered in manufacturing engineering, manufacturing systems engineering and computer-aided engineering. Many of these fields have their origins in mechanical engineering.

These courses are concerned with all stages and aspects of the manufacture of goods. They embrace computer-aided design, robotics, the control of machine tools, materials handling and assembly systems. They also deal with the human aspects of manufacturing processes, so they include relevant parts of economics, management, marketing and sociology.

There is considerable overlap between manufacturing engineering and manufacturing systems engineering courses, but the latter give greater emphasis to the integrated nature of manufacturing technology, and are designed more to give a grasp of underlying engineering principles than to teach specific technical skills in a particular area.

Product design engineering   Product design engineering is the name given to the process by which new or improved products reach the user. There are two main strands in design: engineering design, which concentrates on the function and dimensions of the product, and industrial design, which is concerned with aesthetics and the emotional appeal of the product. Historically, these were often treated as separate disciplines in the UK. However, courses in product design bring together the engineering and human aspects, and are concerned with the total process from the market research that is carried out to determine the need for a new or improved product, through integrated functional and aesthetic design and manufacture, to the marketing of a reliable high-quality and saleable product.

Aeronautical engineering   Aeronautical engineering is concerned primarily with the problems of flight. It developed as a branch of mechanical engineering but is now a highly specialised branch of engineering dealing with aircraft, spacecraft and missiles, together with their power plants, turbines and rocket motors. As the vehicles become more complex, much detailed design work in aerospace industries is conducted by electrical, mechanical and systems engineers, but it is the aeronautical engineer who usually takes overall responsibility for a project.

This Guide also includes courses in avionics, which is concerned with the instrumentation used in aircraft for navigation, control and communications.

Automotive engineering   Automotive engineering covers all aspects of the design, manufacture and testing of a wide range of motor vehicles, including cars, vans, trucks, race cars and off-road vehicles, and their components. Automotive engineering is concerned with aerodynamics, vehicle suspension and dynamics, engine combustion, powertrain engineering, and, increasingly, electronic control systems. Automotive engineers also need an understanding of manufacturing processes and organisation. Despite the fact that there are no longer any wholly UK-owned volume car manufacturers, the UK still has a thriving automotive industry, including foreign-owned volume car manufacturers and many specialist manufacturers. UK companies also play a major role in the design and construction of the race cars used in motorsport throughout the world. All this means that there is still a growing demand for highly qualified graduate automobile engineers.

Agricultural engineering   Agricultural engineering is firmly based on mechanical engineering. Farming is a highly mechanised industry and there is a continuing need for the design, development and maintenance of specialist equipment. Agricultural engineers are also employed in field engineering, land reclamation, drainage and irrigation, land use planning and management, crop storage and processing, and the selection of systems of mechanisation for specific crops. The specialist agricultural engineer is an essential member of one of Britain's most important industries. There is also a need for people with a good knowledge of agricultural engineering coupled with economics and marketing, and possessing the necessary management skills to deal with the many aspects of this rapidly changing industry.

Marine engineering   Marine engineering is concerned with the provision, management and maintenance of engineering systems and equipment associated with anything that works in a marine environment. Principally, this means ships, offshore platforms and undersea vehicles.

In the case of ships, this covers the plant and equipment necessary for the propulsion and safety of the hull, the care and handling of cargoes, and the care, comfort and safety of passengers and crew. On offshore platforms, particular concerns are the supply of electrical power, the pumping systems, and the life support and comfort systems for the crew. For undersea vehicles, areas of interest are the provision of a suitable energy source for propulsion and control, and for manned vehicles, the safety and life support systems for the crew.

Shore-based industries are also involved in marine engineering. As well as designing and building all the complex machinery required, they are also concerned with research and surveying, which contribute much to the improvement in reliability and performance of the machinery.

Naval architecture   Naval architecture is the art of designing craft and structures used at sea, including ships, boats, offshore platforms, submersibles and submarines. Naval architects, while requiring creative imagination, must also possess considerable engineering skill since they are concerned with all aspects of the design and construction of marine vehicles. In addition to considering the structural aspects of vehicle design and the problems of fluid flow around the vehicle, the naval architect is also very much concerned with efficiency and the economics of performance, and, above all, with issues of safety. The environmental consequences of marine vehicle design are also an important part of the naval architect's brief.

Other disciplines   The divisions between the various branches of mechanical engineering, never well defined, are becoming even more blurred. While there is a very great need for engineers with specialist knowledge, the type of specialism required today differs from that needed a generation or more ago. There is considerable overlap between the work of the mechanical and the electrical engineer, and a number of degree courses in mechanical engineering offer specialist options in electronics, instrumentation and systems engineering (see the *Integrated Engineering and Mechatronics* part of this Guide, page 29, for courses in mechatronics).

Careers in mechanical engineering   Most mechanical engineering graduates go straight into employment, mainly in one of a variety of engineering roles. Almost all production, construction and manufacturing depends on mechanical engineering skills, so a graduate in mechanical engineering can follow a career in a wide variety of industries. In addition, many non-engineering professions also recruit numerate and articulate science-based graduates such as mechanical engineers.

The engineering industry (including allied industries such as the automotive, railways and aerospace industries) remains the biggest recruiter of mechanical engineering graduates. Engineering companies design and manufacture equipment for use in manufacturing, the process industries, energy and power generation, consumer products and transportation.

A large number of mechanical engineers are also employed by companies making non-engineering products such as paper or food, and in industries such as power generation, oil, gas and chemicals. The computing industry also employs some mechanical engineers. All these industries build and operate very high-value installations, which may be located anywhere in the world. Mechanical engineers are heavily involved in the design, commissioning and maintenance of large plant, including power stations, oil platforms, pipelines, refineries and chemical works, and they also engage in research and development work for a range of other products used in machinery, such as fuels, lubricants and coolants.

Contractors and consultants   Contracting and consulting organisations are an important sector of the engineering industry, and provide good opportunities for the mechanical engineer. They maintain large teams of engineering professionals to handle work for clients, reducing the workload of in-house engineers and helping them develop their expertise. For many engineers the challenge of project work, travel and change compensates for periods of occasional uncertainty brought about by fluctuations in the economic climate.

The public sector and related public agencies   In the public sector major recruiters include the civil service, particularly the Defence Engineering and Science Group (DESG), and the armed services. A small number of engineers move into regulatory bodies such as the Health and Safety Executive, having gained experience in government agencies or industry. The website www.careers.civil-service.gov.uk carries details of opportunities.

Careers in manufacturing engineering

With its roots in mechanical engineering, manufacturing engineering brings together aspects of a number of other disciplines that bear on manufacturing processes. These include electronics and computer control systems, robotics and methods of automatic weighing, counting, labelling and filling. There is great emphasis on management efficiency, quality, safety and planned maintenance to avoid an unexpected shutdown of production plant. Where there are non-stop production processes, some engineers may work shifts to cover the whole 24 hours. High-quality manufacturing engineering graduates are currently in short supply and consequently in great demand.

The pattern of employment is very similar to that of mechanical engineers. Many graduates start their careers in the engineering and allied industries, including automotive and aerospace companies. Other recruiters include chemicals and defence companies, and several organisations in the energy and extractive industry sector. Manufacturers of finished goods, including instruments, electrical goods, engines, pumps, consumer products and more expensive items, are showing an increasing interest in graduates in manufacturing engineering. Most of these manufacturers make individual products in batches or on an assembly line, rather than bulk products made in continuous process plants.

Contracting and consulting companies supply equipment and services to manufacturing industry, including whole factories and individual systems, and some of these also recruit manufacturing engineers.

Other opportunities for mechanical and manufacturing engineers

Besides roles in engineering research, design and development, and contracting and commissioning, mechanical and manufacturing engineers can also consider more commercial jobs. Commercial engineers are involved in technical sales and marketing, customer service and technical and sales support. Some of these are roles for experienced engineers, while others may provide openings for graduate trainees. Many courses now include elements of business studies, which can give commercially minded engineers an insight into this side of operations.

Many mechanical and manufacturing engineers occupy general management roles, which require not only an understanding of the principles and design of the equipment used, but the skills needed to organise and communicate with a variety of other employees. Some graduates begin their careers as management trainees, but the majority of engineers move into management roles as they gain in experience.

For more information on career opportunities, see the Institution of Mechanical Engineers website at www.imeche.org.uk.

Chapter 2: The courses

TABLE 2a lists the specialised and combined courses at universities and colleges in the UK that lead to the award of an honours degree in mechanical or manufacturing engineering, or a closely related subject. When the table was compiled it was as up to date as possible, but sometimes new courses are announced and existing courses withdrawn, so before you finally fill in your application you should check the UCAS website, www.ucas.com, to make sure the courses you plan to apply for are still on offer.

See Chapter 2 in the *Introduction* part of this Guide (page 6) for advice on how to use TABLE 2a and for an explanation of what the various columns mean. To help you locate and compare courses, TABLE 2a (as well as TABLE 2b, and the first two tables in Chapter 3) is divided into three sections: mechanical engineering; manufacturing engineering and product design (including computer-aided engineering and courses in mechanical and manufacturing engineering); and engineering related to a number of specific industries – aeronautical (including avionics), agricultural, automotive and marine engineering. However, the boundaries between these groups are not sharp, so you would be wise to check what courses there are in other groups before focusing on those in a single group that particularly interests you.

Table 2a — First-degree courses in **Mechanical and Manufacturing Engineering**

Institution / Course title	①②③ see combined subject list – Table 2b	Degree	Duration / Number of years	Foundation year (● at this institution; ○ at franchised institution; ◐ second-year entry)	Modes of study (● full time; ◕ part time; ○ time abroad; ◉ sandwich)	Modular scheme ✿	Course type (● specialised; ◐ combined)	No of combined courses
Mechanical Engineering								
Aberdeen www.abdn.ac.uk								
Mechanical engineering		BEng/MEng	4, 5	◐	● ◕ ○		● ◐	5
Aston www.aston.ac.uk								
Mechanical engineering		BEng/MEng	3, 4, 5	●	● ◉		●	0
Bath www.bath.ac.uk								
Innovation and engineering design ①		MEng	4, 5	●	● ○ ◉	✿	● ◐	2
Mechanical engineering ②		MEng	4, 5	●	● ○ ◉	✿	● ◐	2
Belfast www.qub.ac.uk								
Mechanical engineering		BEngMEng	3, 4	●	● ◉		● ◐	1
Birmingham www.bham.ac.uk								
Mechanical engineering		BEng/MEng	3, 4	●	● ○ ◉		● ◐	2
Blackpool and The Fylde C www.blackpool.ac.uk								
Mechanical and production engineering		BEng	3	●	● ◕		●	0
Bolton www.bolton.ac.uk								
Mechanical engineering		BEng	3, 4	●	● ◕ ○		●	0
Bradford www.bradford.ac.uk								
Mechanical engineering		BEng/MEng	3, 4, 5	●	● ○ ◉		●	0
Brighton www.brighton.ac.uk								
Mechanical engineering		BEng/MEng	3, 4, 5		● ◉		●	0

First-degree courses in **Mechanical and Manufacturing Engineering**

Institution / Course title	Degree	Duration (Number of years)	Foundation year: at this institution	at franchised institution	second-year entry	Modes of study: full time / part time	time abroad	sandwich	Modular scheme	Course type (specialised / combined)	No of combined courses
Bristol www.bris.ac.uk											
Mechanical engineering	BEng/MEng	3, 4, 5				●	O			●	0
Bristol UWE www.uwe.ac.uk											
Mechanical engineering	BEng/MEng	3, 4, 5	●	O		● ◗	O	◕	✿	●	0
Brunel www.brunel.ac.uk											
Mechanical engineering	BEng/MEng	3, 4, 5	●			●	O	◕		●	0
Mechanical engineering with building services	BEng/MEng	3, 4, 5	●			●	O	◕		●	0
Cambridge www.cam.ac.uk											
Mechanical engineering	BA/MEng	3, 4				●			✿	●	0
Cardiff www.cardiff.ac.uk											
Mechanical engineering	BEng/MEng	3, 4, 5	●			●	O	◕		●	0
City www.city.ac.uk											
Mechanical engineering	BEng/MEng	3, 4, 5		O		●	O	◕		●	0
Coventry www.coventry.ac.uk											
Mechanical engineering	BEng/MEng	3, 4	●			●	O	◕		●	0
Mechanical engineering design	BEng/MEng	3, 4				●		◕		●	0
Dundee www.dundee.ac.uk											
Mechanical engineering	BEng	4			◑	● ◗	O			●	0
Durham www.durham.ac.uk											
Mechanical engineering	MEng	4				●	O	◕		●	0
Edinburgh www.ed.ac.uk											
Mechanical engineering	BEng/MEng	4, 5			◑	●	O			● ◕	3
Exeter www.exeter.ac.uk											
Mechanical engineering	BEng/MEng	3, 4				●	O		✿	●	0
Glamorgan www.glam.ac.uk											
Mechanical engineering	BEng	3, 4	●	O		● ◗	O	◕		● ◕	1
Glasgow www.gla.ac.uk											
Mechanical design engineering	BEng/MEng	4, 5			◑	●	O			●	0
Mechanical engineering	BEng/MEng	4, 5			◑	●	O			●	0
Glasgow Caledonian www.gcal.ac.uk											
Computer-aided mechanical engineering	BSc	4			◑	●			✿	●	0
Greenwich www.gre.ac.uk											
Mechanical engineering	BEng	3, 4	●			● ◗		◕		●	0
Mechanical engineering technology	BEng	3, 4	●			●		◕		●	0
Heriot-Watt www.hw.ac.uk											
Mechanical engineering	BEng/MEng	4, 5			◑	●	O	◕		● ◕	2
Hertfordshire www.herts.ac.uk											
Mechanical engineering	BEng	3, 4		O		●	O	◕		●	0
Huddersfield www.hud.ac.uk											
Engineering design: mechanical	BEng/MEng	3, 4, 5	●			●	O	◕		●	0
Mechanical engineering	BEng/MEng	3, 4, 5	●			●	O	◕		●	0
Hull www.hull.ac.uk											
Mechanical engineering	BEng	3		O		● ◗			✿	●	0
Imperial College London www.imperial.ac.uk											
Mechanical engineering	MEng	4				●	O	◕		●	0
King's College London www.kcl.ac.uk											
Mechanical engineering	BEng/MEng	3, 4				●				●	0
Kingston www.kingston.ac.uk											
Mechanical engineering	BEng/MEng	3, 4	●			●		◕		●	0

(continued) Table 2a

First-degree courses in **Mechanical and Manufacturing Engineering**

Institution / Course title	Degree	Duration (Number of years)	Foundation year	Modes of study	Modular scheme	Course type	No of combined courses
Kingston (continued)							
Mechanical engineering design	BEng	3, 4	●	● ◑		●	0
Lancaster www.lancs.ac.uk							
Engineering (mechanical)	BEng/MEng	3, 4		● ◑		●	0
Leeds www.leeds.ac.uk							
Mechanical engineering	BEng/MEng	3, 4	●	● ○		●	0
Leicester www.le.ac.uk							
Mechanical engineering	BEng/MEng	3, 4, 5	●	● ○ ◑		●	0
Liverpool www.liv.ac.uk							
Mechanical engineering ①	BEng/MEng	3, 4	●	● ○		● ◑	1
Mechanical systems and design engineering	BEng/MEng	3, 4	●	● ○		●	0
Liverpool John Moores www.ljmu.ac.uk							
Mechanical engineering	BEng	3, 4	● ○	● ◦ ◑		●	0
London South Bank www.sbu.ac.uk							
Mechanical engineering	BEng	3, 4	●	● ◦ ○ ◑		●	0
Loughborough www.lboro.ac.uk							
Mechanical engineering	BEng/MEng	3, 4, 5	●	● ○ ◑		●	0
Manchester www.man.ac.uk							
Mechanical engineering	BEng/MEng	3, 4, 5	●	● ○ ◑		● ◑	1
Manchester Metropolitan www.mmu.ac.uk							
Mechanical engineering	BSc/BEng	3, 4	● ○	● ◦ ○ ◑		●	0
Napier www.napier.ac.uk							
Mechanical engineering	BEng	4	◑	●		●	0
Newcastle www.ncl.ac.uk							
Mechanical and design engineering	MEng	4	●	● ○		●	0
Mechanical and railway engineering	MEng	4	●	●		●	0
Mechanical engineering ①	BEng/MEng	3, 4, 5	●	● ○		● ◑	4
North East Wales I www.newi.ac.uk							
Mechanical engineering	BEng	3	●	●		● ◑	1
Northumbria www.northumbria.ac.uk							
Mechanical engineering	BEng	3, 4	●	● ◑		● ◑	1
Nottingham www.nottingham.ac.uk							
Mechanical engineering	BEng/MEng	3, 4	●	● ○		●	4
Oxford www.ox.ac.uk							
Mechanical engineering	MEng	4	●	●		●	0
Oxford Brookes www.brookes.ac.uk							
Computer-aided mechanical engineering	BSc	3, 4	●	● ◑		●	0
Mechanical engineering	BSc	3, 4	●	●		●	0
Paisley www.paisley.ac.uk							
Mechanical engineering	BEng	4, 5	◑	● ◦ ◑	✿	●	0
Plymouth www.plymouth.ac.uk							
Mechanical engineering	BEng/MEng	3, 4, 5	● ○	● ◦ ○ ◑		● ◑	2
Portsmouth www.port.ac.uk							
Mechanical engineering	BEng/MEng	3, 4	●	● ○ ◑		●	0
Queen Mary www.qmul.ac.uk							
Mechanical engineering	BEng/MEng	3, 4	●	● ○ ◑	✿	● combined	1
Robert Gordon www.rgu.ac.uk							
Mechanical engineering	BEng/MEng	4, 5	◑	● ◦		● ◑	4

First-degree courses in **Mechanical and Manufacturing Engineering**

Institution / Course title	①②③ see combined subject list – Table 2b	Degree	Duration (No. of years)	Foundation year: at this institution ● / at franchised institution ○ / second-year entry ◐	Modes of study: full time ● / part time ▬	Modes: time abroad ○	Modes: sandwich ◑	Modular scheme	Course type: specialised ● / combined ◑	No of combined courses
Sheffield www.sheffield.ac.uk										
Mechanical engineering ①	①	BEng/MEng	3, 4	●	●	○			● ◑	5
Mechanical systems engineering ②	②	BEng/MEng	3, 4	●	●				● ◑	2
Sheffield Hallam www.shu.ac.uk										
Mechanical engineering		BEng/MEng	3, 4, 5	● ○	● ▬		◑		● ●	1
Southampton www.soton.ac.uk										
Acoustical engineering		BEng/MEng	3, 4	○	●				●	0
Aeronautics and astronautics		BEng/MEng	3, 4	●	●				●	0
Mechanical engineering ①	①	BEng/MEng	3, 4	●	●	○	◑		● ◑	4
Southampton Solent www.solent.ac.uk										
Mechanical design		BEng	3, 4	●	● ▬				●	0
Staffordshire www.staffs.ac.uk										
Mechanical engineering		BEng/MEng	4, 5	● ○	● ▬	○	◑		●	0
Mechanical systems design		BEng/MEng	4, 5	● ○	● ▬	○	◑		●	0
Strathclyde www.strath.ac.uk										
Mechanical engineering		BEng/MEng	4, 5	◐ ●	●	○			● ◑	5
Sunderland www.sunderland.ac.uk										
Engineering design and manufacture		BEng	3, 4		●	○	◑		●	0
Mechanical engineering with design		BEng	3, 4		●	○	◑	✿	●	0
Surrey www.surrey.ac.uk										
Mechanical engineering		BEng/MEng	3, 4, 5	●	●	○	◑		●	0
Sussex www.sussex.ac.uk										
Mechanical engineering		BEng/MEng	3, 4	●	●	○			● ◑	1
Swansea www.swan.ac.uk										
Mechanical engineering		BSc/BEng/MEng	3, 4, 5	●	●	○	◑		● ◑	1
Swansea IHE www.sihe.ac.uk										
Mechanical and manufacturing engineering		BEng	3	●	●				●	0
Teesside www.tees.ac.uk										
Mechanical engineering		BEng	3, 4	●	● ▬	○	◑		●	0
UCE Birmingham www.uce.ac.uk										
Mechanical and automotive design		BEng	3	●	●				●	0
Mechanical engineering ①	①	BEng	3, 4	●	● ▬		◑		● ◑	1
UCL www.ucl.ac.uk										
Mechanical engineering		BEng/MEng	3, 4, 5		●	○	◑		●	0
Ulster www.ulster.ac.uk										
Mechanical engineering		BEng/MEng	3, 4, 5		●	○	◑		●	0
Wales (UWIC) www.uwic.ac.uk										
Manufacturing systems engineering		BSc	3		●				●	0
Warwick www.warwick.ac.uk										
Mechanical engineering		BEng/MEng	3, 4, 5		●	○			●	0
Wolverhampton www.wlv.ac.uk										
Mechanical engineering		BEng	3, 4		●		◑	✿	●	0

Manufacturing Engineering and Product Design

Institution / Course title	①②③	Degree	Duration	Foundation year	Modes: full/part time	Modes: abroad	Modes: sandwich	Modular	Course type	No of combined courses
Anglia Ruskin www.anglia.ac.uk										
Computer-aided product design		BSc	3		●	○			●	0
Aston www.aston.ac.uk										
Engineering product design		BSc	3, 4	●	●		◑		●	0
Industrial product design		BSc	3, 4	●	●		◑		●	0

(continued) Table 2a — First-degree courses in **Mechanical and Manufacturing Engineering**

Legend:
- ⊖ ⊗ ⊘ see combined subject list – Table 2b
- **Foundation year:** ● at this institution · ○ at franchised institution · ◑ second-year entry
- **Modes of study:** ● full time · ◗ part time · ○ time abroad · ◐ sandwich
- **Modular scheme:** ✿
- **Course type:** ● specialised · ◐ combined

Institution / Course title	Degree	Duration (years)	Foundation: this institution	Foundation: franchised	Foundation: 2nd-year entry	Modes: full time	Modes: part time	Modes: time abroad	Modes: sandwich	Modular scheme	Course type: specialised	Course type: combined	No of combined courses
Aston (continued)													
Product design and management	BSc	3, 4	●			●			◐		●		0
Sustainable product design	BSc	3, 4	●			●			◐		●		0
Bath www.bath.ac.uk													
Manufacturing	MEng	4, 5	●			●		○	◐	✿	●	◐	2
Belfast www.qub.ac.uk													
Manufacturing engineering	BEng	3, 4	●			●			◐		●	◐	1
Bolton www.bolton.ac.uk													
Computer-aided product design	BSc	3, 4	●			●	◗	○			●		0
Bournemouth www.bournemouth.ac.uk													
Computer-aided product design	BSc	4							◐		●		0
Design engineering	BSc	4		○			◗	○	◐		●		0
Product design	BSc	4		○				○	◐		●		0
Bradford www.bradford.ac.uk													
Industrial design	BSc	3, 4	●						◐		●		0
Brighton www.brighton.ac.uk													
Product design	BSc	3, 4	●			●	◗	○	◐		●		0
Brunel www.brunel.ac.uk													
Industrial design	BSc	3, 4	●			●		○	◐		●		0
Product design	BSc	3, 4	●			●		○	◐		●		0
Product design engineering	BSc	3, 4				●		○	◐		●		0
Buckinghamshire Chilterns UC www.bcuc.ac.uk													
Product design	BSc	3	●			●					●		0
Cambridge www.cam.ac.uk													
Manufacturing engineering	BA/MEng	3, 4	●			●				✿	●		0
Cardiff www.cardiff.ac.uk													
Manufacturing engineering	BEng/MEng	3, 4	●			●			◐		●		0
Coventry www.coventry.ac.uk													
Industrial product design	BSc/MDes	3, 4		○		●		○	◐		●		0
Durham www.durham.ac.uk													
Design, manufacture and management	MEng	4				●		○	◐		●		0
East London www.uel.ac.uk													
Product design	BSc	3, 4	●			●		○	◐		●		0
Glasgow www.gla.ac.uk													
Product design engineering	BEng/MEng	4, 5			◑	●		○			●		0
Glasgow Caledonian www.gcal.ac.uk													
Manufacturing systems engineering	BEng	4, 5			◑	●	◗	○	◐		●	◐	1
Greenwich www.gre.ac.uk													
Engineering product design	BSc	3, 4				●			◐		●		0
Industrial automation	BEng	3, 4	●			●	◗		◐		●		0
Manufacturing systems engineering	BEng	3, 4	●			●			◐		●		0
Harper Adams UC www.harper-adams.ac.uk													
Engineering design and development	BSc	3, 4				●			◐		●		0
Hertfordshire www.herts.ac.uk													
Manufacturing engineering	BEng	3, 4		○		●	◗	○	◐		●		0
Huddersfield www.hud.ac.uk													
Manufacturing and operations management	BSc	3, 4	●			●	◗	○	◐		●		0
Product innovation, design and development	BSc	3, 4	●			●		○	◐		●		0

Institution / Course title	①②③ see combined subject list – Table 2b	Degree	Duration (No. of years)	Foundation year: at this institution ●	at franchised institution ○	second-year entry ◐	Modes: full time ●	part time ↩	time abroad ○	sandwich ◐	Modular scheme ✿	Course type: specialised ●	combined ◐	No. of combined courses
London South Bank www.sbu.ac.uk														
Engineering product design		BSc	3, 4	●			●		○	◐		●		0
Loughborough www.lboro.ac.uk														
Innovative manufacturing technology		MEng	4				●		○	◐		●		0
Manufacturing engineering and management ①		BEng	3, 4	●			●		○	◐			◐	1
Product design and manufacture		BEng/MEng	3, 4, 5	●			●		○	◐		●		0
Manchester Metropolitan www.mmu.ac.uk														
Automation and control		BEng	3, 4	●			●					●		0
Napier www.napier.ac.uk														
Product design engineering		BSc	4		◐		●					●		0
Newcastle www.ncl.ac.uk														
Mechanical and manufacturing engineering		MEng	4	●			●		○			●		0
Northumbria www.northumbria.ac.uk														
Computer-aided product design		BSc	3, 4				●			◐		●		0
Manufacturing systems engineering		BEng	3, 4	●			●	↩		◐		●		0
Product design and technology		BSc	4				●			◐		●		0
Nottingham www.nottingham.ac.uk														
Manufacturing engineering and management ①		BEng/MEng	3, 4	●			●		○	◐		●	◐	6
Mechanical design, materials and manufacture ②		BEng/MEng	3, 4	●			●			◐	✿		◐	1
Product design and manufacture		BEng	3				●					●		0
Nottingham Trent www.ntu.ac.uk														
Computer-aided product design		BSc	3, 4				●			◐		●		0
Product design		BSc	3, 4				●			◐		●		0
Paisley www.paisley.ac.uk														
Product design and development		BSc	4, 5				●			◐		●		0
Plymouth www.plymouth.ac.uk														
Mechanical design and manufacture		BSc	3, 4	●	○		●	↩	○	◐		●	◐	1
Portsmouth www.port.ac.uk														
Computer-aided product design		BSc	3, 4	●			●		○	◐		●		0
Mechanical and manufacturing engineering		BEng	3, 4	●			●		○	◐		●		0
Product design and innovation		BSc	3, 4	●			●		○	◐		●		0
Product design and modern materials		BSc	3, 4	●			●		○	◐		●		0
Strathclyde www.strath.ac.uk														
Manufacturing engineering and technology ①		BEng/MEng	4, 5		◐		●		○	◐	✿	●	◐	2
Product design engineering ②		BEng/MEng	4, 5		◐		●		○	◐	✿	●	◐	2
Sunderland www.sunderland.ac.uk														
Product design		BSc	3, 4				●		○	◐		●		0
Sussex www.sussex.ac.uk														
Product design		BSc	3, 4				●			◐		●		0
Swansea www.swan.ac.uk														
Product design engineering		BEng/MEng	3, 4				●		○	◐		●		0
Product design technology		BEng	3, 4				●		○	◐		●		0
Swansea IHE www.sihe.ac.uk														
Manufacturing systems engineering		BEng	3	●			●	↩				●		0
Teesside www.tees.ac.uk														
Computer-aided design engineering		BEng	3, 4	●			●	↩	○	◐		●		0
UCE Birmingham www.uce.ac.uk														
Engineering product design		BSc	3, 4	●			●	↩	○	◐		●		0

(continued) Table 2a

First-degree courses in **Mechanical and Manufacturing Engineering**

Institution / Course title	①②③ see combined subject list – Table 2b	Degree	Duration Number of years	Foundation year ● at this institution	○ at franchised institution	◐ second-year entry	Modes of study ● full time; ▬ part time	○ time abroad	◐ sandwich	Modular scheme	Course type ● specialised; ◕ combined	No of combined courses
Wales (Newport) www.newport.ac.uk												
Automation systems		BEng	3	●	○		● ▬	○	◐		●	0
Wales (UWIC) www.uwic.ac.uk												
Mechanical systems engineering		BSc	3				●				●	0
Warwick www.warwick.ac.uk												
Manufacturing and mechanical engineering		BEng/MEng	3, 4, 5				●	○			●	0
Aeronautical, Agricultural, Automotive and Marine Engineering												
Aston www.aston.ac.uk												
Automotive product design		BSc	3, 4	●			●		◐		●	0
Medical product design		BSc	3, 4	●			●		◐		●	0
Bath www.bath.ac.uk												
Aerospace engineering ①		MEng	4, 5	●			●	○	◐	✿	● ◕	2
Automotive engineering ②		MEng	4, 5	●			●	○	◐	✿	● ◕	2
Belfast www.qub.ac.uk												
Aeronautical engineering		BEng/MEng	3, 4, 5	●					◐	✿	●	0
Birmingham www.bham.ac.uk												
Mechanical and automotive engineering		BEng/MEng	3, 4	●			●	○	◐		●	0
Bolton www.bolton.ac.uk												
Automobile engineering		BEng	3	●			● ▬				●	0
Automotive product design		BSc	3, 4	●			● ▬	○			●	0
Bradford www.bradford.ac.uk												
Automotive design technology		BSc	3, 4	●			●		◐		●	0
Mechanical and automotive engineering		BEng/MEng	3, 4, 5	●			●	○	◐		●	0
Vehicle technology		BSc	3				●				●	0
Brighton www.brighton.ac.uk												
Automotive engineering		BEng	3, 4				●		◐		●	0
Bristol www.bris.ac.uk												
Aeronautical engineering		MEng	4				●	○			●	0
Avionic systems		BEng/MEng	3, 4				●	○			●	0
Bristol UWE www.uwe.ac.uk												
Aerospace manufacturing engineering		BEng/MEng	3, 4, 5	●	○		● ▬	○	◐		●	0
Aerospace systems engineering		BEng/MEng	3, 4, 5	●	○		●	○	◐		●	0
Motorsport engineering		BEng	3				●				●	0
Brunel www.brunel.ac.uk												
Aerospace engineering		BEng/MEng	3, 4, 5	●			●		◐		●	0
Aviation engineering		BEng/MEng	3, 4, 5	●			●		◐		●	0
Mechanical engineering with aeronautics		BEng/MEng	3, 4, 5	●			●	○	◐		●	0
Mechanical engineering with automotive design		BEng/MEng	3, 4, 5	●			●	○	◐		●	0
Motorsport engineering		BEng/MEng	3, 4, 5	●			●	○	◐		●	0
Space engineering		BEng/MEng	3, 4, 5	●			●	○	◐		●	0
Cambridge www.cam.ac.uk												
Aerospace and aerothermal engineering		BA/MEng	4				●			✿	●	0
Central Lancashire www.uclan.ac.uk												
Motorsports engineering		BEng	3, 4				● ▬		◐		●	0
City www.city.ac.uk												
Aeronautical engineering		BEng/MEng	3, 4, 5		○		●	○	◐		●	0
Air transport engineering		BEng/MEng	3, 4, 5	○	○		●	○	◐		●	0
Automotive and motor sport engineering		BEng/MEng	3, 4, 5		○		●	○	◐		●	0

First-degree courses in **Mechanical and Manufacturing Engineering**

Institution / Course title	①②③ see combined subject list – Table 2b	Degree	Duration (Number of years)	Foundation year: ● at this institution; ○ at franchised institution; ◐ second-year entry			Modes of study: ● full time; ➤ part time; ○ time abroad; ◐ sandwich			✧ Modular scheme	Course type: ● specialised; ◑ combined	No of combined courses
Coventry www.coventry.ac.uk												
Aerospace systems engineering		BEng	3, 4	●			●	○	◐		●	0
Aerospace technology		BEng	3, 4	●			●	○	◐		●	0
Automotive engineering		BEng/MEng	3, 4	●	○		●	○	◐		●	0
Automotive engineering design		BEng	3, 4	●			●	○	◐		●	0
Avionics technology		BEng	3, 4				● ➤				●	0
Motorsport and motorcycle engineering		BEng	3, 4				●		◐		●	0
Motorsport and powertrain engineering		BEng	3, 4				●		◐		●	0
Motorsport engineering		BEng	3, 4	●			●		◐		●	0
Durham www.durham.ac.uk												
Aeronautics		MEng	4				●				●	
Farnborough CT www.farn-ct.ac.uk												
Aeronautical engineering		BSc	3	●			● ➤				●	0
Glamorgan www.glam.ac.uk												
Aerospace engineering		BEng	3, 4	●	○		●	○	◐		●	0
Aircraft maintenance engineering		BSc	3				●				●	0
Glasgow www.gla.ac.uk												
Aeronautical engineering		MEng	4, 5			◐	●	○			●	0
Avionics		MEng	4, 5			◐	●	○			●	0
Mechanical engineering with aeronautics		BEng/MEng	4, 5			◐	●	○			●	0
Greenwich www.gre.ac.uk												
Automotive engineering with management		BEng	3				●				●	0
Marine engineering technology		BSc	3, 4				●		◐		●	0
Harper Adams UC www.harper-adams.ac.uk												
Agricultural engineering ①		BSc/MEng	3, 4, 5	●			● ➤	○	◐		● ◑	2
Agriculture and mechanisation		BSc	4						◐		●	0
Off-road vehicle design		BSc/MEng	3, 4, 5	●				○	◐		●	0
Heriot-Watt www.hw.ac.uk												
Automotive engineering		BEng	4				●				●	0
Hertfordshire www.herts.ac.uk												
Aerospace engineering		BEng/MEng	3, 4, 5		○		●	○	◐		●	0
Aerospace systems engineering		BEng/MEng	3, 4, 5		○		●	○	◐		●	0
Aerospace technology with management/pilot studies		BSc	3, 4				●		◐		●	0
Automotive engineering		BEng/MEng	3, 4, 5		○		●	○	◐		●	0
Automotive engineering with motorsport		BEng/MEng	3, 4, 5		○		●	○	◐		●	0
Motorsport technology		BSc	3				●	○			●	0
Huddersfield www.hud.ac.uk												
Automotive design		BEng/MEng	4, 5	●			●	○	◐		●	0
Automotive design and technology		BSc	3, 4	●			●	○	◐		●	0
Automotive engineering		BEng/MEng	3, 4, 5	●			●	○	◐		●	0
Automotive product innovation		BSc	3, 4	●			●				●	0
Automotive technology		BSc	3, 4	●			●	○	◐		●	0
Mechanical and automotive design		BEng/MEng	3, 4, 5	●			●	○	◐		●	0
Motorsport technology		BSc	3, 4	●			●	○	◐		●	0
Imperial College London www.imperial.ac.uk												
Aeronautical engineering		MEng	4				●	○			●	0
Kingston www.kingston.ac.uk												
Aerospace engineering		BEng/MEng	3, 4, 5	●			●	○	◐		●	0
Aerospace engineering and astronautics		BEng/MEng	3, 4, 5	●			●	○	◐		●	0

(continued) Table 2a

First-degree courses in **Mechanical and Manufacturing Engineering**

Institution / Course title	① ② ③ see combined subject list – Table 2b	Degree	Duration (Number of years)	Foundation year (● at this institution, ○ at franchised institution, ◑ second-year entry)			Modes of study (● full time, ▼ part time, ○ time abroad, ◑ sandwich)				Modular scheme	Course type (● specialised; ◑ combined)	No of combined courses
Kingston (continued)													
Aerospace engineering design		BEng	3, 4	●			●		○	◑		●	0
Motorcycle engineering design		BEng	3, 4	●			●		○	◑		●	0
Motorsport engineering		BEng	3, 4	●			●			◑		●	0
Leeds www.leeds.ac.uk													
Aeronautical and aerospace engineering		BEng/MEng	3, 4	●			●		○			●	0
Automotive engineering		BEng/MEng	3, 4	●			●		○			●	0
Liverpool www.liv.ac.uk													
Aerospace engineering		BEng/MEng	3, 4				●		○	◑		●	0
Aerospace engineering with pilot studies		BEng/MEng	3, 4				●					●	0
Avionic systems		BEng/MEng	3, 4	◑			●					●	0
Avionic systems with pilot studies		BEng/MEng	3, 4				●					●	0
Liverpool John Moores www.ljmu.ac.uk													
Automotive engineering		BEng	3, 4	●	○		●			◑		●	0
Mechanical and marine engineering		BEng	3, 4	●	○		●			◑		●	0
Loughborough www.lboro.ac.uk													
Aeronautical engineering		BEng/MEng	3, 4, 5	●			●			◑		●	0
Automotive engineering		BEng/MEng	3, 4, 5	●			●			◑		●	0
Manchester www.man.ac.uk													
Aerospace engineering		BEng/MEng	3, 4, 5	●			●		○	◑		● ◑	1
Manchester Metropolitan www.mmu.ac.uk													
Automotive engineering		BEng	3, 4	●			●					●	0
Newcastle www.ncl.ac.uk													
Marine engineering		BEng/MEng	3, 4	●			●		○		◆	●	0
Marine technology		BEng/MEng	3, 4	●			●		○		◆	●	0
Mechanical and automotive engineering		MEng	4	●			●		○			●	0
Naval architecture		BEng/MEng	3, 4	●			●		○			●	0
Offshore engineering		BEng/MEng	3, 4	●			●		○		◆	●	0
Small craft technology		BEng/MEng	3, 4	●			●		○		◆	●	0
North East Wales I www.newi.ac.uk													
Aeronautical engineering ①		BEng	3	●			●					● ◑	2
Performance car technology		BSc	3	●			●					●	0
Nottingham www.nottingham.ac.uk													
Mechanical engineering (aerospace)		MEng	4	●			●			◑		●	0
Mechanical engineering (automotive)		MEng	4	●			●			◑		●	0
Mechanical engineering (bioengineering)		BEng/MEng	3, 4	●			●					●	0
Oxford Brookes www.brookes.ac.uk													
Automotive engineering		BEng/MEng	3, 4, 5	●			●			◑		●	0
Motorsport engineering		BEng/MEng	3, 4, 5	●			●			◑		●	0
Motorsport technology		BSc	3, 4	●			●			◑		●	0
Plymouth www.plymouth.ac.uk													
Marine sports technology		BSc	3	●			●					●	0
Marine technology		BEng	3, 4	●	○		●	▼	○	◑		●	0
Portsmouth www.port.ac.uk													
Marine sports technology		BSc	3, 4	●			●		○	◑		●	0
Queen Mary www.qmul.ac.uk													
Aerospace engineering		BEng/MEng	3, 4	●			●		○	◑	◆	●	0
Avionics		BEng	3	●			●		○	◑	◆	●	0
Sports engineering		MEng	4				●					●	0

First-degree courses in **Mechanical and Manufacturing Engineering**

Institution / Course title	① ② ③ see combined subject list – Table 2b	Degree	Duration (Number of years)	Foundation year (● at this institution; ○ at franchised institution; ◑ second-year entry)	Modes of study (● full time; ➳ part time; ○ time abroad; ◐ sandwich)	Modular scheme	Course type (● specialised; ◑ combined)	No of combined courses
Robert Gordon www.rgu.ac.uk								
Mechanical and offshore engineering		BEng	4, 5	◑	● ➳		●	0
Salford www.salford.ac.uk								
Aeronautical engineering		BSc/BEng/MEng	3, 4, 5	● ○	● ○ ◐		●	0
Aircraft engineering with pilot studies		BEng	3, 4		● ◐		●	0
Aviation technology with pilot studies		BSc	3, 4		● ◐		●	0
Mechanical engineering (aerospace)		BSc/BEng/MEng	3, 4, 5	●	● ○ ◐		●	0
Sheffield www.sheffield.ac.uk								
Aerospace engineering		BEng/MEng	3, 4	●	● ◐		●	0
Aerospace engineering with private pilot instruction		BEng/MEng	3, 4	●	●		●	0
Sheffield Hallam www.shu.ac.uk								
Automotive technology		BSc	3, 4		● ◐		●	0
Southampton www.soton.ac.uk								
Ship science ①		BEng/MEng	3, 4	●	● ○		● ◑	4
Ship science/yacht and small craft		MEng	4	●	● ○		●	0
Southampton Solent www.solent.ac.uk								
Yacht and powercraft design		BEng	3	● ○	●		●	0
Staffordshire www.staffs.ac.uk								
Aeronautical design technology		BSc	3, 4, 5	● ○	● ○ ◐		●	0
Automotive design technology		BSc	3, 4	● ○	● ○ ◐		●	0
Strathclyde www.strath.ac.uk								
Aeromechanical engineering		BEng/MEng	4, 5	◑	● ○		●	0
Naval architecture and marine engineering		BEng/MEng	4, 5	◑	●		●	0
Naval architecture and ocean engineering		BEng/MEng	4, 5	◑	●		●	0
Naval architecture and small craft engineering		BEng/MEng	4, 5	◑	●		●	0
Sunderland www.sunderland.ac.uk								
Automotive design and technology		BSc	3, 4		● ○ ◐		●	0
Automotive engineering with design		BEng	3, 4		● ○ ◐		●	0
Automotive product design		BSc	3, 4		● ○ ◐		●	0
Surrey www.surrey.ac.uk								
Aerospace engineering		BEng/MEng	3, 4, 5	●	● ○ ◐		●	0
Sussex www.sussex.ac.uk								
Automotive engineering		BEng/MEng	3, 4		●		●	0
Swansea www.swan.ac.uk								
Aerospace communications		BEng/MEng	3, 4, 5		● ○ ◐		●	0
Aerospace engineering ①		BEng/MEng	3, 4, 5		● ○ ◐		● ◑	4
Aerospace engineering with propulsion		BEng/MEng	3, 4, 5		● ○ ◐		●	0
Swansea IHE www.sihe.ac.uk								
Automotive electronics systems		BEng	3	●	● ➳		●	0
Automotive engineering		BEng	3	●	● ➳		●	0
Automotive manufacturing		BEng	3	●	●		●	0
Motorcycle engineering		BEng	3	●	●		●	0
Motorsport engineering and design		BEng	3	●	● ➳		●	0
Motorsport manufacturing engineering		BEng	3	●	● ➳		●	0
UCE Birmingham www.uce.ac.uk								
Automotive engineering		BEng	3, 4	●	● ➳ ○ ◐		●	0
UCL www.ucl.ac.uk								
Naval architecture and marine engineering		BEng/MEng	3, 4, 5		● ○ ◐		●	0

Mechanical and Manufacturing Engineering

(continued) Table 2a	First-degree courses in **Mechanical and Manufacturing Engineering**									
Institution Course title	① ② ③ see combined subject list – Table 2b	**Degree**	**Duration** Number of years	**Foundation year** ● at this institution ○ at franchised institution ◑ second-year entry	**Modes of study** ● full time; ▼ part time ○ time abroad ◑ sandwich	**Modular scheme** ❀	**Course type** ● specialised; ◐ combined	**No of combined courses**
Warwick www.warwick.ac.uk								
Automotive engineering		BEng/MEng	3, 4		● ○		●	0
Wolverhampton www.wlv.ac.uk								
Automotive system engineering		BEng	3		●	❀	●	0
York www.york.ac.uk								
Avionics		MEng	4	●	● ○		●	0

Subjects available in combination with mechanical or manufacturing engineering

TABLE 2b shows those subjects that can make up at least one-third of your degree programme when taken in combination with mechanical or manufacturing engineering in the combined degrees listed in TABLE 2a. See Chapter 2 in the *Introduction* (page 6) for general information about combined courses and for an explanation of how to use TABLE 2b.

Table 2b	Subjects to combine with **Mechanical or Manufacturing Engineering**

Mechanical Engineering

Aeronautics   North East Wales I, Strathclyde
Artificial intelligence   Robert Gordon
Automobile engineering   Strathclyde
Business studies   Birmingham, Sheffield②, Sussex
Computer-aided engineering   Aberdeen, Heriot-Watt, Plymouth, Sheffield Hallam
Control engineering   Aberdeen
Electrical engineering   Edinburgh, Robert Gordon, Strathclyde
Energy studies   Edinburgh
Energy technology   Heriot-Watt, Robert Gordon, Southampton①
Engineering management   Southampton①
Environmental engineering   Robert Gordon
European studies   Aberdeen
Finance   Strathclyde

French   Bath① ②, Nottingham, Sheffield①
German   Bath① ②, Nottingham, Sheffield①
Italian   Sheffield①
Management studies   Aberdeen, Edinburgh, Liverpool①, Manchester, Newcastle①, Sheffield① ②, Swansea
Manufacturing engineering   Belfast, Glamorgan
Materials engineering   Birmingham, Newcastle①
Materials science   Aberdeen, Plymouth, Queen Mary, Southampton①, Strathclyde
Mathematics   Newcastle①, Northumbria, Nottingham
Mechatronics   Newcastle①, Southampton①
Product design   UCE Birmingham①
Spanish   Nottingham, Sheffield①

Subjects to combine with **Mechanical or Manufacturing Engineering**	
Manufacturing Engineering	

Business studies   Glasgow Caledonian, Plymouth
Enterprise/entrepreneurship   Strathclyde① ②
French   Bath, Nottingham①
German   Bath, Nottingham①
Italian   Nottingham①

Japanese   Nottingham①
Management studies   Loughborough①, Nottingham①, Strathclyde① ②
Materials engineering   Nottingham②
Mechanical engineering   Belfast
Spanish   Nottingham①

Aeronautical, Agricultural, Automotive and Marine Engineering

Electronic engineering   North East Wales I①
Electronics   Swansea①
Engineering management   Southampton①
French   Bath① ②
German   Bath① ②
Management studies   Harper Adams UC①, Manchester, Swansea①
Marine engineering   Southampton①

Marketing   Harper Adams UC①
Materials science   Southampton①, Swansea①
Mechanical engineering   North East Wales I①
Naval architecture   Southampton①
Power engineering   Swansea①

Other courses that may interest you

You may find courses in the *Integrated Engineering and Mechatronics* part of this Guide (page 29) and the applied physics courses in the *Physics and Chemistry* Guide of interest. The following courses are closely related to the courses in this part of the Guide but do not fit neatly into any of the Guides in the series.

- Sports engineering (Bath and Sheffield)
- Automotive mechatronics (Sunderland)
- Mechanical and medical engineering (Hull).

Chapter 3: The style and content of the courses

Industrial experience and time abroad Many courses provide a range of opportunities for spending a period of industrial training in the UK or abroad, or of study abroad. TABLE 3a gives information about the possibilities for the courses in this Guide. See Chapter 3 in the *Introduction* (page 11) for further information.

Table 3a — Time abroad and sandwich courses

Institution / Course title ① ②: see notes after table	Named 'international' variant of the course	Location: ● Europe; ○ North America; ◐ industry; ◑ academic institution				Maximum time abroad (months)	Time abroad assessed	Language study: ○ optional; ● compulsory; * contributes to assessment	Socrates–Erasmus	Sandwich courses: ● thick; ○ thin	Arranged by: ● institution; ○ student
Mechanical Engineering											
Aberdeen											
Mechanical engineering	●	●		◐	◑	12	●	○*	●		
Aston											
Mechanical engineering ① ②	●	●	○	◐	◑	15	●	○		●	●
Bath											
Innovation and engineering design ③	●	●	○	◐	◑	12	●			●	●
Mechanical engineering ③	●	●	○	◐	◑	12	●	○*	●	●	●
Belfast											
Mechanical engineering ③ ⑤				◐		12		○		●	● ○
Birmingham											
Mechanical engineering ① ③		●	○	◐	◑	12	●	○*	●	●	●
Bolton											
Mechanical engineering						15		○*			
Bradford											
Mechanical engineering ① ③ ⑤		●		◐		12		○	●	● ○	●
Bristol											
Mechanical engineering ③		●			◑	12	●	●*	●		
Bristol UWE											
Mechanical engineering ③		●	○	◐	◑	12	●	○*		●	● ○
Brunel											
Mechanical engineering ③		●	○	◐	◑	18		○*	●	●	● ○
Mechanical engineering with building services ③		●	○	◐	◑	18		○*	●	●	● ○
Cambridge											
Mechanical engineering ③								○*			
Cardiff											
Mechanical engineering ③	●	●		◐	◑	12		○	●	●	○
City											
Mechanical engineering ③		●	○	◐	◑	12	●	○	●	●	●
Coventry											
Mechanical engineering ① ② ③										●	● ○
Dundee											
Mechanical engineering ③			○		◑	12	●				
Durham											
Mechanical engineering ③	●	●	○	◐	◑	12	●	○			

Time abroad and sandwich courses

Institution / Course title ① ②: see notes after table	Named 'international' variant of the course	Location: ● Europe; ○ North America; ⊸ industry; ◐ academic institution	Maximum time abroad (months)	Time abroad assessed	Language study: ○ optional; ● compulsory; * contributes to assessment	Socrates–Erasmus	Sandwich courses: ● thick; ○ thin	Arranged by: ● institution; ○ student
Exeter								
Mechanical engineering ③		● ○ ◐	6	●	●*	●		
Glamorgan								
Mechanical engineering		● ⊸	12		○		● ●	●
Glasgow								
Mechanical design engineering ③		● ○ ⊸ ◐	12	●	○*	●		
Mechanical engineering ③	●	● ○ ⊸ ◐	12	●	○*	●		
Glasgow Caledonian								
Computer-aided mechanical engineering						●		
Greenwich								
Mechanical engineering ②							●	●
Heriot-Watt								
Mechanical engineering ③		● ○ ⊸ ◐	12		○*			
Hertfordshire								
Mechanical engineering ③		● ○ ⊸ ◐	12	●	○*		● ●	● ○
Huddersfield								
Engineering design: mechanical		● ○ ⊸	12	●	○		● ●	●
Mechanical engineering		● ○ ⊸	12	●	○		● ●	●
Hull								
Mechanical engineering ① ③		◐	3	●	○*	●	● ●	○
Imperial College London								
Mechanical engineering ②	●	● ⊸ ◐	9	●	○*	●	● ●	○
Kingston								
Mechanical engineering ③							● ●	●
Mechanical engineering design ③							● ●	●
Leeds								
Mechanical engineering ③		● ○ ◐	12	●	○*	●		
Leicester								
Mechanical engineering ③	●	● ○ ⊸ ◐	12	●	○		● ●	○
Liverpool								
Mechanical engineering ③					○			
Mechanical systems and design engineering ③					○			
Liverpool John Moores								
Mechanical engineering ③					○		● ●	●
London South Bank								
Mechanical engineering ①		● ⊸ ◐	24	●	○		● ●	● ○
Loughborough								
Mechanical engineering ①		● ⊸ ◐	12	●	○*		● ●	● ○
Manchester								
Mechanical engineering ③	●	● ⊸	12	●	○		● ●	●
Manchester Metropolitan								
Mechanical engineering ①	●	● ⊸ ◐	12	●	○*		● ●	●
Napier								
Mechanical engineering							○ ●	
Newcastle								
Mechanical and design engineering ① ③		● ◐	9	●	○*	●		
Mechanical and railway engineering ① ③		● ◐		●	○*	●		
Mechanical engineering ① ③	●	● ◐	9	●	○*	●		

Mechanical and Manufacturing Engineering

229

(continued) Table 3a — Time abroad and sandwich courses

① ② : see notes after table

Institution / Course title	Named 'international' variant of the course	Location: ● Europe; ○ North America; ◖ industry; ◑ academic institution	Maximum time abroad (months)	Time abroad assessed	Language study: ○ optional; ● compulsory; * contributes to assessment	Socrates–Erasmus	Sandwich courses: ● thick; ○ thin	Arranged by: ● institution; ○ student
Northumbria								
Mechanical engineering ③							●	● ○
Nottingham								
Mechanical engineering ③ ④		● ◑	6	●	○*	●	●	○
Oxford								
Mechanical engineering ③					○*			
Oxford Brookes								
Computer-aided mechanical engineering							●	● ○
Mechanical engineering ③ ⑤			12		○*		●	● ○
Paisley								
Mechanical engineering						●	● ○	●
Plymouth								
Mechanical engineering ③		● ○ ◖ ◑	12		○	●	●	● ○
Portsmouth								
Mechanical engineering ③	●	● ○ ◖ ◑	12	●		●	●	● ○
Queen Mary								
Mechanical engineering ③	●	● ○	12	●	○	●	○	●
Robert Gordon								
Mechanical engineering ③								
Sheffield								
Mechanical engineering	●	● ○	12	●	○*	●		
Mechanical systems engineering					○*			
Sheffield Hallam								
Mechanical engineering	●	● ◖ ◑	9	●	○*	●	●	●
Southampton								
Acoustical engineering ③	●	● ○ ◖ ◑	2	●	●*	●		
Mechanical engineering		● ◑	12	●	○*	●		
Southampton Solent								
Mechanical design ③					○			
Staffordshire								
Mechanical engineering ③		● ◖ ◑	12	●	○*		●	●
Mechanical systems design ③		● ◖ ◑	12	●	○*		●	● ○
Strathclyde								
Mechanical engineering ③	●	● ○ ◖ ◑	6	●	○*	●		
Sunderland								
Engineering design and manufacture ③		● ○ ◖ ◑	12	●			●	● ○
Mechanical engineering with design ③		● ○ ◖ ◑	10	●			●	● ○
Surrey								
Mechanical engineering		● ○ ◖ ◑	12	●		●	●	●
Sussex								
Mechanical engineering ③	●	● ○ ◖ ◑	10	●	○*			
Swansea								
Mechanical engineering ① ③	●	● ○ ◖ ◑	12	●		●		
Teesside								
Mechanical engineering ③		● ◖ ◑	12	●		●	●	● ○
UCE Birmingham								
Mechanical engineering ③		● ◖ ◑					●	●
UCL								
Mechanical engineering ① ② ③		◖	12		○*		●	○

Time abroad and sandwich courses

Institution / Course title ①②: see notes after table	Named 'international' variant of the course	Location: ● Europe; ○ North America; ▶ industry; ◖ academic institution	Maximum time abroad (months)	Time abroad assessed	Language study: ○ optional; ● compulsory; * contributes to assessment	Socrates–Erasmus	Sandwich courses: ● thick; ○ thin	Arranged by: ● institution; ○ student
Ulster								
Mechanical engineering		● ▶	12	●	○*	●	●	●
Warwick								
Mechanical engineering ③		● ○ ◖	12	●	○*	●	●	○
Wolverhampton								
Mechanical engineering							●	○

Manufacturing Engineering and Product Design

Institution / Course title	Named 'international' variant	Location	Max time abroad	Time abroad assessed	Language study	Socrates–Erasmus	Sandwich courses	Arranged by
Anglia Ruskin								
Computer-aided product design		● ▶	12	●	○*	●		
Aston								
Engineering product design							●	● ○
Industrial product design							●	● ○
Product design and management							●	● ○
Sustainable product design							●	○
Bath								
Manufacturing ③	●	● ○ ▶ ◖	12	●	○*	●	●	●
Belfast								
Manufacturing engineering ③		▶	12		○		●	● ○
Bolton								
Computer-aided product design		● ○ ◖	15			●		
Bradford								
Industrial design ③							●	● ○
Brighton								
Product design		● ▶ ◖	24		○	●	●	○
Brunel								
Industrial design		● ○ ▶ ◖	12	●	○*	●	●	● ○
Product design ③		● ○ ▶ ◖	12	●	○○*	●	●	● ○
Product design engineering ③		● ▶ ◖	12	●	○*	●	●	● ○
Cambridge								
Manufacturing engineering ③					●*			
Cardiff								
Manufacturing engineering						●		
Coventry								
Industrial product design ③		● ▶					●	● ○
Durham								
Design, manufacture and management ③	●	● ○ ▶ ◖	12	●	○			
East London								
Product design		● ◖	6			●	●	○
Glasgow								
Product design engineering ③		● ▶ ◖	12	●	○*	●		
Glasgow Caledonian								
Manufacturing systems engineering ③		● ○ ▶ ◖	14	●	○	●	●	●
Greenwich								
Engineering product design							●	● ○
Industrial automation ③							●	●
Manufacturing systems engineering ③								
Hertfordshire								
Manufacturing engineering ③		● ○ ▶ ◖	12	●	○*	●	●	○

(continued) Table 3a

Time abroad and sandwich courses

Key:
- ① ②: see notes after table
- Location: ● Europe; ○ North America; ⏷ industry; ◑ academic institution
- Maximum time abroad (months)
- Language study: ○ optional; ● compulsory; * contributes to assessment
- Sandwich courses: ● thick; ○ thin
- Arranged by: ● institution; ○ student

Institution / Course title	Named 'international' variant of the course	Location	Maximum time abroad (months)	Time abroad assessed	Language study	Socrates–Erasmus	Sandwich courses	Arranged by
Huddersfield								
Manufacturing and operations management		● ○ ⏷ ◑	12	●	○		● ●	●
Product innovation, design and development		● ⏷ ◑	12	●			● ●	●
London South Bank								
Engineering product design ① ③		● ⏷ ◑	12	●			● ●	● ○
Loughborough								
Innovative manufacturing technology ②		● ⏷ ◑	12	●	○*		● ●	●
Manufacturing engineering and management ①		● ⏷ ◑	12	●	○*		● ●	● ○
Product design and manufacture ① ③		● ⏷ ◑	12	●	○*		● ●	● ○
Newcastle								
Mechanical and manufacturing engineering ① ③		● ◑			○*	●		
Northumbria								
Manufacturing systems engineering ③							●	● ○
Product design and technology								● ○
Nottingham								
Manufacturing engineering and management ④	●	● ⏷ ◑	12		○*	●		○
Mechanical design, materials and manufacture ③		● ⏷ ◑			○			
Plymouth								
Mechanical design and manufacture ③		● ○ ⏷ ◑	12	●	○	●	● ●	○
Portsmouth								
Computer-aided product design ③		● ○ ⏷ ◑	12	●	○*		● ●	●
Mechanical and manufacturing engineering ③		● ○ ⏷ ◑	12	●	○*	●	● ●	●
Product design and innovation ③		● ○ ⏷ ◑	12	●	○*		● ●	● ○
Product design and modern materials ③		● ○ ⏷ ◑	12	●	○*		● ●	● ○
Strathclyde								
Manufacturing engineering and technology ① ③ ⑤		● ○ ⏷ ◑	12	●	○*	●		● ○
Product design engineering ① ③ ⑤		● ○ ⏷ ◑	12	●	○*	●		● ○
Swansea								
Product design engineering ① ③	●	● ○ ⏷ ◑	12	●		●	● ●	○
Teesside								
Computer-aided design engineering ③		● ⏷ ◑	12	●			● ●	● ○
UCE Birmingham								
Engineering product design ③		● ⏷ ◑	12				●	● ●
Wales (Newport)								
Automation systems ⑤		● ⏷	12		○		●	● ○
Warwick								
Manufacturing and mechanical engineering ③		● ○ ◑	12	●	○*	●	● ●	○

Aeronautical, Agricultural, Automotive and Marine Engineering

Institution / Course title	Named 'international' variant of the course	Location	Maximum time abroad (months)	Time abroad assessed	Language study	Socrates–Erasmus	Sandwich courses	Arranged by
Aston								
Automotive product design							●	● ○
Medical product design							●	● ○
Bath								
Aerospace engineering ③	●	● ○ ⏷ ◑	12	●	○*	●	● ●	●
Automotive engineering ③	●	● ○ ⏷ ◑	12	●	○*	●	● ●	●
Belfast								
Aeronautical engineering ② ③		● ○ ⏷ ◑	12	●	○*	●	○	○
Birmingham								
Mechanical and automotive engineering ①		● ○ ⏷ ◑	12	●		●	● ●	●

Time abroad and sandwich courses

Institution — Course title (①②: see notes after table)	Named 'international' variant of the course	Location: ● Europe; ○ North America; ◖ industry; ◑ academic institution	Maximum time abroad (months)	Time abroad assessed	Language study: ○ optional; ● compulsory; * contributes to assessment	Socrates–Erasmus	Sandwich courses: ● thick; ○ thin	Arranged by: ● institution; ○ student
Bolton								
Automobile engineering ③								
Automotive product design		● ○ ◖ ◑	15		○*	●		
Bradford								
Automotive design technology ③						●	● ○	● ○
Mechanical and automotive engineering ① ③ ⑤		● ◖	12			●	● ○	●
Bristol								
Aeronautical engineering ① ② ③ ④	●	● ○ ◑	12		○*	●		○
Avionic systems ① ③		● ○ ◑		●	○*	●		
Bristol UWE								
Aerospace manufacturing engineering ① ③		● ○ ◖ ◑	12	●	○*	●	●	● ○
Aerospace systems engineering ① ③		● ○ ◖ ◑	12	●		●	●	● ○
Brunel								
Aerospace engineering ③						●	●	● ○
Mechanical engineering with aeronautics ③		● ○ ◖ ◑	18		○*	●	●	● ○
Mechanical engineering with automotive design ③		● ○ ◖ ◑	18		○*	●	●	● ○
Motorsport engineering ③		● ○ ◖ ◑				●	●	○
Cambridge								
Aerospace and aerothermal engineering ③					○*			
Central Lancashire								
Motorsports engineering						●		○
City								
Aeronautical engineering ③		● ○ ◖ ◑	12	●	○	●	●	
Air transport engineering ③		● ○ ◖ ◑	12	●	○	●	●	
Coventry								
Aerospace systems engineering		◖	12				●	● ○
Aerospace technology		◖					●	● ○
Automotive engineering ① ② ③		◖					●	● ○
Automotive engineering design	●	◖	15				●	○
Avionics technology		◖					●	● ○
Motorsport engineering		◖					●	○
Durham								
Aeronautics ③	●	● ○ ◖ ◑	12	●	○			
Glasgow								
Aeronautical engineering		● ◖ ◑	5	●	○*	●		
Avionics		● ◖ ◑	5	●	○*	●		
Mechanical engineering with aeronautics ③		● ○ ◖ ◑	12	●	○*	●		
Greenwich								
Marine engineering technology		◖	12	●			●	●
Harper Adams UC								
Agricultural engineering ③		● ○ ◖	12		○		●	● ○
Off-road vehicle design		● ○ ◖	12		○		●	●
Hertfordshire								
Aerospace engineering ③		● ○ ◖ ◑	12	●	○*	●	●	○ ○
Aerospace systems engineering ③		● ○ ◖ ◑	12	●	○*	●	●	○ ○
Automotive engineering ③		● ○ ◖ ◑	15	●	○*	●	●	○ ○
Automotive engineering with motorsport		● ○ ◖ ◑	15	●		●	●	○ ○
Motorsport technology		● ○ ◖ ◑	12	●		●	●	○

Mechanical and Manufacturing Engineering

Mechanical and Manufacturing Engineering

Time abroad and sandwich courses

Institution — Course title (① ②: see notes after table)

Location: ● Europe; ○ North America; ◗ industry; ◖ academic institution

Language study: ○ optional; ● compulsory; * contributes to assessment

Sandwich courses: ● thick; ○ thin

Arranged by: ● institution; ○ student

Course title	Named 'international' variant	Europe	N. America	Industry	Academic inst.	Max time abroad (months)	Time abroad assessed	Language study	Socrates–Erasmus	Sandwich courses	Arranged by
Huddersfield											
Automotive design		●	○	◗		12	●			●	●
Automotive design and technology ③		●	○	◗		12	●		●	●	●
Automotive engineering ③		●	○	◗		12	●		●	●	●
Automotive product innovation		●	○	◗		12	●		●	●	●
Automotive technology		●	○	◗	◖	12	●		●	●	●
Mechanical and automotive design		●	○	◗		12	●		●	●	●
Motorsport technology		●	○	◗		12		○		●	●
Imperial College London											
Aeronautical engineering	●	●	○	◗	◖	4	●	○*			
Kingston											
Aerospace engineering ③		●	○	◗		12	●	○	●	●	● ○
Aerospace engineering and astronautics ③		●	○	◗		12	●	○	●	●	● ○
Aerospace engineering design ③		●	○	◗	◖	12	●		●	●	● ○
Motorcycle engineering design ③		●	○	◗		12	●		●	●	○
Leeds											
Automotive engineering ③			○		◖	12	●	○*	●	●	●
Liverpool											
Aerospace engineering ① ③		●	○	◗		12		○		●	○
Aerospace engineering with pilot studies ①											
Liverpool John Moores											
Automotive engineering ③									●	●	●
Mechanical and marine engineering								○	●	●	○
Loughborough											
Aeronautical engineering ③									●	●	●
Automotive engineering ③									●	●	●
Manchester											
Aerospace engineering ③ ④	●	●		◗	◖	9	●	●*			
Newcastle											
Marine engineering ③								○	●		
Marine technology ③								○	●		
Mechanical and automotive engineering ① ③		●			◖	9	●	○*	●		
Naval architecture ③								○	●		
Offshore engineering ③								○	●		
Small craft technology ③								○	●		
Oxford Brookes											
Automotive engineering								○		●	● ○
Portsmouth											
Marine sports technology ③										●	● ○
Queen Mary											
Aerospace engineering ② ③		●	○	◗	◖	12	●	○*	●	●	○
Avionics ② ③		●	○	◗	◖	12	●	○*	●	●	○
Robert Gordon											
Mechanical and offshore engineering ③											
Salford											
Aeronautical engineering ③		●		◗		12	●	○	●	●	●
Mechanical engineering (aerospace) ① ② ③	●	●	○	◗	◖	12	●	○*	●	● ○	●
Sheffield											
Aerospace engineering ③											

Time abroad and sandwich courses

Institution — Course title ① ②: see notes after table	Named 'international' variant of the course	Location: ● Europe; ○ North America; ▶ industry; ◐ academic institution	Maximum time abroad (months)	Time abroad assessed	Language study: ○ optional; ● compulsory; * contributes to assessment	Socrates–Erasmus	Sandwich courses: ● thick; ○ thin	Arranged by: ● institution; ○ student
Sheffield (continued)								
Aerospace engineering with private pilot instruction ③								
Southampton								
Ship science ③	●	● ◐	12	●	○*	●	●	○
Ship science/yacht and small craft ③		● ◐	12	●	○*	●	●	○
Staffordshire								
Aeronautical design technology ③		● ▶ ◐	12	●	○*		●	● ○
Automotive design technology ③		● ▶ ◐	12	●	○*		●	● ○
Strathclyde								
Aeromechanical engineering ③	●	● ○ ▶ ◐	6	●	○*	●		
Naval architecture and marine engineering		● ○ ◐	12	●	○	●		
Naval architecture and small craft engineering		● ○ ◐	12	●	○			
Sunderland								
Automotive design and technology ③		● ○ ▶ ◐	12				●	
Automotive engineering with design ③		● ○ ▶ ◐	12				●	● ○
Automotive product design ③		● ○ ▶ ◐	12				●	● ○
Surrey								
Aerospace engineering		● ○ ▶ ◐	12			●	●	●
Swansea IHE								
Automotive engineering ③								
Motorsport engineering and design ③								
UCE Birmingham								
Automotive engineering ③		● ▶	12				●	●
UCL								
Naval architecture and marine engineering ① ② ③		▶	12	●	○*		●	○
Wolverhampton								
Automotive system engineering							●	○
York								
Avionics ③		● ○ ▶ ◐	12	●	○	●	● ○	○

① A year of industrial experience before the course starts is recommended for students on non-sandwich courses

② A year of industrial experience before the course starts is recommended for students on sandwich courses

③ Students on non-sandwich courses can take a year out for industrial experience

④ Other patterns of sandwich course : Bristol Aeronautical engineering 3–1–1; Manchester Aerospace engineering 1st half of year 4 (MEng); Nottingham Manufacturing engineering and management 1–3–1; 1–4–1; 2–1–1–1; 2–1–2–1 Mechanical engineering 1–3–1

⑤ Minimum period of vacation industrial experience for non-sandwich students (weeks) : Belfast Manufacturing engineering 12 Mechanical engineering 12; Bradford Mechanical and automotive engineering 52 Mechanical engineering 52; Cambridge Aerospace and aerothermal engineering 8 Manufacturing engineering 8 Mechanical engineering 8; Oxford Brookes Mechanical engineering 2; Southampton Acoustical engineering 20; Strathclyde Manufacturing engineering and technology 10 Product design engineering 10; Wales (Newport) Automation systems 12

Introductory period The early stages of all the engineering courses covered in this part of the *Engineering* Guide have the same basic content. For example, many of the courses in aeronautical engineering share their first year with a mechanical engineering course. Indeed, many institutions begin an even wider range of their engineering courses with a common introductory period: see TABLE 3b for where this is the case and at what point specialisation occurs.

Basic content   The courses almost always include the following basic subjects (the titles may differ between institutions):

<u>Mathematics and statistics</u>   The development of appropriate mathematical tools for the analysis and solution of engineering problems

<u>Computing</u>   Programming, and the use of computers for design, analysis and report writing, including the use of commercial software packages for these activities

<u>Mechanics</u>   The analysis of motion (velocity and acceleration) and forces in moving parts, such as linkage mechanisms and turbine rotors, and the relationship between force and motion (inertia effects)

<u>Structures</u>   The analysis of static forces and the calculation of stresses and strains, for example in pressure vessels, pipes, cranes and dams

<u>Properties of materials</u>   Explanations of why materials are strong or weak, ductile or brittle, and methods of improving these properties

<u>Thermodynamics</u>   The theoretical analysis of heat transfer (for example in boilers), and the performance and efficiency of heat engines, such as internal combustion engines, and steam and gas turbines

<u>Fluid mechanics</u>   The analysis of how gases and liquids flow around or through fixed or moving objects

<u>Electricity</u>   Circuit theory, motors and control devices

<u>Manufacturing</u>   The generation of simple and complex shapes to accurate dimensions in various materials by casting, moulding, plastic deformation and the use of machine tools

<u>Engineering drawing (design)</u>   You will first learn how to define the shape and size of components, and then how to communicate ideas and information through drawings and sketches; this knowledge is used as a basis for the design process. In most situations there will be more than one solution, and it is at this stage that the engineer must use judgement and creative skills to arrive at the most satisfactory design.

Specialised content: Mechanical engineering   The range of subjects covered in the first two years of most mechanical engineering and related courses is broadly similar. The first two years are concerned with the basic principles underlying the theory and practice of mechanical engineering. It is not until specialisation occurs towards the end of a course that significant differences appear. The later years of mechanical engineering courses develop the basic themes and add control, information technology, business and management studies, and quality assurance.

Considerable emphasis is placed on workshop and practical work. In the workshops and laboratory, you gain experience in manufacturing processes, experimental method and practical skills. You also learn to appreciate the practical relevance of what you have been taught in lectures, and how to write reports. These activities are usually fully integrated within the total curriculum.

Manufacturing engineering　　The initial years of courses in manufacturing engineering include many of the technology topics common to a range of engineering courses. They also introduce and emphasise subjects of particular interest to the manufacturing engineer, such as manufacturing technology, production processes, materials and materials science, management, economics, human factors, automation and quality engineering. There is also an increasing emphasis on computer-aided manufacturing systems (CAM), computer-integrated manufacturing (CIM) and robotics.

These core subjects are required as a foundation for final-year topics, which come under two broad headings: technology and management. The proportion under each of these headings, the range of subjects offered and the amount of choice vary from course to course.

Aeronautical engineering　　In the later years of aeronautical engineering courses, the major subjects studied include mathematics, aerodynamics, aircraft structures, aircraft stability and control, and aerospace engineering design.

Engineering and management　　There are a number of courses with titles involving engineering and management. These cover both the technological and management elements of engineering or manufacturing systems, and highlight the interrelationships between these elements.

Structure of final year(s)　　After the first two years, most MEng courses require a further two years of academic study, while BEng courses require just one. There is considerable variation in the way the final years are structured. At some institutions all students follow a common course with little choice, while at others there is a free choice of subjects from a wide range of available options. Between these two extremes there are a large number of institutions in which the final years contain a core curriculum but allow considerable freedom of choice for the remainder.

All engineering courses feature a major project in their final year, typically occupying about 25% of the study time: see Chapter 4 for more details.

Specialised topics TABLE 3b shows you where specific subjects are available as compulsory (●) or optional (○) components, or where there are both compulsory and optional components of a topic (◐). The table also indicates when a subject is available only on the MEng variant of a course (◡) – see TABLE 3c for more information on MEng courses. Note that the titles given to topics in the table may not correspond exactly with those used in individual prospectuses.

Table 3b — Final-year course content

Institution / Course title — ● compulsory; ○ optional; ◐ compulsory + options; ◡ MEng only; ① ②: see notes after table. Time of specialisation: S = semester; T = term; Y = year

Institution / Course title	Time	CAD	CAM/CIM	Manufacturing systems	Control engineering/automation	Quality control	NC machine tools	Acoustics/vibration	Energy resources	Engine technology	Optical methods	Stress analysis methods	Machine design	Electronics	Mechatronics	Avionics	Fluid dynamics	Planned maintenance	Materials	Advanced materials	Polymers	Systems engineering	Environmental engineering	Work design/ergonomics	Operational research	Artificial intelligence	Business/management
Mechanical Engineering																											
Aberdeen — Mechanical engineering	Y3	○		○									○						○	○	○		○				●
Aston — Mechanical engineering ①	Y2	●	○	○	●		○	●						◡			○		●	○							●
Bath — Innovation and engineering design ①	Y3	●		●	●					○										○				○			●
Bath — Mechanical engineering ②	Y3	○	○	○	●	○			○	○										○			○	○			●
Belfast — Mechanical engineering ①	Y2	○	○	●	●	○	○	○	○	●				●	●				●								
Birmingham — Mechanical engineering ①			○	○	○	○	○						○												○		○
Bolton — Mechanical engineering ①	Y1	●	○	○	●	○		●					○	●			●	●	●	●				○			
Bradford — Mechanical engineering ①	Y3	○	○	○	○														○	○							○
Bristol — Mechanical engineering			○	○	●				○	○			○				○	○	○	○				○	○	○	○
Bristol UWE — Mechanical engineering ①	Y2	○	○	○	○				○	○							○		○	○							●
Brunel — Mechanical engineering ①	Y3	○		●	●	○			○	○	○		●	●		○	●		○	○			○	●			●
Brunel — Mechanical engineering with building services ②	Y3			●	●				●	●				●						●			●				●
Cambridge — Mechanical engineering ①	Y3	○		○	○		○						○				○		○	○			○		○		○
Cardiff — Mechanical engineering ①	S2	◐		○	○	○			○	○			○				○			○						●	○
City — Mechanical engineering	Y2	○	○	○	○	○			○	○	○		○				○	○	○			○					
Coventry — Mechanical engineering ①		●		○	●				○	○			○				○		○								○
Dundee — Mechanical engineering ①	Y2	●		●						●			●				●		●	●						●	
Durham — Mechanical engineering ①	Y3			○	○			○	◐	○			○	◐			◐										○
Edinburgh — Mechanical engineering ①	Y2	○	○	○	○				○	○				○			○		○	○							○

Final-year course content

Legend:
● compulsory; ○ optional; ◑ compulsory + options; ▼ MEng only
S = semester; T = term; Y = year

Institution / Course title (① ② : see notes after table)	Time of specialisation	CAD	CAM/CIM	Manufacturing systems	Control engineering/automation	Quality control	NC machine tools	Acoustics/vibration	Energy resources	Engine technology	Optical methods	Stress analysis methods	Machine design	Electronics	Mechatronics	Avionics	Fluid dynamics	Planned maintenance	Materials	Advanced materials	Polymers	Systems engineering	Environmental engineering	Work design/ergonomics	Operational research	Artificial intelligence	Business/management
Exeter — Mechanical engineering ①	Y 2	●	○	○	○	○			●	●		●	○				○	○	◑	◑	○				●	○	◑
Glamorgan — Mechanical engineering	Y 1	○			○	●		○					○				○		○						○		
Glasgow — Mechanical design engineering ①			○		○		○	▼	○		○			○					●	●				▼	◑		●
Glasgow — Mechanical engineering ②			○		○		○	▼	○					○					○	▼				▼			●
Glasgow Caledonian — Computer-aided mechanical engineering	Y 2	●	●						●			●															
Greenwich — Mechanical engineering ①	S 2			○	○			○	○				○				○	○									●
Heriot-Watt — Mechanical engineering	Y 3	○	○	○	○	●	●	●	○	○							○	○							○	●	●
Hertfordshire — Mechanical engineering ①	Y 2			●	●	●		●									●										
Huddersfield — Engineering design: mechanical ①	Y 2	●		●	●	●		●				●	●						●								●
Huddersfield — Mechanical engineering ②	Y 2	●		●	●	●		●										●	●	▼							●
Hull — Mechanical engineering	S 2	○	○	○	○	○		○	○			○					○	○	○					○		○	●
Imperial College London — Mechanical engineering ①			○	○	○	○		○	○					○			○	○	○					○		○	
King's College London — Mechanical engineering ①	Y 2		●		●							●	○														○
Kingston — Mechanical engineering	Y 2	◑	◑		◑		◑					◑	◑						◑	◑	◑					◑	◑
Kingston — Mechanical engineering design	Y 2	◑			◑	◑	◑					◑	◑				◑	◑	◑	◑	◑			◑	◑		◑
Leeds — Mechanical engineering	Y 3	○	○	○	○	○		○																	○	○	
Leicester — Mechanical engineering ①	S 4			●	●			○	○			○					○	○	○			○			○	○	●
Liverpool — Mechanical engineering	Y 2	●	▼	▼	●	▼		▼				●	●						●	▼	▼						▼
Liverpool — Mechanical systems and design engineering	Y 2	●	▼	●	●	▼		▼				●	●						●	▼	▼						▼
Liverpool John Moores — Mechanical engineering ①	Y 2	●	○	○	○	○	○	○	○	○		○	●				○		○	○							○
London South Bank — Mechanical engineering	Y 2			◑	◑	◑													◑	◑							▼
Loughborough — Mechanical engineering ①			○	○		○	●		○	○		○	○	○	○			○		○	○	○	○				●
Manchester — Mechanical engineering			○	○	○			○	○			○	○				○	○							○		○
Manchester Metropolitan — Mechanical engineering ①	Y 3	○	○		○	●		○				○					○		●	●							
Napier — Mechanical engineering	Y 2	○	○	○	○									○					○								

(continued) Table 3b

Final-year course content

● compulsory; ○ optional; ◐ compulsory + options; ▬ MEng only
① ② : see notes after table
S = semester; T = term; Y = year

Institution / Course title	Time of specialisation	CAD	CAM/CIM	Manufacturing systems	Control engineering/automation	Quality control	NC machine tools	Acoustics/vibration	Energy resources	Engine technology	Optical methods	Stress analysis methods	Machine design	Electronics	Mechatronics	Avionics	Fluid dynamics	Planned maintenance	Materials	Advanced materials	Polymers	Systems engineering	Environmental engineering	Work design/ergonomics	Operational research	Artificial intelligence	Business/management
Newcastle																											
Mechanical and design engineering ①		●	●	●	●							●	●	●	●		●		●	●							●
Mechanical and railway engineering ②		●	●	●	●				●						●				●								●
Mechanical engineering ③		●	●	●	●				●			●	●		●				●								●
North East Wales I																											
Mechanical engineering	S 2	●	●	●	●	●	●	●	●	●		●	●	●	●	●	●		●	●	○		○	○	○	○	○
Northumbria																											
Mechanical engineering ①					◐			●									○										○
Nottingham																											
Mechanical engineering ①		○			◐			○	○				○	○			○			○	○						◐
Oxford																											
Mechanical engineering ①	Y 3	○	○	○	○		○	○	○				○	○	○				○	○	○		○	○	○		
Oxford Brookes																											
Mechanical engineering ①		○	○		●	○		◐													●						●
Plymouth																											
Mechanical engineering ①	Y 3	●		●	○	○											○		●	○				●	●		●
Portsmouth																											
Mechanical engineering		●	●	●	●	●			●			●			○				●	○	○			●	●		●
Queen Mary																											
Mechanical engineering	Y 2	○	○	●	●		○										○								○	●	●
Robert Gordon																											
Mechanical engineering ①	S 2			●								●						●	●		●						●
Sheffield																											
Mechanical engineering		○	●	●	○		○	○	○	▬		●	○		●			▬	●	●	○		▬	▬			●
Sheffield Hallam																											
Mechanical engineering	S 2	●		○	○	○											●		●								●
Southampton																											
Acoustical engineering								●																			
Mechanical engineering ①		○	○		○	●		○	○		▬	○	●	○	○		○		○	○	○		○	○			●
Southampton Solent																											
Mechanical design	S 2	●	●										●						●	●							●
Staffordshire																											
Mechanical engineering	Y 2	●		●	○			○				○							○		○	○					○
Mechanical systems design	Y 2	●	●	○				○	○			○							○		●	○					○
Strathclyde																											
Mechanical engineering		●		○	◐	○	○	◐	○	○		●	●	●	○		●		◐	◐	◐	●	◐				◐
Sunderland																											
Engineering design and manufacture ①	Y 2	●		●									●				●										●
Mechanical engineering with design		●		●								●							●		●				●		●
Surrey																											
Mechanical engineering ①	Y 2	●						○				○	○				○		▬	▬	▬		○				●
Sussex																											
Mechanical engineering	Y 2	○		○		○	○		○																		●
Swansea																											
Mechanical engineering		●	●	●	●	●	●	●	●			●							●								▬

Final-year course content

Legend: ● compulsory; ○ optional; ◐ compulsory + options; ▼ MEng only
Time of specialisation: S = semester; T = term; Y = year

Institution / Course title	Time of specialisation	CAD	CAM/CIM	Manufacturing systems	Control engineering/automation	Quality control	NC machine tools	Acoustics/vibration	Energy resources	Engine technology	Optical methods	Stress analysis methods	Machine design	Electronics	Mechatronics	Avionics	Fluid dynamics	Planned maintenance	Materials	Advanced materials	Polymers	Systems engineering	Environmental engineering	Work design/ergonomics	Operational research	Artificial intelligence	Business/management
Swansea IHE — Mechanical and manufacturing engineering ①				●	●							●								●							
Teesside — Mechanical engineering ①	Y 2	●	●				●					●	●		○		●		●								●
UCL — Mechanical engineering ①			●			○	○	▼	○			○		○			○		○	○		○					
Ulster — Mechanical engineering ①		●	●	●	●		●	●			●								○	○		○	●	○			
Warwick — Mechanical engineering ①	Y 2		○		▼		▼	○			▼	▼	●		○		▼			▼		▼	▼	▼			▼
Wolverhampton — Mechanical engineering	Y 2			●	●																						●

Manufacturing Engineering and Product Design

Institution / Course title	Time of specialisation	CAD	CAM/CIM	Manufacturing systems	Control engineering/automation	Quality control	NC machine tools	Acoustics/vibration	Energy resources	Engine technology	Optical methods	Stress analysis methods	Machine design	Electronics	Mechatronics	Avionics	Fluid dynamics	Planned maintenance	Materials	Advanced materials	Polymers	Systems engineering	Environmental engineering	Work design/ergonomics	Operational research	Artificial intelligence	Business/management
Anglia Ruskin — Computer-aided product design	Y 2	●	●									●							●					●			
Aston — Engineering product design ①			○	○	○	○		○	○				○	○					○					○			●
Industrial product design			◐	◐		◐	◐	◐	◐				◐	◐					◐				◐	◐			◐
Product design and management ②			◐	◐	◐				◐	◐		◐	◐	◐	◐				◐	◐		◐		◐			◐
Sustainable product design ③			◐	○					●	◐			○	◐				◐	◐	◐		●	◐				◐
Bath — Manufacturing ①	Y 3	○	●	●		○													○	○			○		○		
Belfast — Manufacturing engineering ①	Y 2	○	●	●	●	●	●	○	○			●							●					●			●
Bolton — Computer-aided product design ①	Y 2	●	●								○	○							●	●	●						○
Bournemouth — Design engineering		●	●	●								●							●	●							
Product design	Y 2	●	○		○	●	○		●	○			●	○				○	●	●		○	●				
Bradford — Industrial design ①			○	○	○	○	○					○	○						○					○			○
Brighton — Product design				●																	○		●				●
Brunel — Industrial design	Y 3	●	●		○	○						●	●	●									●				●
Product design		●	●									●	●	●									●				●
Product design engineering		●	●									●	●	●									●				●
Buckinghamshire Chilterns UC — Product design ①	Y 2			●									●										●				
Cardiff — Manufacturing engineering	Y 2	○	◐	●		○	○							○	○				○	○					●	●	○	●
Coventry — Industrial product design			○																								●

(continued) Table 3b — Final-year course content

● compulsory; ○ optional; ① ② : see notes after table
◑ compulsory + options; ▶ MEng only
S = semester; T = term; Y = year

Institution / Course title	Time of specialisation	CAD	CAM/CIM	Manufacturing systems	Control engineering/automation	Quality control	NC machine tools	Acoustics/vibration	Energy resources	Engine technology	Optical methods	Stress analysis methods	Machine design	Electronics	Mechatronics	Avionics	Fluid dynamics	Planned maintenance	Materials	Advanced materials	Polymers	Systems engineering	Environmental engineering	Work design/ergonomics	Operational research	Artificial intelligence	Business/management
Durham																											
Design, manufacture and management ①	Y 3	●	●		○		●	○				○			○										●	○	
East London																											
Product design		●																									
Glasgow																											
Product design engineering ①		●			○			○					○	●					●	●				●			●
Glasgow Caledonian																											
Manufacturing systems engineering	Y 3	●	●	●	●	●	●												●	●							●
Greenwich																											
Engineering product design ①		●		●																							●
Industrial automation ②	Y 2			○	○	●											○		○								●
Manufacturing systems engineering	S 2	○	○	○						○	○				○				○								●
Hertfordshire																											
Manufacturing engineering ①			●																								
Huddersfield																											
Manufacturing and operations management	Y 2	●	●	●	○	●	●				●		●	●	●				●								●
Product innovation, design and development ①		●	●			●									●				●								●
London South Bank																											
Engineering product design		●																	●					●			
Loughborough																											
Innovative manufacturing technology ①		○	○	●	○	○	●						○						○	○	○		○				●
Manufacturing engineering and management ②		○	○	●		○	○								○			○	○	○	○	○	○	○			●
Product design and manufacture ③		○	○	○	○	○	○					○							○	○	○		○	○			●
Newcastle																											
Mechanical and manufacturing engineering ①		●	●	●	●	●	●					●	●	●			●		●	●						●	●
Northumbria																											
Manufacturing systems engineering			●	●	●														○								○
Product design and technology ①				●																							●
Nottingham																											
Manufacturing engineering and management		●	●	●	○	○	○						○	○			○		○	○	○		○	○		○	●
Mechanical design, materials and manufacture		◑		◑	◑	◑									◑				●	◑	◑	◑					●
Plymouth																											
Mechanical design and manufacture ①		●	●	●		○																			○		
Portsmouth																											
Computer-aided product design ①		●	●																								●
Mechanical and manufacturing engineering		●	●	●	●	●	●		●			●							●								●
Product design and innovation ②																				●	●			●			●
Product design and modern materials ③																				●				●			●

Final-year course content

Legend: ● compulsory; ○ optional; ◑ compulsory + options; ▼ MEng only; S = semester; T = term; Y = year
① ② : see notes after table

Institution / Course title	Time of specialisation	CAD	CAM/CIM	Manufacturing systems	Control engineering/automation	Quality control	NC machine tools	Acoustics/vibration	Energy resources	Engine technology	Optical methods	Stress analysis methods	Machine design	Electronics	Mechatronics	Avionics	Fluid dynamics	Planned maintenance	Materials	Advanced materials	Polymers	Systems engineering	Environmental engineering	Work design/ergonomics	Operational research	Artificial intelligence	Business/management
Strathclyde																											
Manufacturing engineering and technology	Y 2	●	●	●	●	●	●	○					○	○	○				○			○			○	○	○
Product design engineering	Y 2	●	○	○	○	○	○						○	○	○			○	●	●	●			●	○	●	○
Swansea																											
Product design engineering			●	●	●	●						●							●	●			●				●
Swansea IHE																											
Manufacturing systems engineering ①				●	●	●													●								
Teesside																											
Computer-aided design engineering ①	Y 2	●	●	●			●	●						●					●					●			
Wales (Newport)																											
Automation systems ①	Y 2	●	●	●	●	●									○											○	
Warwick																											
Manufacturing and mechanical engineering ①	Y 2	●	●	●	●	●	▼		▼	▼	▼	○		●					▼					▼	▼	●	▼

Aeronautical, Agricultural, Automotive and Marine Engineering

Institution / Course title	Time of specialisation	CAD	CAM/CIM	Manufacturing systems	Control engineering/automation	Quality control	NC machine tools	Acoustics/vibration	Energy resources	Engine technology	Optical methods	Stress analysis methods	Machine design	Electronics	Mechatronics	Avionics	Fluid dynamics	Planned maintenance	Materials	Advanced materials	Polymers	Systems engineering	Environmental engineering	Work design/ergonomics	Operational research	Artificial intelligence	Business/management
Aston																											
Automotive product design			●	●	●	○	○			○			○	○	○		○	○							○	○	●
Medical product design ①		◑	◑	◑		◑	◑						◑		◑		◑	◑							◑	◑	◑
Bath																											
Aerospace engineering ①	Y 3		●		●		○									○	●		●								●
Automotive engineering ②	Y 3	○	○		●	○	●								○				●				○				●
Belfast																											
Aeronautical engineering ①			○		●	●	▼			●						○	●		●	○							●
Birmingham																											
Mechanical and automotive engineering							●			●														●			
Bolton																											
Automobile engineering ①	Y 1	●		○		○		○	○	●		○							●	○	○			○			
Automotive product design ②	Y 1	●													○				●	●	●			○			○
Bradford																											
Automotive design technology ①		○	○	○	○	○						○	○						○	○							○
Mechanical and automotive engineering	Y 3	○	○	○	○	○							○						○	○							○
Bristol																											
Aeronautical engineering ①			○	○	●		○					○		○					●						○	○	○
Avionic systems ②				○	◑		○						◑	◑	◑		◑								○	○	
Bristol UWE																											
Aerospace manufacturing engineering	Y 2	●	●	●	○	●	●						○	○	○		●		●	●			○	○		●	●
Aerospace systems engineering	Y 2	●	●	●	○	●	●						○		●	●	●		●							●	●
Brunel																											
Mechanical engineering with aeronautics ①	Y 3			●	●			●		●				●	●		●		●			●				●	●
Mechanical engineering with automotive design ②	Y 3			●				●		●				●			●		●				●			●	●

(continued) Table 3b — Final-year course content

● compulsory; ○ optional; ◐ compulsory + options; ▼ MEng only
S = semester; T = term; Y = year

① ② : see notes after table

Institution / Course title	Time of specialisation	CAD	CAM/CIM	Manufacturing systems	Control engineering/automation	Quality control	NC machine tools	Acoustics/vibration	Energy resources	Engine technology	Optical methods	Stress analysis methods	Machine design	Electronics	Mechatronics	Avionics	Fluid dynamics	Planned maintenance	Materials	Advanced materials	Polymers	Systems engineering	Environmental engineering	Work design/ergonomics	Operational research	Artificial intelligence	Business/management
Brunel (continued)																											
Motorsport engineering										●																	●
Cambridge																											
Aerospace and aerothermal engineering	Y 3	○		○	○			○						○			○		○	○				○		○	○
Central Lancashire																											
Motorsports engineering ①		●											●														
City																											
Aeronautical engineering ①	Y 2	○	○					○		○						●	○	○									●
Air transport engineering ②	Y 2		○		○			○		○						●	○	○									●
Coventry																											
Aerospace systems engineering ①														●		●	●		●	●		●					●
Aerospace technology	Y 2	○		●	●								●	●		●	●		●	○							●
Automotive engineering ②	Y 2	●	●	●		○	○											○		●	○				○		●
Automotive engineering design ③	Y 2	○	○	○				○																			●
Avionics technology	Y 2	○		○	○									●		●	○				●						●
Motorsport engineering		●	●	●						●							●			○							●
Farnborough CT																											
Aeronautical engineering ①														○		○	○										●
Glasgow																											
Aeronautical engineering ①	Y 3	●			○					○							○										●
Avionics	Y 2				●									●		●	●									○	
Mechanical engineering with aeronautics ②		○						○		▼		○					○		○	▼	▼						●
Greenwich																											
Automotive engineering with management		○	○		○	●			●	●			●														●
Marine engineering technology ①	Y 2			○				●		●			●				●		●								●
Harper Adams UC																											
Agricultural engineering ①		●			●									●	●				●					●			
Off-road vehicle design ②		●													●				●								●
Hertfordshire																											
Aerospace engineering ①	Y 2			●	●	●											●		●								●
Aerospace systems engineering ②	Y 2															●	●		●								●
Automotive engineering ③	Y 2			○						●									●						●		●
Automotive engineering with motorsport ④	Y 2																										●
Motorsport technology ⑤					●					●			●														●
Huddersfield																											
Automotive design ①	Y 2	●			●			●		●		●								●				●			●
Automotive design and technology	Y 2	●	●	●				●		●		●						●		●				●			●
Automotive engineering ②	Y 2	●			●			●		●							●			●							●
Automotive product innovation	Y 2	●						●		●					●					●		●					●
Automotive technology ③	Y 2	●			●			●		●										●							●
Mechanical and automotive design ④	Y 2	●			●							●	●							▼							●
Motorsport technology	Y 2	●	●	●	○		●	●		●			●	●					●					●			●
Imperial College London																											
Aeronautical engineering ①																○			○								○

Final-year course content

Legend: ● = compulsory; ○ = optional; ◑ = compulsory + options; ▼ = MEng only. ①②: see notes after table. S = semester; T = term; Y = year.

Institution / Course title	Time of specialisation	CAD	CAM/CIM	Manufacturing systems	Control engineering/automation	Quality control	NC machine tools	Acoustics/vibration	Energy resources	Engine technology	Optical methods	Stress analysis methods	Machine design	Electronics	Mechatronics	Avionics	Fluid dynamics	Planned maintenance	Materials	Advanced materials	Polymers	Systems engineering	Environmental engineering	Work design/ergonomics	Operational research	Artificial intelligence	Business/management
Kingston																											
Aerospace engineering	Y 2	◑	◑		◑	◑						◑		◑					◑	◑	◑						◑
Aerospace engineering and astronautics	Y 2	◑	◑		◑	◑						◑		◑					◑	◑	◑						◑
Aerospace engineering design	Y 2					●				●		●						●	●								●
Motorcycle engineering design ①	Y 2	◑	◑	◑	◑		◑			◑							◑		◑	◑	◑			◑			◑
Leeds																											
Automotive engineering		○	○	○	○	○	○		○																	○	○
Liverpool																											
Aerospace engineering			○	○	○	○	○							●					○	●	○					○	●
Aerospace engineering with pilot studies			○	○	○	○	○							●					○	●	○						●
Liverpool John Moores																											
Automotive engineering ①	Y 2	●	○	○						○		○	○				○		○	○							○
Mechanical and marine engineering	Y ?	○			○	○			○	●		○	○			○	○		○	○							○
Loughborough																											
Aeronautical engineering ①						○				●			○			○	○			○				▼	▼		▼
Automotive engineering ②						○				○							○			○				▼	▼		▼
Manchester																											
Aerospace engineering ①				●	●											○	●		●	●							●
Newcastle																											
Marine engineering ①	Y 2	◑	◑	◑	◑	▼		◑	◑				◑	◑			◑	◑	◑	◑			◑	▼	▼	◑	◑
Marine technology ②	Y 2	◑	◑	◑	◑	◑		◑	◑				◑	◑			◑	◑	◑	◑		◑	◑	◑	◑	◑	◑
Mechanical and automotive engineering ③			●	●	●	●				●		●	●	●			●		●	●							●
Naval architecture ④	Y 2	◑	◑	◑	◑	◑		◑	◑				◑				◑	◑	◑	◑		◑	◑	◑	◑	◑	◑
Offshore engineering ⑤	Y 2	◑	◑	◑	◑	▼		◑	◑				◑	◑			◑	◑	◑	◑		◑	◑	◑	◑	◑	◑
Small craft technology ⑥	Y 2	◑	◑	◑	◑	▼		▼	◑				◑				◑	◑	◑	◑		◑	◑	◑	◑	◑	◑
North East Wales I																											
Aeronautical engineering	S 2	○	○	○	○	○	○	○	○	○			○	○	○	○	○	○	○	○		○	○		○	○	○
Oxford Brookes																											
Automotive engineering ①	T 6	●	○		○		●												●	●							●
Plymouth																											
Marine technology ①	Y 3	○								○									○								
Portsmouth																											
Marine sports technology ①		●							●	●				●			●		●	●	●		●				●
Queen Mary																											
Aerospace engineering ①	Y 2			◑										◑						○					◑		●
Avionics	Y 2			◑					○					●						○					○	○	
Sports engineering															○					○							
Robert Gordon																											
Mechanical and offshore engineering ①	S 2			●		●	●		●								●		●								●
Salford																											
Aeronautical engineering	Y 2									●				○	●												
Mechanical engineering (aerospace) ①	Y 2		○	○																			○	○	○	○	
Sheffield																											
Aerospace engineering ①			○	○							○		○			○	○		○	○		○					●

(continued) Table 3b — Final-year course content

Institution / Course title
①②: see notes after table
● compulsory; ○ optional; ◑ compulsory + options; ▶ MEng only
S = semester; T = term; Y = year

Institution / Course title	Time of specialisation	CAD	CAM/CIM	Manufacturing systems	Control engineering/automation	Quality control	NC machine tools	Acoustics/vibration	Energy resources	Engine technology	Optical methods	Stress analysis methods	Machine design	Electronics	Mechatronics	Avionics	Fluid dynamics	Planned maintenance	Materials	Advanced materials	Polymers	Systems engineering	Environmental engineering	Work design/ergonomics	Operational research	Artificial intelligence	Business/management
Sheffield (continued)																											
Aerospace engineering with private pilot instruction ②				○	○					●		●		○					○	○		●					●
Southampton																											
Ship science ①				●				●		●							●	●	●	●		○	○	○	○		●
Ship science/yacht and small craft ②					●			○		●							●	●	●	○		○	○	○	○		●
Southampton Solent																											
Yacht and powercraft design ①		●																●	●	● ●					●		●
Staffordshire																											
Aeronautical design technology ①	Y 2	●			●		●			●		○	○	○	○		●		●								●
Automotive design technology ②	Y 2	●					○			●		○			●		●		●		○						●
Strathclyde																											
Aeromechanical engineering	Y 2	●		○	◑	○	○	○	◑	○	◑	●	●	●				●	◑	◑	◑	◑	●		◑		◑
Naval architecture and small craft engineering ①	Y 3																										○
Sunderland																											
Automotive design and technology ①		●								●									●			●					●
Automotive engineering with design ②	Y 2		●							●									●			●		●			●
Automotive product design ③		●								●									●								●
Surrey																											
Aerospace engineering ①	Y 2				○							○	○			●	○		○	○							●
Swansea IHE																											
Automotive electronics systems ①														●													
Automotive engineering ②		●			●		●			●									● ●								
Motorcycle engineering ③																			● ●								
Motorsport engineering and design ④		●								●									● ●								
Motorsport manufacturing engineering ⑤																			●								●
UCE Birmingham																											
Automotive engineering ①																											
UCL																											
Naval architecture and marine engineering ①		●		○	○		▶ ▶			●		○					●		○	○			○				●
Wolverhampton																											
Automotive system engineering	Y 2		○							○	○						○										○
York																											
Avionics ①		●		◑ ●								●		● ●							○				○	●

Mechanical Engineering

Aston ①Thermodynamics; product design; formula student racing car project

Bath ①Manufacturing processes ●; global design ●; innovation ●; human resources ●; supply management ○ ②Structural mechanics ●; thermofluids ●; global design ○; innovation ○; biomechanics ○; fluid power ○; vehicle ride and handling ○

Belfast ①Welding technology; plastics technology; design for assembly; strengthening mechanisms in metals

Birmingham ① Automotive engineering; thermofluids; advanced mechanics; language; CAE
Bolton ① Composite materials
Bradford ① Automotive engineering
Bristol UWE ① Aerofluids; mechanics of materials and structures; finite element analysis
Brunel ① Biomechanics; bioenergetics; biofluid mechanics; building services ② Electrical services and lighting design
Cambridge ① Advanced tribology
Cardiff ① Energy management; process engineering; machines and tribology; thermodynamics; solid mechanics; risk and hazards
Coventry ① Dynamics; thermodynamics; vehicle aerodynamics; chassis engineering; finite element analysis; automotive engines and transmissions; aerospace structures and materials, propulsion systems and aerodynamics
Dundee ① Solid mechanics
Durham ① Tribology; aerodynamics
Edinburgh ① Finite element analysis; project management
Exeter ① Mechanical reliability; life cycle engineering
Glasgow ① Applied design systems; design; thermodynamics; process engineering; lasers and optical systems; built environment; dynamics ② Lasers and optical systems; process engineering; built environment; dynamics; thermodynamics; design
Greenwich ① Choice of stream: mechanical process engineering; 'conventional' mechanical engineering; advanced manufacturing engineering
Hertfordshire ① Mechanical engineering design
Huddersfield ① Finite element analysis; design ② Finite element analysis
Imperial College London ① Advanced fracture mechanics; tribology; nuclear reactor technology; mechanical transmissions technology; welding, joining and adhesives; combustion
King's College London ① Thermofluids; numerical methods
Leicester ① Instrumentation; reliability; tribology
Liverpool John Moores ① Engineering design; engineering analysis

Loughborough ① Heat and mass transfer; internal combustion engines; finite element methods; machine dynamics; microprocessors; turbomachinery; welding and joining; digital processing; lasers; fracture mechanics
Manchester Metropolitan ① Structural design; fluid and aerodynamics; heat transfer; dynamics
Newcastle ① Joining technology; mechanical power transmissions; manufacturing technology; management of new product introduction ② Joining technology; mechanical power transmissions; railway operation (signalling and electrical systems; environmental management) ③ Bioengineering; mechanical power transmission; method; thermo-fluid dynamics; energy management; joining technology
Northumbria ① Mechanics; energy studies; condition monitoring
Nottingham ① Thermal systems engineering; component analysis and failure assessment; dynamics of mechanical systems
Oxford ① Turbomachinery; biomedical engineering; power transmission
Oxford Brookes ① Joining technology; advanced dynamics; thermal power systems; strength of components
Plymouth ① Thermal engineering
Robert Gordon ① Engineering analysis; failure analysis; systems analysis; project and operations management; plant performance
Southampton ① Bioengineering; electro-mechanical sustainable energy
Sunderland ① Modern design methodologies
Surrey ① Energy methods in stress analysis; propulsion; design; applied fracture mechanics
Swansea IHE ① Finite element analysis; condition monitoring
Teesside ① Design; computer-aided analysis; mechanics of solids; dynamics
UCL ① Heat transfer; applied mechanics; mathematics; thermodynamics; power transmission; language
Ulster ① Plant engineering; engineering design; systems reliability
Warwick ① Appropriate technology; automotive engineering; instrumentation; robotics; sustainability

Manufacturing Engineering and Product Design

Aston ① Product design and realisation; design of thermo/fluid systems ② Formula student racing car project; aesthetics ③ Environmental impact assessment; recycling of products; sustainability; ethics
Bath ① Manufacturing processes ●; geometric modelling ○; biomechanics ○; tribology ○; supply management ○
Belfast ① Welding technology; plastics technology; design for assembly; strengthening mechanisms in metals
Bolton ① Visualisation technology
Bradford ① Project management; product design; ergonomics

Buckinghamshire Chilterns UC ① Design
Durham ① Microprocessor systems
Glasgow ① Product design engineering; applied design systems; lasers; design and technology
Greenwich ① Product development; future-based product design ② Thermal power plant and heat transfer
Hertfordshire ① Manufacturing strategy; engineering decision support systems; reliability engineering
Huddersfield ① Innovation; product liability; creativity

Loughborough ① Rapid prototyping; reverse engineering; rapid manufacturing; structural integrity; sports engineering ② Rapid protptyping; sports engineering ③ Product design; innovation; reverse engineering; rapid prototyping; rapid manufacturing; CNC technology
Newcastle ① Joining technology; manufacturing technology
Northumbria ① Modelling for manufacture
Plymouth ① Underwater engineering; composites manufacture

Portsmouth ① Personal and professional development ●; product management systems ●
③ Marketing; product design management
① Product design management; marketing
Swansea IHE ① Finite element analysis
Teesside ① Design analysis project; computer modelling
Wales (Newport) ① Project management
Warwick ① Automotive engineering; sustainability; robotics

Aeronautical, Agricultural, Automotive and Marine Engineering

Aston ① Design of medical applications (eg surgical instruments, wheelchairs); ergonomics; materials
Bath ① Aircraft stability and control ●; aircraft performance and design ●; integrated manufacture and data management ●; aerodynamics ○; engineering plasticity ○; fluid power ○ ② Stuctural mechanics ●; thermofluids ●; power transmission ●; engineering plasticity ○; fluid power ○
Belfast ① Advanced aircraft structures; finite element analysis
Bolton ① Vehicle ergonomics; vehicle electronics; composite materials; applied numerical methods; quality and reliability ② Motor vehicle studies; modelling and visualisation
Bradford ① Vehicle testing; vehicle engineering
Bristol ① Computational aerodynamics; turbulence; multivariable control; non-linear dynamics and chaos; computational structural analysis; helicopter dynamics and aerodynamics ② Computer graphics; non-linear dynamics and chaos; networks and protocols; advanced computer architecture; image processing; multi-stage decision-making; optical communications; speech processing; neural networks; concurrent and distributed systems; formal methods
Brunel ① Flight mechanics; vehicle structures ② Thermofluid dynamics; vehicle structures and dynamics
Central Lancashire ① Formula student teams responsible for design, development and manufacture of race car
City ① Mathematics; structural dynamics ② Gas and turbine engineering
Coventry ① Aerostructures; aerodynamics ② Total quality management; marketing; finance; project management ③ Chassis engineering; automotive engines and transmissions; vehicle aerodynamics
Farnborough CT ① Aircraft structural design; air transport management; trials management; design for reliability and maintainability; software engineering; advanced aerodynamics
Glasgow ① Aeroelasticity; aero structures; space systems ② Lasers; thermodynamics; process engineering; built environment; dynamics; design; aeronautics
Greenwich ① Materials in service; vessel and fleet technical management; ship management

Harper Adams UC ① Off-road vehicle design; product development and testing; field engineering; farm and industrial buildings ② Product development and test; off-road vehicle design
Hertfordshire ① Aerospace engineering design; aerospace performance, propulsion and design; mechanics and properties of materials; stability and control of aircraft; aerospace structural design and analysis; aerodynamics ② Logistics engineering; aerospace engineering design ③ Vehicle structural analysis and manufacture; vehicle engineering design; vehicle dynamics ④ Motorsport engineering; aerodynamics and engine design for motorsport; vehicle engineering design; vehicle dynamics; vehicle structural analysis and manufacture; mechanics and properties of materials ⑤ Motorsport engineering
Huddersfield ① ② Finite element analysis; vehicle design ③ Project management; product liability; vehicle design ④ Finite element analysis; design
Imperial College London ① Structural dynamics; wing design; applications of fluid dynamics; computational mechanics; systems analysis; turbulence and turbulence modelling; helicopter dynamics; finite elements
Kingston ① Motorcycle engineering; motorcycle design
Liverpool John Moores ① Automotive design
Loughborough ① Aircraft design ② Vehicle design; advanced vehicle engineering; vehicle dynamics
Manchester ① Aerodynamics; finite element analysis; structural dynamics; aircraft structures; helicopters
Newcastle ① Marine engineering ② Marine-related subjects ③ Internal combustion engines; mechanical power transmissions; manufacturing technology; automotive applications; thermal power and propulsive systems ④ Naval architecture ⑤ Drilling engineering ⑥ Marine-related subjects
Oxford Brookes ① Thermodynamics; stress; dynamics; power train engineering; chassis engineering
Plymouth ① Naval architecture; marine and offshore engineering
Portsmouth ① Marine equipment; materials; marine leisure development; sports law; marine

manufacture; environmental law and policy; sports and globalisation

Queen Mary ① Aerospace structures; spacecraft design

Robert Gordon ① Engineering analysis; project and operations management; offshore engineering; plant performance

Salford ① Finite element analysis

Sheffield ① ② Aerodynamics and flight mechanics

Southampton ① ② Wide range of topics in marine and small craft design, structure, propulsion, safety; marine law; materials; finite element analysis; languages

Southampton Solent ① Structural design; aero-hydrodynamics; computational mechanics; materials and production for specialised craft

Staffordshire ① Flight technology; human factors ② Automobile systems design

Strathclyde ① Dynamics of marine vehicles; marine systems design; small craft aero-hydrodynamics; small craft structures; small craft systems and manufacture; risk management and reliability

Sunderland ① Automotive systems technology; modern design methodologies ② Vehicle suspension systems; vehicle electrical systems ③ Automotive systems technology; modern design methodologies

Surrey ① Numerical methods; propulsion; spacecraft systems; spacecraft dynamics

Swansea IHE ① Vehicle control systems; computer networks ② Automotive control and diagnostic systems; engine design; engine management systems; computational methods ③ Computer methods; engine structural design; structure and materials; vehicle control systems ④ Automotive control and diagnostic systems; engine design; structures; engine management systems; computational methods ⑤ Finite element analysis; engine design; control technology

UCE Birmingham ① Power train systems; body chassis engineering

UCL ① Naval architecture; language

York ① Radar systems; navaids; flight control systems; digital signal processing; transducers and instrumentation; parallel computing; real-time systems; software engineering

General studies and additional subjects

Specialised engineering courses run the risk of focusing the student's attention entirely on technical subjects. To prevent this, some institutions include topics on liberal, complementary or general studies in their specialised courses. TABLE 3a contains information about language teaching; you will need to look at prospectuses or ask institutions direct for information about other subjects.

Where a course is part of a modular degree scheme, there will probably be more opportunities to take a wide range of other subjects, though the requirements for professional qualification may mean that the choice is more restricted than it would be if you were following a course in a non-engineering subject. TABLE 2a shows where a course is part of a modular degree scheme allowing a relatively free choice of modules from a large number of subjects.

BEng/MEng courses

TABLE 2a lists many courses that can lead to the award of either a BEng or an MEng degree. For these courses, most of the tables in this Guide give information specifically for the BEng stream (the tables show information for MEng courses that are MEng only). Much of the information will also apply to the MEng stream, but TABLE 3c shows you where there are differences for the MEng stream. It shows at what point the MEng course separates from the BEng (for those courses where they are not separate from the start), and what proportion of students are expected to leave with an MEng degree.

Table 3c

MEng course differences

Institution	Course title	MEng separates from BEng	Students receiving MEng %	More engineering	More management	More languages	Other differences
Aberdeen	Mechanical engineering	Year 3	30	●	●		More extensive group project, engineering analysis and methods, project management
Aston	Mechanical engineering	Year 3	30	●	●		Advanced projects
Belfast	Aeronautical engineering						Business, management and economics studied in more detail
	Mechanical engineering	Year 3	25				Greater emphasis on mechanical engineering design, professional studies and manufacturing engineering
Birmingham	Mechanical and automotive engineering	Year 3	40	●			Year 3 project
	Mechanical engineering	Year 3	40	●			Project in year 3
Bradford	Mechanical and automotive engineering	Year 4		●			Dissertation in industry or university
	Mechanical engineering	Year 4		●			Enhanced academic-study; dissertation in industry or university
Bristol	Aeronautical engineering						Group design project; larger research project; more options
	Avionic systems	Year 3	60				Major group design project year 3; larger research project in final year; more options
	Mechanical engineering	Year 4	85				Major project
Bristol UWE	Aerospace manufacturing engineering	Year 3	10				Year 3: industrial case studies; year 4: major dissertation; broader choice of advanced level modules
	Aerospace systems engineering	Year 3	20	●			
	Mechanical engineering	Year 3	10	●			Research dissertation; broader study
Brunel	Aerospace engineering	Year 3	30	●	●		
	Mechanical engineering	Year 3	30	●	●		Major group design project
	Mechanical engineering with aeronautics	Year 3	50	●	●	●	Major group design project
	Mechanical engineering with automotive design	Year 3	50	●	●	●	Major group design project
	Mechanical engineering with building services	Year 3	30	●	●		Major group design project
	Motorsport engineering	Year 3	30	●	●		Major group design project
Cambridge	Aerospace and aerothermal engineering	Year 3	100				Exit after 3 years with BA exceptional
	Manufacturing engineering	Year 3	100	●			Exit after 3 years with BA exceptional
	Mechanical engineering		100				Exit after 3 years with BA exceptional
Cardiff	Manufacturing engineering	Start	30	●			Greater depth and breadth; more project work (especially groups)
	Mechanical engineering	Year 3	50	●	●		Additional group and individual project activities
City	Aeronautical engineering	Year 3	20				Individual project over 2 years (instead of 1)
	Air transport engineering	Year 3	20				Individual and group design projects over 2 years (instead of 1)
	Mechanical engineering	Year 3	20				Multidisciplinary projects; language options; management studies; individual and group design projects over 2 years (instead of 1)
Edinburgh	Mechanical engineering	Year 4	75				Industrial/European placement; group working exercise; advanced study modules
Exeter	Mechanical engineering	Year 2	40	●			Year 4: interdisciplinary group project

MEng course differences

Institution	Course title	MEng separates from BEng	Students receiving MEng %	More engineering	More management	More languages	Other differences
Glasgow	Aeronautical engineering	Year 3	10				
	Avionics	Year 4	10				
	Mechanical design engineering	Year 4	50	●	●		More project work
	Mechanical engineering	Year 4	50	●	●		More project work
	Mechanical engineering with aeronautics	Year 4	50	●	●		More project work
	Product design engineering	Year 4	50				More project work
Harper Adams UC	Agricultural engineering	Year 3	25	●	●		Broader base
	Off-road vehicle design	Year 3	30	●	●		
Heriot-Watt	Mechanical engineering	Year 3	10	●			Extensive project work; more options; deeper study of chosen subjects
Hertfordshire	Aerospace engineering	Year 4	5	●	●		Independent study; teamwork
	Aerospace systems engineering	Year 4	5	●	●		Independent study; teamwork
	Automotive engineering	Year 4	5	●	●		Independent study; teamwork
	Automotive engineering with motorsport			●	●		Independent study; teamwork
Huddersfield	Automotive design	Year 4	5	●	●		Major group project
	Automotive engineering	Year 4	5	●	●		Major group project
	Engineering design: mechanical	Year 4	5	●	●		
	Mechanical and automotive design	Year 4	5	●	●		Major group project
	Mechanical engineering	Year 4	5	●	●		
King's College London	Mechanical engineering	Year 2	15	●			
Kingston	Aerospace engineering	Year 2	25	●	●		Different modules studied
	Aerospace engineering and astronautics	Year 3	25	●	●		Different modules studied
Leeds	Automotive engineering	Year 2	90				
	Mechanical engineering	Year 2	50				
Leicester	Mechanical engineering	Year 3	25	●	●		Further options; major group design project
Liverpool	Aerospace engineering	Year 3	50	●			Options for some specialisation in flight technology, structure and materials, manufacturing, and avionic systems
	Aerospace engineering with pilot studies	Year 3	50	●			Options for some specialisation in flight technology, structure and materials, manufacturing, and avionic systems
	Mechanical engineering	Year 3	30	●	●		Longer project period; options in final year
	Mechanical systems and design engineering	Year 3	30	●	●		Longer period spent on project work; options in final year
Loughborough	Aeronautical engineering	Year 2	50	●	●	●	Advanced specialised modules, wide range of options, work with research teams and study overseas
	Automotive engineering	Year 2	50	●	●	●	Advanced specialised modules, wide range of options, work with research teams and study overseas
	Mechanical engineering	Year 3		●	●	●	Greater depth; business and language opportunities; more substantial projects; higher standard for progression
	Product design and manufacture	Year 3	20	●	●	●	Specialist modules in final year; optional study in Singapore; major professional development project
Manchester	Mechanical engineering	Year 3	40	●	●		

(continued) Table 3c — MEng course differences

Institution	Course title	MEng separates from BEng	Students receiving MEng %	More engineering	More management	More languages	Other differences
Newcastle	Marine engineering	Year 2	25		●		
	Marine technology	Year 2	25		●		
	Mechanical engineering	Year 3	60	●			
	Naval architecture	Year 2	25				
	Offshore engineering	Year 2			●		
	Small craft technology	Year 2	25		●		
Northumbria	Manufacturing systems engineering	Year 3					Greater breadth and depth; practical group projects
	Mechanical engineering	Year 3					Breadth and depth; organisational issues
Nottingham	Manufacturing engineering and management	Year 3					
	Mechanical design, materials and manufacture	Year 3	50	●	●		More project work
	Mechanical engineering	Year 3		●			
	Mechanical engineering (bioengineering)	Year 3			●		
Oxford Brookes	Automotive engineering	Year 3					
Portsmouth	Mechanical engineering	Year 4	20				MEng made up from elements of postgraduate MSc in Mechanical Engineering
Queen Mary	Aerospace engineering	Year 3					
	Mechanical engineering	Year 2					Specific courses in year 4; individual and group projects; computational mechanics; aircraft propulsion; advanced project management
Robert Gordon	Mechanical and offshore engineering	Year 3	10				Major multidisciplinary project, year 5
	Mechanical engineering	Year 3	10	●	●		Major multidisciplinary project, year 5
Salford	Aeronautical engineering		20				
	Mechanical engineering (aerospace)		30				
Sheffield	Aerospace engineering	Year 3	67	●			Group design project; choice of 8 career streams
	Mechanical engineering	Year 3	80	●	●		Group industrial design and project work; Master's level modules in year 4; larger individual project
	Mechanical systems engineering	Year 2					
Southampton	Acoustical engineering	Year 3	50	●	●		More industrial focus
	Mechanical engineering	Year 3	70	●	●		Group design and multidisciplinary projects
	Ship science	Year 3	90	●	●		Group and individual design projects
	Ship science/yacht and small craft			●	●		Group and individual design projects
Staffordshire	Mechanical engineering	Year 4	15				
	Mechanical systems design	Year 3	15	●			
Strathclyde	Aeromechanical engineering	Year 4	70	●	●		Group project
	Manufacturing engineering and technology	Year 3	40	●	●		Experience in research, industry and group work
	Mechanical engineering	Year 4	70	●	●	●	Advanced group project
	Naval architecture and small craft engineering	Year 4			●		Wider specialisation
	Product design engineering	Year 3	50	●	●		Experience in research, industry and group work
Surrey	*Both courses*	Year 3		●	●		Greater breadth and depth
Sussex	Mechanical engineering	Year 3					

Institution	Course title	MEng separates from BEng	Students receiving MEng %	More engineering	More management	More languages	Other differences
Swansea	Mechanical engineering	Level 3	20		●		More depth in technical subjects; bigger, industrially-related projects
UCL	Mechanical engineering	Year 3	40	●			Second class honours level of performance required for MEng entry; year 4 group design project
	Naval architecture and marine engineering	Year 3	40	●			Second class honours level of performance required for MEng entry; year 4 group ship design project
Warwick	Manufacturing and mechanical engineering	Year 3	60	●			Optional year's study abroad
	Mechanical engineering	Year 2		●			Optional year's study abroad; multi-disciplinary group project

MEng course differences

See Chapter 4 in the *Introduction* (page 16) for general information about teaching and assessment methods used in all types of engineering course, including those covered in this part of the Guide. It also explains how to interpret TABLE 4, which gives information about projects and assessment methods used on individual courses.

Mechanical and Manufacturing Engineering

Table 4 — Assessment methods

Institution	Course title	Key for frequency of assessment column: ● term; ◐ semester; ○ year	Frequency of assessment	Years of exams contributing to final degree (years of exams not contributing to final degree)	Coursework: minimum/maximum %	Project/dissertation: minimum/maximum %	Time spent on projects in: first/intermediate/final years %			Group projects: ● compulsory; ○ optional	Orals: ◐ if borderline; ● everyone / ○ for projects; ● everyone
Aberdeen	Mechanical engineering		◐	(1),(2),3,4,5	15/20	39/42	20	20	50	●	○
Anglia Ruskin	Computer-aided product design			3	10	90	80	80	80		◐○
Aston	Automotive product design		○	1,2,3,4	20/20	20/20	20	30	30	●	◐○
	Engineering product design		○	1,2,3,4	20/30	30/35	20	30	30	●	◐○
	Industrial product design		◐	(1),2,3,4	20/20	20/20	20	30	30	●	◐○
	Mechanical engineering		○	1,2,3,4,5	20/20	20/20	40	15	30	●	◐○
	Medical product design			1,2,3,4	20/20	20/20	20	20	40		
	Product design and management		○	1,2,3,4	20/20	20/20	20	20	40		◐○
	Sustainable product design		○	1,2,3,4	20/20	20/20	20	20	40		◐○
Bath	Aerospace engineering		◐	(1),2,3,4	0/50	40/40	10	15	50	●	○
	Automotive engineering		◐	(1),2,3,4	0/50	40/40	10	15	50	●	○
	Innovation and engineering design		◐	(1),2,3,4	0/50	40/40	10	15	50	●	○
	Manufacturing		◐	(1),2,3,4	0/50	40/40	10	15	50	●	○
	Mechanical engineering		◐	(1),2,3,4	0/50	40/40	10	15	50	●	○
Belfast	Aeronautical engineering		◐	(1),2,3,4	10/20	20/30	15	20	30	●	○
	Manufacturing engineering		◐	(1),2,3	10/15	25/30	10	25	25	○	◐○
	Mechanical engineering		◐	(1),2,3	10/15	25/30	10	25	25	○	◐○
Birmingham	Mechanical and automotive engineering		◑	(1),2,3,4	30/40	14/14	30	35	50	●	○
	Mechanical engineering		◐	(1),2,3,4	30/40	14/14	30	35	50	●	○
Blackpool and The Fylde C	Mechanical and production engineering		◐	1,2,3		20/33	0	33	33	●	
Bolton	Automobile engineering		○	(1),2,3	20/25	20/25	10	10	25		◐○
	Automotive product design		◐	(1),2,3	40/50	40/50	25	40	50	●	◐○
	Computer-aided product design		◐	(1),3	0/20	60/80	30	50	50		○
	Mechanical engineering		○	(1),2,3	71/81	17/25	10	30	30	●	○
Bournemouth	Design engineering		○	(1),2,(3),4	24	40	20	20	50	●	○●
	Product design		◑	(1),2,(3),4	10/10	50/60	12	15	80	●	◐○
Bradford	Automotive design technology		◐	(1),2,3,4					33		
	Industrial design		◐	(1),2,3,4					33		
	Mechanical and automotive engineering		◐	(1),2,3,4,5	10/20	17/17	5	25	25		○
	Mechanical engineering		◐	(1),2,3,4,5	10/20	17/17	5	10	25	○	○
Brighton	Product design				50	50	30	30	60		
Bristol	Aeronautical engineering		○	(1),2,3,4	20/20	27/27	5	25	33	●	◐○
	Avionic systems		○	(1),2,3,4	5/10	25/30	8	20	33	●	◐○
	Mechanical engineering		○	(1),2,3,4	7/10	30/40	20	32	42	●	○

Assessment methods

Institution	Course title	Frequency of assessment (Key: ◑ term; ◐ semester; ○ year)	Years of exams contributing to final degree (years of exams not contributing to final degree)	Coursework: minimum/maximum %	Project/dissertation: minimum/maximum %	Time spent on projects in: first/intermediate/final years %			Group projects: ● compulsory; ○ optional	Orals: ◑ if borderline; ● for projects; ○ everyone
Bristol UWE	Aerospace manufacturing engineering	◑	(1),**2,3**	25/**50**	20/**30**	20	20	20	●	◑○
	Aerospace systems engineering	◑	(1),**2,3**	25/**50**	20/**30**	10	15	25		○
	Mechanical engineering	◑	(1),**2,3**	10	25	10	25	30	●	○
Brunel	Aerospace engineering	○	(1),**2,3,4**							○●
	Industrial design	○	(1),**2,3**	50/**65**	20/**20**					○●
	Mechanical engineering	○	(1),**2,3,4**	27/**27**	14/**14**	29	17	58	●	○●
	Mechanical engineering with aeronautics	○	(1),**2,3,4,5**	27/**27**	14/**14**	29	17	58	●	○●
	Mechanical engineering with automotive design	◑	(1),**2,3,4,5**	27/**27**	14/**14**	29	17	58	●	○●
	Mechanical engineering with building services	○	(1),**2,3,4,5**	27/**27**	14/**14**	29	17	50	●	◑●
	Motorsport engineering	○	(1),**2,3,4**	27/**27**	14/**14**	29	17	58	●	○●
	Product design	○	(1),**2,3**	50/**65**	20/**20**					○●
	Product design engineering	○	(1),**2,3**	50/**65**	20/**20**					
Buckinghamshire Chilterns UC	Product design	◑	(1),**2,3**	40/**70**	30/**60**	60	60	70	●	○
Cambridge	Aerospace and aerothermal engineering	○	(1),(2),**3,4**	0/**20**	50/**50**	5	25	50	●	○
	Manufacturing engineering	○	(1),(2),**3,4**	**20**	50/**50**	5	20	50	●	○
	Mechanical engineering	○	(1),(2),**3,4**	0/**20**	50/**50**	5	20	50	●	○
Cardiff	Manufacturing engineering	◑	(1),**2,3,4**	10/**15**	25/**25**	10	10	25	●	◑○●
	Mechanical engineering	◑	(1),**2,3,4,5**	8/**8**	25/**25**	25	25	25	●	◑○●
Central Lancashire	Motorsports engineering	○	**2,3**	50	33	20	30	66	●	◑○●
City	Aeronautical engineering	○	(1),**2,3**	20/**25**	10/**15**	10	15	25	●	◑○
	Air transport engineering	○	(1),**2,3,4**	20/**25**	10/**15**	10	15	25	●	◑○
	Mechanical engineering	○	(1),**2,3**	20/**25**	10/**15**	10	15	25	●	◑○
Coventry	Aerospace systems engineering	○	(1),**2,3**	30/**40**	25/**25**	25	25	25	●	◑○
	Aerospace technology	◐	(1),**2,3**	23/**23**	50/**50**	40	40	50	●	○
	Automotive engineering		(1),**2,3**	30/**45**	30/**40**	20	20	40	●	◑○
	Automotive engineering design	◐	(1)	0/**70**	30/**100**	30	100	100	●	◑○
	Avionics technology	◐	(1),**2,3**	30/**40**	30/**45**	25	30	30		◑○
	Industrial product design	◐	(1)		100	30	40	100		◑○
	Mechanical engineering	○	(1),**2,3**	15/**20**	40/**40**	25	25	50	●	◑○
	Motorsport engineering	◐	(1),**2,3**	30/**40**	30/**45**	25	30	30	●	◑○
Dundee	Mechanical engineering	◑	(1),(2),**3,4**	10/**25**	20/**30**	15	25	30	●	
Durham	Aeronautics	○	(1),**2,3,4**	5/**10**	25/**25**	5	10	50	●	◑○
	Design, manufacture and management	○	(1),**2,3,4**	5/**10**	25/**25**	5	10	50	●	◑○
	Mechanical engineering	○	(1),**2,3,4**	5/**10**	25/**25**	5	10	50	●	◑○
East London	Product design	◑	(1),**2,3**	40/**40**	60/**60**	40	50	60	●	○
Edinburgh	Mechanical engineering	◑	(1),(2),**3,4,5**	22/**22**	26/**26**	0	15	30	●	◑○
Exeter	Mechanical engineering	◑	(1),**2,3,4**	20/**35**	25/**30**		25	25	◐	○●
Farnborough CT	Aeronautical engineering	◑	(1),**2,3**	50	**50**	25	25	25	●	○
Glamorgan	Mechanical engineering	◑	(1),**2,4**	40/**40**	40/**40**	15	25	40	●	○
Glasgow	Aeronautical engineering	◐	(1),(2),**3,4,5**	5/**5**	20/**20**		10	20	●	○
	Avionics		(1),(2),**3,4,5**	10/**10**	20/**20**		10	20	●	○

(continued) Table 4 — Assessment methods

Institution	Course title	Key for frequency of assessment column: ● term, ◑ semester, ○ year	Frequency of assessment	Years of exams contributing to final degree (years of exams not contributing to final degree)	Coursework: minimum/maximum %	Project/dissertation: minimum/maximum %	Time spent on projects in: first/intermediate/final years %			Group projects: ● compulsory; ○ optional	Orals: ◑ if borderline; ● for projects; ○ everyone
Glasgow (continued)	Mechanical design engineering		◑	(1),(2),**3,4,5**	0/**10**	20/**35**	5	10	40	●	○
	Mechanical engineering		◑	(1),(2),**3,4,5**	0/**10**	20/**35**	5	10	40	●	○
	Mechanical engineering with aeronautics		◑	(1),(2),**3,4,5**	0/**10**	20/**35**	5	10	40	●	○
	Product design engineering		◑	(1),(2),**3,4,5**	15/**15**	50/**50**	20	35	60	●	○
Glasgow Caledonian	Computer-aided mechanical engineering		◑	(1),(2),(3),**4**	30/**70**	30/**30**	12	17	25		
	Manufacturing systems engineering		◑	(1),(2),**3**,(4),**5**	20/**30**	33		10	33		◑○
Greenwich	Automotive engineering with management		◐	(1),**2,3**	20/**20**	25/**25**	15	15	25	●	○
	Engineering product design		◑	(1),**2,3**		50/**50**	25	25	50	●	◑○
	Industrial automation		◐	(1),**2,3**	20/**20**	25/**25**	12	12	25	●	○
	Manufacturing systems engineering		◐	(1),**2,3**	20/**20**	25/**25**	15	15	25	●	◑○
	Marine engineering technology		◐	(1),**2,3**	20/**20**	25/**25**	12	12	25		○
	Mechanical engineering		◐	(1),**2,3**	20/**20**	25/**25**	12	12	25		◑○
Harper Adams UC	Agricultural engineering		◑	(1),**2,4,5**	30/**30**	25/**25**	30	20	50	●	◑○●
	Off-road vehicle design		◑	(1),**2,4,5**	10/**15**	20/**25**	20	20	35	●	○
Heriot-Watt	Mechanical engineering		◐	(1),(2),**3,4,5**	20/**20**	25/**25**	20	15	50		○
Hertfordshire	Aerospace engineering		◐	(1),**2,4,5**	20/**25**	35/**40**	10	20	25	●	○●
	Aerospace systems engineering		◐	(1),**2,3**	20/**25**	35/**40**	10	20	35	●	○●
	Automotive engineering		◐	(1),**2,4,5**	20/**25**	35/**40**	10	20	35		○
	Automotive engineering with motorsport		◐	(1),**2,4,5**	20/**25**	35/**40**	10	20	35		○
	Manufacturing engineering		◐	(1),**2**,(3),**4**	15/**50**	30/**85**	40	20	40		○
	Mechanical engineering		◐	(1),**2,4,5**	20/**25**	35/**40**	10	20	35		○●
Huddersfield	Automotive design		○	(1),**2,4**	30/**50**	15/**15**	50	50	60	●	○●
	Automotive design and technology		○	(1),**2,3,4**	25/**40**	25/**30**	25	25	35		○●
	Automotive engineering		○	(1),**2,4**	30/**50**	15/**15**	50	50	60	●	○●
	Automotive product innovation		○	(1),**2,3,4**	30/**40**	25/**30**	20	20	35		○●
	Automotive technology		○	(1),**2,3**	20/**50**	25/**30**	25	25	30		○●
	Engineering design: mechanical		○	(1),**2,4**	30/**50**	15/**15**	50	50	60	●	○●
	Manufacturing and operations management		○	(1),**2,3,4**	20/**50**	25/**30**	30	30	30		○●
	Mechanical and automotive design		○	(1),**2,4**	30/**50**	15/**15**	50	50	60		○●
	Mechanical engineering		○	(1),**2,4**	30/**50**	15/**15**	50	50	60	●	○
	Motorsport technology		○	(1),**2,3,4**	50/**60**	25/**30**	15	15	30	●	○●
	Product innovation, design and development		○	(1),**2,3,4**	20/**50**	25/**30**	30	30	30		◑○
Hull	Mechanical engineering		◑	(1),**2,3,4**	50	20/**35**	10	30	50	●	◑○●
Imperial College London	Aeronautical engineering		○	**1,2,3,4**	14/**21**	21/**32**	2	17	60	●	
	Mechanical engineering		◐	**1,2,3,4**	15/**20**	15/**20**	5	10	25	●	
King's College London	Mechanical engineering		◑	**1,2,3,4**	32/**38**	21/**21**	6	10	30	●	
Kingston	Aerospace engineering		◑	(1),**2,3,4,5**	30/**50**	25/**25**	10	20	35	●	○
	Aerospace engineering and astronautics		◑	(1),**2,3,4,5**	**75**	**25**	10	20	35		○
	Aerospace engineering design		◑	(1),**2,3**	30/**30**	50/**60**	10	20	50		○
	Mechanical engineering		◑	(1),**2,3,4,5**	**75**	**25**	10	20	25	●	○

Assessment methods

Institution	Course title	Frequency of assessment	Years of exams contributing to final degree (years of exams not contributing to final degree)	Coursework: minimum/maximum %	Project/dissertation: minimum/maximum %	Time spent on projects in: first/intermediate/final years %			Group projects: ● compulsory; ○ optional	Orals: ◑ if borderline; ● for projects; ● everyone
Kingston (continued)	Mechanical engineering design	◑	(1),2,3,4	75	25	10	20	25		○
	Motorcycle engineering design	◑	(1),2,3	75	25	10	20	25	●	○
Leeds	Automotive engineering	◑	(1),2,3,4	10/15	20/30	10	20	33		
	Mechanical engineering	◑	(1),2,3,4	10/15	20/30	10	20	33	●	◑○
Leicester	Mechanical engineering	◑	1,2,3,4	10/20	10/20	10	20	35	●	○
Liverpool	Aerospace engineering	◑	(1),2,3,4	30/30	15/15	5	10	15		◑○
	Aerospace engineering with pilot studies	◑	(1),2,3,4	30/30	15/15	5	10	15		◑○
	Mechanical engineering	◑	(1),2,3,4	0/10	18	5	10	20	○	○
	Mechanical systems and design engineering	◑	(1),2,3,4	0/10	18	5	10	20	●	○
Liverpool John Moores	Automotive engineering	◑	(1),2,3	20/50	30/30	20	20	30	●	○
	Mechanical and marine engineering	◑	(1),2,3	20/50	30/30	20	20	30	○	○
	Mechanical engineering	◑	(1),2,3	20/50	30/30	20	20	30	●	○
London South Bank	Engineering product design	◑	(1),2,3	20/30	50/70	15	50	70	○	○○●
	Mechanical engineering	◑	(1),2,3	20/20	10/25	10	40	40	●	◑
Loughborough	Aeronautical engineering	◑	(1),2,3,4	16/23	35/35	0	10	35	●	○
	Automotive engineering	◑	(1),2,3	16/23	35/35	0	10	35	●	○
	Innovative manufacturing technology	◑	(1),2,3,4	15/25	25/25	15	30	40	●	◑○
	Manufacturing engineering and management	◑	(1),2,3,4	20/20	30/30	10	20	25	●	◑○
	Mechanical engineering	◑	(1),2,3,4,5	15/25	30/30	20	30	45	●	◑○
	Product design and manufacture	◑	(1),2,3,4,5	20/20	35/40	10	20	25	●	◑○
Manchester	Aerospace engineering	◑	1,2,3,4	30/30	12/12		10	25	●	◑○
	Mechanical engineering	○	(1),2,3,4	10/20	25/30	15	10	30	●	○
Manchester Metropolitan	Mechanical engineering	◐●	(1),2,3	48/60	23	3	5	20	○	○
Napier	Mechanical engineering	◑	(1),(2),3,4	30/30	25/25	10	15	25		○
Newcastle	Marine engineering	◑	1,2,3		20	5	10	20	●	◑
	Marine technology	◑	1,2,3,4		20	5	10	20	●	◑
	Mechanical and automotive engineering	◑	(1),2,3,4	7/7	24/24	5	25	33	●	○
	Mechanical and design engineering	◑	(1),2,3,4	7/7	24/24	5	25	33	●	○
	Mechanical and manufacturing engineering	◑	(1),2,3,4	7/7	24/24	5	25	33	●	○
	Mechanical and railway engineering	◑	(1),2,3,4	7/7	24/24	5	25	33	●	○
	Mechanical engineering	◑	(1),2,3,4	7/7	18/24	5	25	33	●	○
	Naval architecture	◑	1,2,3,4		20/25	5	10	20	●	◑
	Offshore engineering	◑	1,2,3		20/25	5	20	20	●	◑
	Small craft technology	◑	1,2,3		20	5	10	20	●	◑
North East Wales I	Aeronautical engineering	◑	1,2,3	40/60	40/60	15	30	60		
	Mechanical engineering	◑	1,2,3	25/35	25/35	20	30	35		
Northumbria	Manufacturing systems engineering	◑	(1),2,3	14/16	34/34	17	21	42		◑○
	Mechanical engineering	◑	(1),2,3,4	14/16	34/34	17	21	42		◑○
Nottingham	Manufacturing engineering and management	◑	1,2,3,4	20/50	18/18	10	10	25	○	◑○
	Mechanical design, materials and manufacture	◑	(1),2,3	20/30	20/30	10	20	30	●	◑

Mechanical and Manufacturing Engineering

Mechanical and Manufacturing Engineering

Assessment methods

Institution	Course title	Frequency of assessment	Years of exams contributing to final degree (years of exams not contributing to final degree)	Coursework: minimum/maximum %	Project/dissertation: minimum/maximum %	Time spent on projects in: first/intermediate/final years %			Group projects: ● compulsory; ○ optional	Orals: ◑ if borderline; ● everyone / ○ for projects
Nottingham (continued)	Mechanical engineering	◑	(1),**2,3,4**	5/**10**	20/**25**	10	10	33	●	◑ ○
	Mechanical engineering (bioengineering)	◑	(1),**2,3,4**							
Oxford	Mechanical engineering		(1),**3,4**	13/**13**	25/**25**	10	15	50	●	◑
Oxford Brookes	Automotive engineering	◑	(1),**2,3,4**	**30**				35		
	Computer-aided mechanical engineering	◑	(1),**2,3,4**							
	Mechanical engineering	◑	(1),**2,3**	36/**51**	11/**11**	25	25	35	●	◑ ○
Paisley	Mechanical engineering	◑	(1),(2),**3,4**		15			15		◑ ○
Plymouth	Marine technology	◑	(1),**2,3**	**60**	**40**	40	40	50	●	◑ ○
	Mechanical design and manufacture	◑	(1),**2,3**	**60**	**40**	40	40	50	●	◑ ○
	Mechanical engineering	◑	(1),**2,3**	**60**	**40**	40	40	50	●	◑ ○
Portsmouth	Computer-aided product design	◑	(1),**2,3,4**	25/**50**	25/**50**	15	30	50	●	◑
	Marine sports technology	◑	(1),**2,3,4**	29/**31**	33/**33**	10	30	33	●	◑ ○ ●
	Mechanical and manufacturing engineering	◑	(1),**2,3,4,5**	25/**60**	20/**30**	15	30	45	●	◑
	Mechanical engineering	◑	(1),**2,3,4,5**	24/**70**	15/**25**	15	30	60	●	◑
	Product design and innovation	◑	(1),**2,3,4**	25/**50**	25/**50**	15	30	50	●	◑
	Product design and modern materials	◑	(1),**2,3,4**	25/**50**	25/**50**	15	30	50	●	◑
Queen Mary	Aerospace engineering	○	**1,2,3,4**	20/**25**	10/**15**	0	20	50	●	◑ ○ ●
	Avionics	◑	**1,2,3**	20/**25**	10/**15**	0	0	50	●	◑ ○ ●
	Mechanical engineering	◑	**1,2,3**	10	20	0	0	25	○	◑ ○
	Sports engineering	○	**1,2,3,4**			25	25	75	●	
Robert Gordon	Mechanical and offshore engineering	◑	(1),(2),**3,4,5**	**20**	**25**	33	20	25	●	◑ ○ ●
	Mechanical engineering	◑	(1),(2),**3,4,5**	**20**	**25**	33	20	25	●	◑ ○ ●
Salford	Aeronautical engineering	◑	(1),**2,3,4**	20	20	20	20	35	●	
	Mechanical engineering (aerospace)	◑	(1),**2,3,4**	40/**50**	17/**17**	10	20	30	●	○
Sheffield	Aerospace engineering	◑	(1),**2,3,4**	10/**15**	25/**25**	5	15	33	●	◑ ○
	Aerospace engineering with private pilot instruction	◑	(1),**2,3**	10/**15**	20/**25**	5	15	25		◑ ○
	Mechanical engineering	◑	(1),**2,3,4**	20/**30**	19/**22**	5	30	33		○
	Mechanical systems engineering	◑	(1),**2,3,4**							
Sheffield Hallam	Mechanical engineering	◑	(1),**2,4**	20/**80**	20/**20**	20	25	25	○	○ ●
Southampton	Acoustical engineering	◑	(1),**2,3,4**	7/**20**	20/**20**	5	25	25	●	◑ ○
	Mechanical engineering	◑	(1),**2,3,4**	27/**35**	17/**32**	4	25	40	○	◑ ○
	Ship science	◑	(1),**2,3,4**	10/**25**	20/**25**	0	25	33	●	○
	Ship science/yacht and small craft	◑	(1),**2,3,4**	10/**25**	20/**25**		25	33		○
Southampton Solent	Mechanical design	◑	(1),**2,3**	40	33	5	10	40	●	◑ ○
	Yacht and powercraft design	◐	(1),**2,3**	40	25/**30**	10	15	25	●	○
Staffordshire	Aeronautical design technology	◑	(1),**2,3**	30/**50**	40/**60**	12	25	40		○ ●
	Automotive design technology	◑	(1),**2,3**	30/**50**	40/**60**	12	25	40	●	○ ●
	Mechanical engineering	◑	(1),**2,3**	20/**30**	40/**50**	12	25	40	●	○ ●
	Mechanical systems design	◑	(1),**2,3**	20/**30**	40/**50**	12	25	40	●	○ ●
Strathclyde	Aeromechanical engineering	◑	(1),**2,3,4,5**	20/**60**	40/**60**	25	25	35		○
	Manufacturing engineering and technology	◑	(1),**2,3,4,5**	30/**50**	30/**70**	15	50	50		○ ●
	Mechanical engineering	◑	(1),**2,3,4,5**	20/**60**	40/**60**	25	25	35	●	○

Key for frequency of assessment column: ◑ term; ◐ semester; ○ year

Assessment methods

Institution	Course title	Frequency of assessment	Years of exams contributing to final degree (years of exams not contributing to final degree)	Coursework: minimum/maximum %	Project/dissertation: minimum/maximum %	Time spent on projects in: first/intermediate/final years %			Group projects: ● compulsory; ○ optional	Orals: ◐ if borderline; ○ for projects; ● everyone
Strathclyde (continued)	Naval architecture and marine engineering	◑	(1),2,3,4,5	20/40	20/30	5	20	40	●	○
	Naval architecture and small craft engineering	◑	(1),2,3,4,5	20/40	20/30	5	20	40	●	○
	Product design engineering	◑	(1),2,3,4,5	30/30	30/70	15	50	50	●	○●
Sunderland	Automotive design and technology	◑	(1),2,3,4	30/40	35/45	20	20	45		●
	Automotive engineering with design	◑	(1),2,3,4	10/20	35/40	20	20	35		○●
	Automotive product design	◑	(1),2,3,4	30/40	50/60	30	40	60	●	●
	Engineering design and manufacture	◑	(1),2,3,4	10/20	35/40	20	20	35		●
	Mechanical engineering with design	◑	(1),2,3,4	10/20	33/40	20	20	35	●	○
Surrey	Aerospace engineering	◑	(1),2,3,4	15/25	15/15	15	10	40	●	○
	Mechanical engineering	◐	(1),2,4	15/25	15/15	15	10	40	●	○
Sussex	Mechanical engineering	○	(1),2,3	15/15	20/20	0	20	20		◐○
Swansea	Mechanical engineering	◑	(1),2,3,4	26/26	21/21	12	47	47	●	○
	Product design engineering	◑	(1),2,3,4	25/25	20/20	10	50	50	●	○
Swansea IHE	Automotive engineering	◑●	(1),2,3	55/55	28/28	0	15	30	●	○
	Motorcycle engineering	◑●	(1),2,3	55/55	28/28	0	15	30	●	○
	Motorsport engineering and design		(1),2,3	55/55	28/28	10	15	30	●	○
Teesside	Computer-aided design engineering	◑	(1),2,3	40/40	30/30	10	25	35	●	◐○●
	Mechanical engineering	◑	(1),2,3	40/50	20/30	10	15	25	○	
UCE Birmingham	Automotive engineering	◑	(1),2,3							○
	Engineering product design	◑	(1),2,3							
	Mechanical engineering	◑	(1),2,3							
UCL	Mechanical engineering	○	1,2,3,4	25/33	20/20	10	10	25		○
	Naval architecture and marine engineering	○	1,2,3,4	25/33	20/20	10	10	25		○
Ulster	Mechanical engineering	◑	(1),(2),3,4	18/29	42/42	25	42	42	●	○●
Wales (Newport)	Automation systems	◑	(1),2,3	20/25	20/25	10	20	25		◐○
Warwick	Manufacturing and mechanical engineering	○	(1),2,3,4	18/24	18/18	12	12	25	●	○●
	Mechanical engineering	○	(1),2,3,4	18/24	18/18	12	12	25	●	○●
Wolverhampton	Automotive system engineering	◑	(1),2,3		25			25		○
	Mechanical engineering	◑	(1),2,3		25			25		○
York	Avionics	◑●	1,2,3,4	80/80	20/20	5	10	60		◐○●

Key for frequency of assessment column: ◐ term; ◑ semester; ○ year

Mechanical and Manufacturing Engineering

See Chapter 5 in the *Introduction* (page 18) for general information about entrance requirements that applies to all types of engineering course, including those covered in this part of the Guide. It also explains how to interpret TABLE 5, which gives information about entrance requirements for individual courses.

Table 5 — Entrance requirements

Institution	Course title	Number of students (includes other courses)	Typical offers (BSc/BEng) UCAS tariff points	A-levels	SCQF Highers	Typical offers (MEng) UCAS tariff points	A-levels	SCQF Highers	● compulsory; ○ preferred A-level Mathematics	A-level Physics
Aberdeen	Mechanical engineering	(170)		CCD	ABBBC		BCC	ABBBC	●	●
Anglia Ruskin	Computer-aided product design		140							
Aston	Automotive product design	15	240–280	BCC	BBBBC				○	
	Engineering product design	20	240–280		BBBBC				○	
	Industrial product design	15	240–280	BCC	BBBBC				○	
	Mechanical engineering	30	240–280	BCC	BBBBC	300	BBB	AABBB	●	○
	Medical product design	3	240–280	BCC	BBBBC				○	
	Product design and management	15	240–280	BCC	BBBBC				○	
	Sustainable product design	3	240–280	BCC	BBBBC				○	
Bath	Aerospace engineering	45					AAB		●	●
	Automotive engineering	15					AAB		●	●
	Innovation and engineering design	12					AAB		●	●
	Manufacturing	10					AAB		●	●
	Mechanical engineering	50					AAB		●	●
Belfast	Aeronautical engineering	40		BBC	BBBC		ABB		●	○
	Manufacturing engineering	20		BBC	BBBC				●	○
	Mechanical engineering	65		BBC	BBBC		BBC	BBBC	●	○
Birmingham	Mechanical and automotive engineering	45	280	ABB	ABBBB	320	ABB	ABBBB	●	○
	Mechanical engineering	80	280	ABB	ABBBB	320	ABB	ABBBB	●	○
Blackpool and The Fylde C	Mechanical and production engineering	20	160						○	○
Bolton	Automobile engineering	50	200						○	○
	Automotive product design	20	200							
	Computer-aided product design	20	200							
	Mechanical engineering	25	200						●	○
Bournemouth	Computer-aided product design		220						○	○
	Design engineering	36	220						○	○
	Product design	70	220						○	○
Bradford	Automotive design technology	(20)	220						○	○
	Industrial design	(20)	200						○	○
	Mechanical and automotive engineering	(60)	200–240			300			○	○
	Mechanical engineering	(60)	200–240			300			○	○
	Vehicle technology		200–240							
Brighton	Automotive engineering		260						●	○
	Mechanical engineering		260						●	○

Entrance requirements

Institution	Course title	Number of students (includes other courses)	Typical offers (BSc/BEng)			Typical offers (MEng)			A-level Mathematics	A-level Physics
			UCAS tariff points	A-levels	SCQF Highers	UCAS tariff points	A-levels	SCQF Highers	● compulsory; ○ preferred	
Brighton (continued)	Product design	30	280						○	○
Bristol	Aeronautical engineering	70					AAB	AAABB	●	●
	Avionic systems	18		ABB	AAABB		ABB	AAABB	●	●
	Mechanical engineering	75	340	AAA	AAAAA	340	AAA	AAAAA	●	●
Bristol UWE	Aerospace manufacturing engineering	60	180–220			240–260			●	○
	Aerospace systems engineering	40	200–240			240–260			●	○
	Mechanical engineering	45	180–220			240–260			●	○
	Motorsport engineering		180–240							
Brunel	Aerospace engineering		280			340			●	●
	Aviation engineering		260			320				
	Industrial design	20	260	BBC					●	●
	Mechanical engineering	(120)	240	CCC	BBBBC	300	BBB	ABBBB	●	●
	Mechanical engineering with aeronautics	(120)	240	CCC	BBBBC	300	BBB	ABBBB	●	●
	Mechanical engineering with automotive design	(120)	240	CCC	BBBBC	300	BBB	ABBBB	●	●
	Mechanical engineering with building services	(120)	240	CCC	BBBBC	320	BBB	ABBBB	●	●
	Motorsport engineering	(120)	280	BBC	BBBBC	340	AAB	ABBBB	●	●
	Product design	20	260–240	BBC	CCCCC				●	○
	Product design engineering	20	260	BBC					●	●
	Space engineering		260			320				
Buckinghamshire Chilterns UC	Product design	20	120–240		CCCC				○	○
Cambridge	*All courses*	(300)		AAA			AAA		●	●
Cardiff	Manufacturing engineering	25	320	BBC		280	ABB		●	○
	Mechanical engineering	70	280	BBC		320	ABB		●	○
Central Lancashire	Motorsports engineering	40	240		BBBB				●	○
City	Aeronautical engineering	60	240			300			●	○
	Air transport engineering	25	240			300			●	○
	Automotive and motor sport engineering	15	240			300				
	Mechanical engineering	15	240			300	ABB		●	○
Coventry	Aerospace systems engineering	30	260						●	●
	Aerospace technology	25	200						○	○
	Automotive engineering	60	260			260			●	○
	Automotive engineering design	24	260						●	●
	Avionics technology	20	200						○	○
	Industrial product design	25	260						○	○
	Mechanical engineering	60	260			260			●	○
	Mechanical engineering design	40	260			260			●	○
	Motorsport and motorcycle engineering		200–240						●	○
	Motorsport and powertrain engineering		200–240						●	○
	Motorsport engineering	20	200–240						●	○
Dundee	Mechanical engineering	30	240			300			●	○

Mechanical and Manufacturing Engineering

Institution	Course title	Number of students (includes other courses)	Typical offers (BSc/BEng) UCAS tariff points	A-levels	SCQF Highers	Typical offers (MEng) UCAS tariff points	A-levels	SCQF Highers	● compulsory; ○ preferred A-level Mathematics	A-level Physics
Durham	Aeronautics	(130)					ABB	AAAA	●	○
	Design, manufacture and management	(140)					ABB	AAAA	●	○
	Mechanical engineering	(140)					ABB	AAAA	●	○
East London	Product design	25	160–200							
Edinburgh	Mechanical engineering	50		BBB	BBBB		BBB	BBBB	●	○
Exeter	Mechanical engineering	40	240			300			●	○
Farnborough CT	Aeronautical engineering	24	80–160						●	●
Glamorgan	Aerospace engineering		140–180							
	Aircraft maintenance engineering		140–180							
	Mechanical engineering	40	140–180						●	○
Glasgow	Aeronautical engineering	85		BBC	ABBBC		BBC	ABBBC	●	●
	Avionics	5		BB	BBBBC		BB	BBBBC	●	●
	Mechanical design engineering	25		CCC	ABBB		BCC	AABB	●	○
	Mechanical engineering	40		CCC	ABBB		BCC	AABB	●	○
	Mechanical engineering with aeronautics	30		CCC	ABBB		BCC	AABB	●	○
	Product design engineering	50		CCC	ABBB		BCC	AABB	●	○
Glasgow Caledonian	Computer-aided mechanical engineering	15		CD	BCC				○	○
	Manufacturing systems engineering	10		BC	BBBC				●	●
Greenwich	Automotive engineering with management		120						○	○
	Engineering product design	15	200	CDD	CCC				○	
	Industrial automation	5	240	CCC	BBCC				○	○
	Manufacturing systems engineering	10	240	CCC	BBCC				●	○
	Marine engineering technology	10	200	DDD	CCCC				○	○
	Mechanical engineering	50	240	CCC	BBCC				●	○
	Mechanical engineering technology		160						○	○
Harper Adams UC	Agricultural engineering	25	240		BBBC	300		AABB	○	○
	Agriculture and mechanisation		160–240							
	Engineering design and development		160–240							
	Off-road vehicle design	20	240		BBBC	300		AABB	○	○
Heriot-Watt	Automotive engineering			CDD	BBBC					
	Mechanical engineering	60		CCD	BBBC		CCD	BBBC	●	○
Hertfordshire	Aerospace engineering	70	260			320			●	○
	Aerospace systems engineering	25	260			320			●	○
	Aerospace technology with management/pilot studies		200							
	Automotive engineering	40	260			320			●	○
	Automotive engineering with motorsport		240			300				
	Manufacturing engineering	(40)	240–280						○	
	Mechanical engineering	30	260						●	○
	Motorsport technology		200							
Huddersfield	Automotive design	30	220		BBBB	260		AABB	●	
	Automotive design and technology	20	220		BBBB				●	
	Automotive engineering	20	220		BBBB	260		AABB	●	○

Entrance requirements

Institution	Course title	Number of students (includes other courses)	Typical offers (BSc/BEng) UCAS tariff points	A-levels	SCQF Highers	Typical offers (MEng) UCAS tariff points	A-levels	SCQF Highers	● compulsory; O preferred — A-level Mathematics	A-level Physics
Huddersfield (continued)	Automotive product innovation	20	220		BBBB					
	Automotive technology	20	180		BBBB					
	Engineering design: mechanical	30	220			260			●	
	Manufacturing and operations management	20	220		BBBB					
	Mechanical and automotive design	15	220		BBBB	260		AABB	●	
	Mechanical engineering	20	220		BBBB	260		AABB	●	O
	Motorsport technology	30	180		RRR					
	Product innovation, design and development	20	220		BBBB					
Hull	Mechanical engineering	30	220–280						●	O
Imperial College London	Aeronautical engineering	70					AAB		●	●
	Mechanical engineering	160					AAB		●	●
King's College London	Mechanical engineering	30		BBB	AABBB		BBB	AABBB	●	O
Kingston	Aerospace engineering	50	240		BBCC	300			●	O
	Aerospace engineering and astronautics	25	240		BBCC	300			●	O
	Aerospace engineering design	30	140						●	O
	Mechanical engineering	20	240		BBCC				●	O
	Mechanical engineering design	20	140		BCC				●	O
	Motorcycle engineering design	35	140		BCC				●	O
	Motorsport engineering		140						●	O
Lancaster	Engineering (mechanical)	(90)		BCC	AABB		BBC	AAAB	●	O
Leeds	Aeronautical and aerospace engineering			AAB	AAABB		AAB	AAABB	●	●
	Automotive engineering	25		BBB	BBBBB		BBB	BBBBB	●	O
	Mechanical engineering	90		BBC	BBBCC		BBC	BBBCC	●	O
Leicester	Mechanical engineering	(90)	260			300			●	O
Liverpool	Aerospace engineering	50	300	BBB	AAAA	340	AAB	AAAAA	●	●
	Aerospace engineering with pilot studies	30	300	BBB	AAAA	340	AAB	AAAAA	●	●
	Avionic systems		280	BBC	AAAB	320	AAB	AAAAB	●	●
	Avionic systems with pilot studies		300	BBB	AAAA	340	AAB	AAAAA	●	●
	Mechanical engineering	60	280	BBC	AAAB	340	AAB	AAAAA	●	●
	Mechanical systems and design engineering	12	280	BBC	AAAB	340	AAB	AAAAA	●	●
Liverpool John Moores	Automotive engineering	15	140–200						●	●
	Mechanical and marine engineering	15	140–200						O	O
	Mechanical engineering	20	140–200						●	●
London South Bank	Engineering product design	30		CC	BBB				O	O
	Mechanical engineering	40		CC	BBB				●	O
Loughborough	Aeronautical engineering	60		ABC	BBB		AAB		●	●
	Automotive engineering	55		DDD			AAB	ABBBB	●	●
	Innovative manufacturing technology	25				300–320	ABB		●	O
	Manufacturing engineering and management	20	260	BCC			ABB		●	O
	Mechanical engineering	120	260	BCC		340	ABB		●	●

Mechanical and Manufacturing Engineering

(continued) Table 5

Entrance requirements

Institution	Course title	Number of students (includes other courses)	Typical offers (BSc/BEng) UCAS tariff points	A-levels	SCQF Highers	Typical offers (MEng) UCAS tariff points	A-levels	SCQF Highers	● compulsory; ○ preferred A-level Mathematics	A-level Physics
Loughborough (continued)	Product design and manufacture	50	260	BCC		300	BBB		●	○
Manchester	Aerospace engineering	(250)	300	ABB		320	AAB		●	●
	Mechanical engineering	(250)	300	ABB		320	AAB		●	○
Manchester Metropolitan	Automation and control		240		BBBB				●	○
	Automotive engineering		240		CCCC					
	Mechanical engineering	50	220–240		CCCC				●	○
Napier	Mechanical engineering	25	220							
	Product design engineering		220							
Newcastle	Marine engineering	(90)		BBB			ABB		●	●
	Marine technology	(90)		BBB	BBBB		ABB	BBBB	●	●
	Mechanical and automotive engineering	10				320	ABB		●	○
	Mechanical and design engineering	10				320	ABB		●	○
	Mechanical and manufacturing engineering	10				320	ABB		●	○
	Mechanical and railway engineering	10				320	ABB		●	○
	Mechanical engineering	75	280	BBB	BBBBB	320	ABB	AABBB	●	○
	Naval architecture	(90)		BBB			ABB		●	●
	Offshore engineering	(90)		BBB			ABB		●	●
	Small craft technology	(90)		BBB			ABB		●	●
North East Wales I	Aeronautical engineering	10	140						○	
	Mechanical engineering	10	140						○	○
	Performance car technology		140							
Northumbria	Computer-aided product design		240		CCCCC					
	Manufacturing systems engineering	25	240		BBBB				●	○
	Mechanical engineering	60	240		CCC				●	○
	Product design and technology		240		BBBB				●	
Nottingham	Manufacturing engineering and management	30		ABB			ABB		●	
	Mechanical design, materials and manufacture	35		ABB			AAB		●	●
	Mechanical engineering	(100)		AAB	AABBB		AAB	AABBB	●	●
	Mechanical engineering (aerospace)	(100)		AAB			AAB		●	●
	Mechanical engineering (automotive)	(100)		AAB			AAB		●	●
	Mechanical engineering (bioengineering)	(100)		AAB			AAB		●	●
	Product design and manufacture	15		ABB					●	
Nottingham Trent	*Both courses*	70	240							
Oxford	Mechanical engineering	(170)					AAA	AAAAB	●	●
Oxford Brookes	Automotive engineering	30		BCC			BCC		●	●
	Computer-aided mechanical engineering			CC						
	Mechanical engineering	35		CCC					●	●
	Motorsport engineering								●	●
	Motorsport technology			CC						
Paisley	Mechanical engineering	30		CD	BBC				●	●

| Institution | Course title | Number of students (includes other courses) | Typical offers (BSc/BEng) | | | Typical offers (MEng) | | | ● compulsory; ○ preferred | |
			UCAS tariff points	A-levels	SCQF Highers	UCAS tariff points	A-levels	SCQF Highers	A-level Mathematics	A-level Physics
Paisley (continued)	Product design and development			DD	BCC					
Plymouth	Marine sports technology		160							
	Marine technology	10	240						●	○
	Mechanical design and manufacture	30	160						○	○
	Mechanical engineering	40	160			240			●	○
Portsmouth	Computer-aided product design	30	200							
	Marine sports technology	20	200	CC					○	○
	Mechanical and manufacturing engineering	40	160						○	○
	Mechanical engineering	40	240			240			●	○
	Product design and innovation	50	200							
	Product design and modern materials	20	200							
Queen Mary	Aerospace engineering	20	260–340		BBBCC	300–340		BBBCC	●	●
	Avionics	5	260–340		BBBCC				●	●
	Mechanical engineering	15	260–340		BBBCC	300–340		BBBCC	●	●
	Sports engineering	5				300–340		BBBCC	●	●
Robert Gordon	Mechanical and offshore engineering	30	220–240		BBCC				●	○
	Mechanical engineering	30		CCD	BBCC		BBB	ABBB	●	○
Salford	Aeronautical engineering	90	240		BBBBC	300		AABBB	●	○
	Aircraft engineering with pilot studies		240						●	○
	Aviation technology with pilot studies		160						●	○
	Mechanical engineering (aerospace)	70	240			300			●	○
Sheffield	Aerospace engineering	70	320			340			●	●
	Aerospace engineering with private pilot instruction	10	320			340	AAB		●	●
	Mechanical engineering	100	320			340			●	○
	Mechanical systems engineering	10		AAB	AABB		AAB	AABB	●	○
Sheffield Hallam	Automotive technology		140							
	Mechanical engineering		160			300			●	○
Southampton	Acoustical engineering	25	320	ABB	AABBB	320	ABB	AABBB	●	●
	Aeronautics and astronautics	85	340	AAB	AAABB	340	AAB	AAABB	●	●
	Mechanical engineering	60	340	AAB		340	AAB		●	●
	Ship science	(40)	320	ABB	AABBB	320	ABB	AABBB	●	●
	Ship science/yacht and small craft	(40)	320	ABB	AABBB	320	AAB BB		●	●
Southampton Solent	Mechanical design	(55)	100						○	○
	Yacht and powercraft design	35	100						●	○
Staffordshire	Aeronautical design technology	100	200–280						○	○
	Automotive design technology	100	200–280						○	○
	Mechanical engineering	50	280			280			●	●
	Mechanical systems design	50	280			280			●	●
Strathclyde	Aeromechanical engineering	45		BBB	ABBBB		AAB	AAAAB	●	●
	Manufacturing engineering and technology			CCC	ABBB		BBB	AABB	●	●
	Mechanical engineering	85		BBB	ABBBB		AAB	AAAAB	●	○
	Naval architecture and marine engineering	(25)		BCC	BBBB		BBB	AAAA	●	○

Mechanical and Manufacturing Engineering

Mechanical and Manufacturing Engineering

Institution	Course title	Number of students (includes other courses)	Typical offers (BSc/BEng)			Typical offers (MEng)			● compulsory; ○ preferred A-level Mathematics	A-level Physics
			UCAS tariff points	A-levels	SCQF Highers	UCAS tariff points	A-levels	SCQF Highers		
Strathclyde (continued)	Naval architecture and ocean engineering	(25)		BCC	BBBB		BBB	AAAA	●	○
	Naval architecture and small craft engineering	(25)		BCC	BBBB		BBB	AAAA	●	○
	Product design engineering	70		CCC	BBBBC		BBB	AABB	●	●
Sunderland	Automotive design and technology	(25)	200–360	BCCC					○	○
	Automotive engineering with design	(25)	200–360	BCCC					○	○
	Automotive product design	(25)	200–360	BCCC					○	○
	Engineering design and manufacture	(25)	200–360	BCCC					○	○
	Mechanical engineering with design	(25)	200	BCCC					○	○
	Product design		200	BCCC					○	○
Surrey	Aerospace engineering	40	260	BCC	BBBBC	300	BBB	BBBBC	●	○
	Mechanical engineering	50	260	BCC	BBBBC	300	BBB	BBBBC	●	○
Sussex	Automotive engineering			BBB	BBBBB		AAB	ABBBB		
	Mechanical engineering			BBB	BBBBB		AAB	ABBBB		
	Product design			BBB						
Swansea	Aerospace communications		260–300			320–380			●	
	Aerospace engineering		260–300			320–380			●	
	Aerospace engineering with propulsion		260–300			320–380			●	
	Mechanical engineering	60	260–300			320–380			●	
	Product design engineering		260			320			●	○
	Product design technology		260–300							
Swansea IHE	Automotive electronics systems		200							
	Automotive engineering	5	200–340						●	○
	Automotive manufacturing		200–340							
	Manufacturing systems engineering		80–340							
	Mechanical and manufacturing engineering		80–340							
	Motorcycle engineering		40–340						○	○
	Motorsport engineering and design	65	200–340						●	○
	Motorsport manufacturing engineering		200–340							
Teesside	*Both courses*	20	180–240						○	○
UCE Birmingham	Automotive engineering		220–240						●	○
	Engineering product design		220–240							
	Mechanical and automotive design		220–240							
	Mechanical engineering		220–240						●	○
UCL	Mechanical engineering	55		BBB			AAB		●	●
	Naval architecture and marine engineering	12		BBB			ABB		●	●
Ulster	Mechanical engineering	36	260			300			●	○
Wales (Newport)	Automation systems	15	160						○	○
Wales (UWIC)	*Both courses*		140							○
Warwick	Automotive engineering			BBB			AAB		○	
	Manufacturing and mechanical engineering	20		BBC			AAB		○	○
	Mechanical engineering	90		BBC			AAB		○	○
Wolverhampton	*Both courses*	20	160–220						○	
York	Avionics	(110)					AAB		●	●

See Chapter 6 in the *Introduction* (page 20) for general information about professional qualification that applies to all types of engineering course, including those covered in this part of the Guide.

TABLE 6 shows courses that have been accredited at Chartered Engineer or Incorporated Engineer level by the various professional institutions, especially the Institution of Mechanical Engineers (IMechE), the Institution of Engineering and Technology (IET) and the Royal Aeronautical Society (RAeroSoc).

Other institutions shown by their initials in TABLE 6 are the Chartered Society of Designers (CSD), the Institution of Civil Engineers (ICE), the Institution of Structural Engineers (IStructE), the Institution of Engineering Designers (IED), the Institution of Agricultural Engineers (IAgrE), the Royal Institution of Naval Architects (RINA), the Institute of Marine Engineering, Science and Technology (IMarEST), the Institute of Materials, Minerals and Mining (IMMM), the Institute of Measurement and Control (InstMC) and the Institute of Operations Management (IOM).

The IMechE accredits courses that can contribute to Chartered Engineer status; the other institutions can accredit courses at both levels, as shown in the table.

Note that where courses are accredited by several institutions, the accreditation by an individual institution may be conditional on your having taken certain options, which may not coincide with those required by another institution. You should also remember that the process of accreditation goes on continuously, so you should check with the professional institutions for the current position on specific courses.

Chapter 7 in the *Introduction* (page 24–26) lists the professional institutions' addresses.

Table 6 — Accreditation by professional institutions

Institution	Course title	IMechE	IET CEng	IET IEng	RAeroSoc CEng	RAeroSoc IEng	Others
Aberdeen	Mechanical engineering	●					
Anglia Ruskin	Computer-aided product design						IED
Aston	Mechanical engineering	●					
Bath	Aerospace engineering				●		
	Automotive engineering	●	●				
	Innovation and engineering design	●	●				
	Manufacturing	●	●				
	Mechanical engineering	●	●				
Belfast	Aeronautical engineering	●			●		
	Manufacturing engineering	●	●				
	Mechanical engineering	●	●				

(continued) Table 6

Accreditation by professional institutions

Institution	Course title	IMechE	IET CEng	IET IEng	RAeroSoc CEng	RAeroSoc IEng	Others
Birmingham	*Both courses*	●					
Bolton	Automotive product design						CSD
	Computer-aided product design						CSD
Bournemouth	Product design						IED
Bradford	Mechanical and automotive engineering	●					
	Mechanical engineering	●					
Brighton	Product design						IED
Bristol	Aeronautical engineering				●		
	Avionic systems		●		●		
	Mechanical engineering	●					
Bristol UWE	Aerospace manufacturing engineering		●				
	Mechanical engineering	●					
Brunel	Aerospace engineering	●					
	Industrial design						CSD; IED
	Mechanical engineering	●					
	Mechanical engineering with aeronautics	●					
	Mechanical engineering with automotive design	●					
	Mechanical engineering with building services	●					
	Motorsport engineering	●					
	Product design						CSD; IED
	Product design engineering						CSD; IED
Cambridge	Aerospace and aerothermal engineering	●			●		
	Manufacturing engineering	●	●				
	Mechanical engineering	●	●		●		
Cardiff	Manufacturing engineering		●				
	Mechanical engineering	●					Energy Institute
City	Aeronautical engineering	●			●		
	Air transport engineering	●			●		
	Mechanical engineering	●					
Coventry	Aerospace systems engineering	●			●		
	Aerospace technology					●	
	Automotive engineering	●					
	Automotive engineering design	●					
	Avionics technology					●	
	Industrial product design						IED
	Mechanical engineering	●					
	Mechanical engineering design	●					
	Motorsport engineering					●	
Dundee	Mechanical engineering	●					
Durham	Design, manufacture and management	●	●				
	Mechanical engineering	●	●				
East London	Product design						IED; CSD
Edinburgh	Mechanical engineering	●					
Exeter	Mechanical engineering	●	●				
Glamorgan	Mechanical engineering	●					

Accreditation by professional institutions

Institution	Course title	IMechE	IET CEng	IET IEng	RAeroSoc CEng	RAeroSoc IEng	Others
Glasgow	Aeronautical engineering	●			●		
	Avionics		●		●		
	Mechanical design engineering	●					
	Mechanical engineering	●					
	Mechanical engineering with aeronautics	●					
	Product design engineering	●					
Glasgow Caledonian	Computer-aided mechanical engineering			●			
	Manufacturing systems engineering	●	●				
Greenwich	Marine engineering technology						IMarEST (IEng)
	Mechanical engineering			●			
Harper Adams UC	Agricultural engineering						IAgrE
	Off-road vehicle design						IAgrE
Heriot-Watt	Mechanical engineering	●					InstE (CEng)
Hertfordshire	Aerospace engineering				●		
	Aerospace systems engineering				●		
	Automotive engineering	●					Institute of Motor Industry
	Automotive engineering with motorsport	●					Institute of Motor Industry
	Mechanical engineering	●					
Huddersfield	Manufacturing and operations management						Institute of Operations Management
Hull	Mechanical engineering	●					
Imperial College London	Aeronautical engineering	●			●		
	Mechanical engineering	●					
King's College London	Mechanical engineering	●					
Kingston	Aerospace engineering	●	●	●	●		
	Aerospace engineering and astronautics	●			●		
	Aerospace engineering design			●	●	●	
	Mechanical engineering	●			●		
	Mechanical engineering design				●		
	Motorcycle engineering design				●		
Lancaster	Engineering (mechanical)	●					
Leeds	Automotive engineering	●					
	Mechanical engineering	●					
Leicester	Mechanical engineering	●					InstMC
Liverpool	Aerospace engineering	●	●		●		
	Aerospace engineering with pilot studies	●	●		●		
	Mechanical engineering	●					
	Mechanical systems and design engineering	●					
Liverpool John Moores	Mechanical and marine engineering	●					IMarEST
London South Bank	Engineering product design						IED
	Mechanical engineering	●					
Loughborough	Aeronautical engineering	●			●		
	Automotive engineering	●					
	Innovative manufacturing technology	●					
	Manufacturing engineering and management	●	●				
	Mechanical engineering	●					
	Product design and manufacture	●	●				

Accreditation by professional institutions

Institution	Course title	IMechE	IET CEng	IET IEng	RAeroSoc CEng	RAeroSoc IEng	Others
Manchester	Aerospace engineering				●		
	Mechanical engineering	●	●				
Manchester Metropolitan	Mechanical engineering	●	●				
Napier	Mechanical engineering	●	●				
Newcastle	Marine engineering						RINA; IMarEST
	Marine technology						RINA; IMarEST
	Mechanical and automotive engineering	●	●				
	Mechanical and design engineering	●	●				
	Mechanical and manufacturing engineering	●	●				
	Mechanical engineering	●	●				
	Naval architecture						RINA; IMarEST
	Offshore engineering						RINA; IMarEST
	Small craft technology						RINA; IMarEST
North East Wales I	Mechanical engineering			●	●		
Northumbria	Manufacturing systems engineering	●	●				
	Mechanical engineering	●	●				
Nottingham	Manufacturing engineering and management		●				
	Mechanical design, materials and manufacture						IMat (CEng)
	Mechanical engineering	●					
	Mechanical engineering (aerospace)	●					
	Mechanical engineering (automotive)	●					
	Mechanical engineering (bioengineering)	●					
Oxford	Mechanical engineering	●	●				
Oxford Brookes	Automotive engineering	●					
	Mechanical engineering	●					
Plymouth	Mechanical design and manufacture				●		
	Mechanical engineering	●					
Portsmouth	Marine sports technology			●	●		
	Mechanical and manufacturing engineering				●		
	Mechanical engineering	●					
Queen Mary	Aerospace engineering				●		
	Avionics				●		
	Mechanical engineering	●					
Robert Gordon	Mechanical and offshore engineering	●					IMarEST
	Mechanical engineering	●					
Salford	Aeronautical engineering				●		
	Mechanical engineering (aerospace)	●					
Sheffield	Aerospace engineering			●	●		IMMM (CEng)
	Aerospace engineering with private pilot instruction			●	●		IMMM (CEng)
	Mechanical engineering	●					
	Mechanical systems engineering			●			InstMC (CEng)
Sheffield Hallam	Mechanical engineering	●	●		●		
Southampton	Acoustical engineering	●					Institute of Acoustics
	Mechanical engineering	●					
	Ship science						RINA; IMarEST
	Ship science/yacht and small craft						RINA; IMarEST

Institution	Course title	IMechE	IET CEng	IET IEng	RAeroSoc CEng	RAeroSoc IEng	Others
Southampton Solent	Yacht and powercraft design						RINA
Staffordshire	Aeronautical design technology				●		
	Automotive design technology				●		
Strathclyde	Aeromechanical engineering	●			●		
	Manufacturing engineering and technology	●	●				
	Mechanical engineering	●					
	Naval architecture and marine engineering						RINA; IMarEST
	Naval architecture and small craft engineering						RINA; IMarEST
	Product design engineering	●	●				
Sunderland	Automotive engineering with design				●		
	Engineering design and manufacture				●		
	Mechanical engineering with design				●		
Surrey	Aerospace engineering	●			●		
	Mechanical engineering	●					
Sussex	Mechanical engineering	●					
Swansea	Mechanical engineering	●					
	Product design engineering	●					
Teesside	Computer-aided design engineering						IED (IEng)
	Mechanical engineering			●			
UCE Birmingham	Automotive engineering	●					
	Mechanical engineering	●					
UCL	Mechanical engineering	●					
	Naval architecture and marine engineering						RINA (CEng); IMarEST (CEng)
Wales (Newport)	Automation systems						InstMC (IEng)
Wales (UWIC)	*Both courses*			●			
Warwick	Manufacturing and mechanical engineering	●	●				InstMC
	Mechanical engineering	●					
York	Avionics			●			

See Chapter 7 in the *Introduction* (page 24) for a list of sources of information that apply to all types of engineering course, including those covered in this part of the Guide. It also lists a number of books that can give some general background to engineering and the work of engineers. Visit the websites of the professional institutions for information on careers, qualifications and course accreditation, as well as information about publications, many of which you can download. Chapter 7 in each of the other parts of the Guide contains further suggestions for background reading in the individual engineering disciplines.

The courses This Guide gives you information to help you narrow down your choice of courses. Your next step is to find out more about the courses that particularly interest you. Prospectuses cover many of the aspects you are most likely to want to know about, but some departments produce their own publications giving more specific details of their courses. University and college websites are shown in TABLE 2a.

You can also write to the contacts listed below.

Aberdeen Student Recruitment and Admissions Service (sras@abdn.ac.uk), University of Aberdeen, Regent Walk, Aberdeen AB24 3FX

Anglia Ruskin Contact Centre (answers@anglia.ac.uk), Anglia Ruskin University, Bishop Hall Lane, Chelmsford CM1 1SQ

Aston Dr Geof Carpenter (g.f.carpenter@aston.ac.uk), School of Engineering and Applied Science, Aston University, Aston Triangle, Birmingham B4 7ET

Bath Admissions Secretary (en-ug-admissions@bath.ac.uk), Department of Mechanical Engineering, University of Bath, Bath BA2 7AY

Belfast Aeronautical engineering Adviser of Studies, Department of Aeronautical Engineering, The Queen's University of Belfast, Belfast BT7 1NN; Manufacturing engineering Mechanical engineering Adviser of Studies, Department of Mechanical Engineering, The Queen's University of Belfast, Ashby Institute, 125 Stranmillis Road, Belfast BT9 5AH

Birmingham Dr R J Cripps (mfg.mech.admissions@bham.ac.uk), School of Engineering, University of Birmingham, Edgbaston, Birmingham B15 2TT

Blackpool and The Fylde C Admissions Office (admissions@blackpool.ac.uk), Blackpool and The Fylde College, Ashfield Road, Bispham, Blackpool FY2 0HB

Bolton Automobile engineering Mechanical engineering Norman Lloyd (nl1@bolton.ac.uk); Automotive product design Admissions Tutor (bp2@bolton.ac.uk); Computer-aided product design Admissions Tutor (sw5@bolton.ac.uk); all at Department of Engineering and Design, Bolton University, Deane Road, Bolton BL3 5AB

Bournemouth Computer-aided product design Product design Bea Dunleavy; Design engineering Mary Faulkner; both Programmes Administrator, Bournemouth University, Studland House, 12 Christchurch Road, Bournemouth BH1 3NA

Bradford Mr Jack Bradley (ug-eng-enquiries@bradford.ac.uk), Admissions Tutor, School of Engineering, Design and Technology, University of Bradford, Bradford BD7 1DP

Mechanical and Manufacturing Engineering

Brighton Automotive engineering Mechanical engineering Admissions Tutor, Department of Mechanical and Production Engineering; Product design Admissions Tutor, School of Engineering; both at University of Brighton, Lewes Road, Brighton BN2 4GJ

Bristol Aeronautical engineering Avionic systems Admissions Tutor (aero-office@bristol.ac.uk), Department of Aerospace Engineering; Mechanical engineering Dr J E Morgan, Admissions Tutor, Department of Mechanical Engineering; both at University of Bristol, Bristol BS8 1TR

Bristol UWE Pat Cottrell (admissions.cems@uwe.ac.uk), Faculty of Computing, Engineering and Mathematics, University of the West of England Bristol, Coldharbour Lane, Frenchay, Bristol BS16 1QY

Brunel Industrial design Product design Product design engineering Dr Blue Ramsay (design.information@brunel.ac.uk); All other courses Ms Petra Godwin (me-ug-admissions@brunel.ac.uk); both Admissions Tutor, School of Engineering and Design, Brunel University, Uxbridge UB8 3PH

Buckinghamshire Chilterns UC Admissions Office (admissions@bcuc.ac.uk), Buckinghamshire Chilterns University College, Queen Alexandra Road, High Wycombe HP11 2JZ

Cambridge Cambridge Admissions Office (admissions@cam.ac.uk), University of Cambridge, Fitzwilliam House, 32 Trumpington Street, Cambridge CB2 1QY

Cardiff Dr T O'Doherty (odoherty@cardiff.ac.uk), School of Engineering, Cardiff University, PO Box 917, Cardiff CF24 1XH

Central Lancashire Dr John Calderbank (jacalderbank@uclan.ac.uk), Department of Technology, University of Central Lancashire, Preston PR1 2HE

City Undergraduate Admissions Office (ugadmissions@city.ac.uk), City University, Northampton Square, London EC1V 0HB

Coventry Aerospace systems engineering Dr S Hargrave (s.hargrave@coventry.ac.uk); Aerospace technology Avionics technology Mr M Basini (m.basini@coventry.ac.uk); Automotive engineering design Dr S Owen (s.owen@coventry.ac.uk); Industrial product design Mr M Evatt (m.evatt@coventry.ac.uk); Motorsport and motorcycle engineering Motorsport and powertrain engineering Motorsport engineering Mr J Baxter (jon.baxter@coventry.ac.uk); All other courses Mr B Dunn (b.dunn@coventry.ac.uk); all at School of Engineering, Coventry University, Priory Street, Coventry CV1 5FB

Dundee Dr M S Pridham (m.s.pridham@dundee.ac.uk), Mechanical Engineering Department, University of Dundee, Dundee DD1 4HN

Durham Dr Tim Short (engineering.admissions@durham.ac.uk), School of Engineering, University of Durham, South Road, Durham DH1 3LE

East London Student Admissions Office (admiss@uel.ac.uk), University of East London, Docklands Campus, 4–6 University Way, London E16 2RD

Edinburgh Undergraduate Admissions Office (sciengug@ed.ac.uk), College of Science and Engineering, University of Edinburgh, The King's Buildings, West Mains Road, Edinburgh EH9 3JY

Exeter Admissions Secretary (eng-admissions@exeter.ac.uk), Department of Engineering, School of Engineering, Computer Science and Mathematics, University of Exeter, North Park Road, Exeter EX4 4QF

Farnborough CT Admissions Administrator (admissions@farn-ct.ac.uk), Farnborough College of Technology, Boundary Road, Farnborough, Hampshire GU14 6SB

Glamorgan Mike Board, School of Design and Advanced Technology, University of Glamorgan, Pontypridd, Mid Glamorgan CF37 1DL

Glasgow Clerk of Faculty of Engineering, Glasgow University, Glasgow G12 8QQ

Glasgow Caledonian Computer-aided mechanical engineering Dr Asou K Roy (a.roy@gcal.ac.uk); Manufacturing systems engineering Angela Geddes (age@gcal.ac.uk); both at School of Engineering, Science and Design, Glasgow Caledonian University, Cowcaddens Road, Glasgow G4 0BA

Greenwich Admissions Co-ordinator (eng-courseinfo@gre.ac.uk), School of Engineering, University of Greenwich, Medway Campus, Pembroke, Chatham Maritime, Kent ME4 4TB

Harper Adams UC Engineering design and development Bill Rowley (browley@harper-adams.ac.uk); All other courses G F D Wakeham (gfdwakeham@harper-adams.ac.uk), Course Manager; both at Harper Adams University College, Newport, Shropshire TF10 8NB

Heriot-Watt Dr P A Kew (p.a.kew@hw.ac.uk), Department of Mechanical and Chemical Engineering, Heriot-Watt University, Riccarton, Edinburgh EH14 4AS

Hertfordshire Dr A Lewis (admissions@herts.ac.uk), Department of Aerospace, Civil and Mechanical Engineering, University of Hertfordshire, College Lane, Hatfield AL10 9AB

Huddersfield Department of Engineering and Technology (engtech@hud.ac.uk), University of Huddersfield, Queensgate, Huddersfield HD1 3DH

Hull Admissions Tutor (engineering-admissions@hull.ac.uk), Department of Engineering, University of Hull, Hull HU6 7RX

Imperial College London Aeronautical engineering Dr K G Woodgate (aero.admissions@imperial.ac.uk), Department of Aeronautics; Mechanical engineering D Robb (d.robb@imperial.ac.uk), Department of Mechanical Engineering; both at Imperial College London, South Kensington Campus, London SW7 2AZ

King's College London Admissions Tutor (ugadmissions.engineering@kcl.ac.uk), Division of Engineering, King's College London, Strand, London WC2R 2LS

Kingston Student Information and Advice Centre, Cooper House, Kingston University, 40–46 Surbiton Road, Kingston upon Thames KT1 2HX

Lancaster Dr R V Chaplin, Engineering Department, Lancaster University, Lancaster LA1 4YR

Leeds Student Support Office (ug-admissions@mech-eng.leeds.ac.uk), School of Mechanical Engineering, University of Leeds, Leeds LS2 9JT

Leicester Mr I M Jarvis (ms263@le.ac.uk), Department of Engineering, University of Leicester, University Road, Leicester LE1 7RH

Liverpool Admissions Tutor (ugeng@liv.ac.uk), Department of Engineering, University of Liverpool, Brownlow Hill, Liverpool L69 3GH

Liverpool John Moores Student Recruitment Team (recruitment@ljmu.ac.uk), Liverpool John Moores University, Roscoe Court, 4 Rodney Street, Liverpool L1 2TZ

London South Bank Admissions Office, London South Bank University, 103 Borough Road, London SE1 0AA

Loughborough Aeronautical engineering Automotive engineering Ms S Boyd (s.boyd@lboro.ac.uk), Department of Aeronautical and Automotive Engineering; All other courses Mrs S C Thorne (mmadmissions@lboro.ac.uk), Admissions Co-ordinator, Wolfson School of Mechanical and Manufacturing Engineering; both at Loughborough University, Loughborough LE11 3TU

Manchester Dr A J Bell (ug.mace@manchester.ac.uk), School of Mechanical, Aerospace and Civil Engineering, University of Manchester, PO Box 88, Manchester M60 1QD

Manchester Metropolitan Dr W J McCann (j.mccann@mmu.ac.uk), Department of Mechanical Design & Manufacture, Manchester Metropolitan University, Chester Street, Manchester M1 5GD

Napier Colin Boswell, School of Engineering, Napier University, 10 Colinton Road, Edinburgh EH10 5DT

Newcastle Marine engineering Marine technology Naval architecture Offshore engineering Small craft technology Admissions Tutor (marinetech.ugadmin@ncl.ac.uk), Department of Marine Technology; All other courses Admissions Office (enquiries@ncl.ac.uk); both at University of Newcastle upon Tyne, Newcastle upon Tyne NE1 7RU

North East Wales I C Fordwhalley, School of Engineering, North East Wales Institute, Mold Road, Wrexham, Clwyd LL11 2AW

Northumbria Admissions (er.educationliaison@northumbria.ac.uk), University of Northumbria, Trinity Building, Northumberland Road, Newcastle upon Tyne NE1 8ST

Nottingham Manufacturing engineering and management Product design and manufacture Dr Catherine Wyken (catherine.wyken@nottingham.ac.uk); All other courses Dr S J Pickering (stephen.pickering@nottingham.ac.uk); both at School of Mechanical, Materials & Manufacturing Engineering, University of Nottingham, University Park, Nottingham NG7 2RD

Nottingham Trent Registry (admissions@ntu.ac.uk), Nottingham Trent University, Burton Street, Nottingham NG1 4BU

Oxford Deputy Administrator (Academic), Department of Engineering Science, Oxford University, Parks Road, Oxford OX1 3PJ

Oxford Brookes Mechanical engineering Mr A Hern; All other courses Mr J Balkwill; both at School of Engineering, Oxford Brookes University, Headington, Oxford OX3 0BP

Paisley J C Watson (wats-mm@paisley.ac.uk), Department of Mechanical and Manufacturing Engineering, University of Paisley, Paisley PA1 2BE

Plymouth Faculty of Technology (d.plane@plymouth.ac.uk), University of Plymouth, Drake Circus, Plymouth PL4 8AA

Portsmouth John R Bishop (john.bishop@port.ac.uk), Admissions Tutor, Faculty of Technology, University of Portsmouth, Anglesea Building, Anglesea Road, Portsmouth PO1 3DJ

Queen Mary Marian Langbridge (m.langbridge@qmul.ac.uk), Department of Engineering, Queen Mary University of London, Mile End Road, London E1 4NS

Robert Gordon Admissions Office (admissions@rgu.ac.uk), The Robert Gordon University, Schoolhill, Aberdeen AB10 1FR

Salford Mr A J White (a.j.white@salford.ac.uk), Department of Aeronautical and Mechanical Engineering, University of Salford, Salford M5 4WT

Sheffield Aerospace engineering Aerospace engineering with private pilot instruction Mrs J Bradbury (j.bradbury@sheffield.ac.uk); Mechanical engineering Undergraduate Admissions Tutor (admit.mech@sheffield.ac.uk); both at Department of Mechanical Engineering; Mechanical systems engineering Admissions (ugacse@sheffield.ac.uk), Department of Automatic Control and Systems Engineering; all at University of Sheffield, Mappin Street, Sheffield S1 3JD

Sheffield Hallam R Crampton, School of Engineering, Sheffield Hallam University, Pond Street, Sheffield S1 1WB

Southampton Acoustical engineering Dr T P Waters (tpw@isvr.soton.ac.uk), Institute of Sound and Vibration Research; Aeronautics and astronautics Mechanical engineering Admissions Tutor (admissions@mech.soton.ac.uk), Mechanical Engineering; Ship science Ship science/yacht and small craft Professor G E Hearn (ugship@ship.soton.ac.uk), Ship Science; all at School of Engineering Sciences, University of Southampton, Southampton SO17 1BJ

Southampton Solent Faculty of Technology Admissions, Southampton Solent University, East Park Terrace, Southampton SO14 0RD

Staffordshire Ann Grainger, School of Engineering and Advanced Technology, Staffordshire University, Beaconside, Stafford ST18 0AD

Strathclyde Aeromechanical engineering Mechanical engineering Dr Jim Wood (jwood@mecheng.strath.ac.uk), Department of Mechanical Engineering; Manufacturing engineering and technology Product design engineering Caroline McGuire (c.maguire@dmem.strath.ac.uk), Department of Design, Manufacture and Engineering Management; both at University of Strathclyde, Glasgow G1 1XJ; All other courses Mr D L Smith, Department of Ship and Marine Technology, University of Strathclyde, 100 Montrose Street, Glasgow G4 0LZ

Sunderland Student Recruitment (student-helpline@sunderland.ac.uk), University of Sunderland, Chester Road, Sunderland SR1 3SD

Surrey Miss Natasha Baines (eng-admissions@surrey.ac.uk), School of Engineering, University of Surrey, Guildford GU2 7XH

Sussex Admissions Tutor (ug.admissions@engineering.sussex.ac.uk), Department of Engineering, University of Sussex, Falmer, Brighton BN1 9QT

Swansea Mrs Lynette Jones (l.j.jones@swan.ac.uk), Department of Mechanical Engineering, University of Wales Swansea, Singleton Park, Swansea SA2 8PP

Swansea IHE Automotive electronics systems Sean McCartan (autoelec@sihe.ac.uk), Faculty of Applied Design and Engineering, Swansea Institute of Higher Education, Mount Pleasant, Swansea SA1 6ED; Automotive engineering Automotive manufacturing Motorsport engineering and design Roger Dowden (auto@sihe.ac.uk); Manufacturing systems engineering Mechanical and manufacturing engineering Dennis Lawlor (june.williams@sihe.ac.uk); Motorcycle engineering Dr Owen Williams (moto@sihe.ac.uk); Motorsport manufacturing engineering Richard Thomas (richard.thomas@sihe.ac.uk); all at Swansea Institute of Higher Education, Townhill Road, Swansea SA2 0UT

Teesside Dr T Shaw, School of Science and Technology, University of Teesside, Middlesbrough TS1 3BA

UCE Birmingham Information Officer (enquiries@tic.ac.uk), Technology Innovation Centre, University of Central England in Birmingham, Millennium Point, Curzon Street, Birmingham B4 7XG

UCL Dr Chris Nightingale (ugadmissions@meng.ucl.ac.uk), Department of Mechanical Engineering, University College London, Torrington Place, London WC1E 7JE

Ulster Mr R H McKeown, School of Electrical and Mechanical Engineering, University of Ulster, Shore Road, Newtownabbey BT37 0QB

Wales (Newport) Admissions (admissions@newport.ac.uk), University of Wales College, Newport, PO Box 101, Newport NP18 3YH

Wales (UWIC) Marketing and Student Recruitment (admissions@uwic.ac.uk), University of Wales Institute, Cardiff, PO Box 377, Western Avenue, Llandaff CF5 2SG

Warwick Director of Undergraduate Admissions, Department of Engineering, University of Warwick, Coventry CV4 7AL

Wolverhampton Admissions Unit (enquiries@wlv.ac.uk), University of Wolverhampton, Compton Road West, Wolverhampton WV3 9DX

York K L Todd, Department of Electronics, University of York, Heslington, York YO10 5DD